Ecological Studies, Vol. 171

Analysis and Synthesis

Ecological Studies

Volumes published since 1998 are listed at the end of this book.

Springer

Berlin
Heidelberg
New York
Hong Kong
London
Milan
Paris
Tokyo

M. L. Martínez N. P. Psuty (Eds.)

Coastal Dunes

Ecology and Conservation

With 108 Figures, 3 in Color, and 33 Tables

Springer

Dr. M. Luisa Martínez
Departamento de Ecología Vegetal
Instituto de Ecología, A.C.
km 2.5 Antigua Carretera a Coatepec 351
Xalapa, Veracruz 91070
México

Dr. Norbert P. Psuty
Institute of Marine and Coastal Sciences
Rutgers University
New Brunswick
New Jersey 08901
USA

Cover illustration: Background: Coastal foredunes migrating inland, Fire Island, New York, USA. The house has since been picked up and moved inland, along with a few others that were in jeopardy of tumbling into the sea (Photo N. Psuty). Upper right hand insert: *Chamaecrista chamaecristoides* a shrubby legume endemic to the Gulf of Mexico with two disjunct populations on the Pacific. It is one of the first colonizers of mobile dunes and facilitates survival and growth of late colonizers (Photo M.L. Martínez). Lower left hand insert: Coastal dune system, La Mancha, in central Gulf of Mexico, Mexico. The photo shows different successional stages from early (mobile dunes) to late (tropical rain forest growing on dunes, in the back) (Photo M.L. Martínez). Lower right hand insert: Production of drinking water in the coastal dune area of Meijendel in the Netherlands. The photo shows an artificial lake where pre-treated water of the river Meuse is infiltrated into the subsoil (Photo K. Tomeï)

ISSN 0070-8356
ISBN 3-540-40829-0 Springer-Verlag Berlin Heidelberg New York

Library of Congress Cataloging-in-Publication Data

Coastal dunes : ecology and conservation / M.L. Martinez, N.P. Psuty.
p. cm. – (Ecological studies ; vol. 171)
Includes bibliographical references and index.
ISBN 3-540-40829-0 (alk. paper)
1. Sand dune ecology. 2. Sand dune conservation. 3. Coastal ecology. I. Martinez, M. L. (M. Luisa), 1963- II. Psuty, Norbert P. III. Ecological studies ; v. 171.

QH541.5.S26C625 2004
577.5'83–dc22 2003063512

Springer-Verlag Berlin Heidelberg New York
Springer-Verlag is a part of Springer Science+Business Media

springeronline.com

Printed in Germany

Production: Friedmut Kröner, 69115 Heidelberg, Germany
Cover design: *design & production* GmbH, 69126 Heidelberg
Typesetting: Kröner, 69115 Heidelberg, Germany

31/3150 YK – 5 4 3 2 1 0 – Printed on acid free paper

The assistance and ideas of Prof. Roy A. Lubke during the early stages of this book are gratefully acknowledged.

Preface

Coastal dunes are characterized by a high ecological diversity, which is the result of a wide set of geomorphological features, environmental heterogeneity, and species variability. These ecosystems have a worldwide distribution covering almost every latitude, from tropical to polar. However, in spite of this global abundance and their ecological and economic relevance, coastal dunes have been substantially altered by human activities, and many are already severely and irreversibly degraded.

Sand dunes have been studied for a long time (as early as 1835). However, there has been strong emphasis on the mid-latitude dune systems and little attention given to low-latitude situations. Unfortunately, it is in these lower latitudes, the tropics, where much of the modern exploitation and coastal development for tourism is occurring. In addition, the modest communication and collaboration among scientists studying coastal dunes in tropical and temperate latitudes have generated a degree of scientific isolation and have limited the occasions of comparing data, performing interdisciplinary studies, and coordinating joint research programs. In an effort to foster scientific dialogue and encourage collaboration, this book brings together coastal dune specialists from tropical and temperate latitudes covering a wide set of topics and experiences.

The concept for this book started at the joint meeting of the XVI International Botanical Congress and the Annual Meeting of the ATB (Association of Tropical Biology) held in St. Louis, Missouri, USA, 1–7 August 1999. During the Symposium titled "Coastal sand dunes: their ecology and restoration", a group of dune specialists made presentations on dune morphology; the roles of species and groups of species in maintaining ecological processes; and specific proposals to promote dune conservation, protection, enhancement, and wise utilization. This meeting between multinational colleagues (mostly tropical and subtropical) led to the opportunity to exchange information and gain new perspectives, and spurred conceptual development of this collection. The theme for the book matured and evolved to include patterns and processes occurring in *both* tropical and temperate latitudes, but with a bias to the neglected tropical areas. The original set of participants was expanded to increase the variety of topics and experiences. In the end, 48 authors from 9 different countries contributed to the book's contents.

A major product of this book is a set of recommendations for future research, identifying some of the most relevant topics of which detailed knowledge is still lacking. It also identifies potential management tools that will promote and maintain the rich diversity of the dune environments, independent of the latitude where they occur. Finally, the paradox of conservation versus increasing coastal development considers the maintenance of the natural dynamics of coastal dunes together with the changes wrought by human activities. That is, a dynamic approach is necessary in order to achieve an enlightened conservation of the coastal environment.

This book was peer-reviewed by many experts, whose comments greatly improved the quality of each chapter: J.M. van Alphen, S.M. Arens, Pieter G.E.F. Augustinus, Michael Barbour, Janusz Blaszkowski, Robert Boyd, Oscar Briones, Ragan Callaway, R.M. Crawford, A.J. Davy, Omar Defeo, Wilfried H.O. Ernst, Alberto González, Rudolf de Groot, A.P. Grootjans, Patrick Hesp, Peter Hietz, Gilles Houlle, A.H.L. Huiskes, R. Karr, Suzanne Koptur, Robert Manson, M. Anwar Maun, Catherine Meur-Ferrec, Roland Paskoff, Edmund Penning-Rousell, Orrin H. Pilkey, Thomas Poulson, Gretel van Rooyen, John Sawyer, Ian Spellerberg, Martyn Sykes, David Sylvia, Guillermo Tell, Leonard B. Thien, and S.E. van Wieren. Specifically, we gratefully acknowledge Martyn Caldwell for his editorial advice and his thorough revision of the book.

Finally, the senior editor, M. Luisa Martinez, who bore much of the editorial workload of the book, would like to thank her colleagues who gave logistic support, namely: Octavio Pérez-Maqueo, Gabriela Vázquez, Rosario Landgrave, M. Luisa Vázquez, Antonio Martínez, Ana Martínez, Alejandra and Carolina Vela, Josefa Vázquez, Nickteh Sánchez, and Araceli Toga. Thanks also to Valeria Pérez-Martínez for the many meaningful moments while we were working on the book, and to Dieter Czeschlik and Andrea Schlitzberger for their constant interest and support throughout the different stages of the book, from the very beginning.

Partial financial support to elaborate the book was provided by the Instituto de Ecología, A.C. (902–17 and 902–17–516).

Xalapa, Ver. (Mexico) — *M. Luisa Martínez*
Highlands, New Jersey (USA) — *Norbert Psuty*
Grahamstown (South Africa) — *Roy Lubke*

Contents

Contributors

AARDE, R. J. VAN

Conservation Ecology Research Unit, Department of Zoology and Entomology, University of Pretoria, Pretoria 0002, South Africa

ADEMA, E.B.

Department of Plant Biology, University of Groningen, P.O. Box 14, 9750 AA Haren, The Netherlands

AHUMADA, B.

Baja California Ecology Office. Via Rápida Oriente No. 1. Int. 6. Centro de Gobierno. Zona Río, México

BAEYENS, G.

Amsterdam Water Supply, Vogelenzangseweg 21, 2114 BA Vogelenzan, The Netherlands

BAKKER, T.W.M.

Dune Water Works of South Holland, Postbox 34, 2270 AA Voorburg, The Netherlands

BEKKER, R.M.

Department of Plant Biology, University of Groningen, P.O. Box 14, 9750 AA Haren, The Netherlands

CORKIDI, L.

Tree of Life Nursery, 33201 Ortega Highway, San Juan Capistrano, California 92693, USA

Cruz, Y.

Sciences School. Baja California University, Km 106 Carr. Tijuana Ensenada, 22800 Ensenada, B.C., México

Cuautle, M.

Departamento de Ecología Vegetal, Instituto de Ecología, A.C., Apdo. 63, Xalapa, Veracruz 91070, México

Díaz Barradas, M.C.

Department of Plant Biology and Ecology, The University of Seville, Ap. 1095, 41080 Sevilla, Spain

Díaz-Castelazo, C.

Departamento de Ecología Vegetal, Instituto de Ecología, A.C., Apdo. 63, Xalapa, Veracruz 91070, México

Espejel, I.

Sciences School. Baja California University, Km 106 Carr. Tijuana Ensenada, 22800 Ensenada, B.C., México

Ferreira, S.

Auckland Conservancy, Department of Conservation, Auckland, New Zealand

Gallego Fernández, J.B.

Department of Plant Biology and Ecology, The University of Seville, Ap. 1095, 41080 Sevilla, Spain

García Mora, R.

Department of Plant Biology and Ecology, The University of Seville, Ap. 1095, 41080 Sevilla, Spain

García Novo, F.

Department of Plant Biology and Ecology, The University of Seville, Ap. 1095, 41080 Sevilla, Spain

García-Franco, J.G.

Departamento de Ecología Vegetal, Instituto de Ecología, A.C., km 2.5 Antigua Carretera a Coatepec No. 301, Xalapa, Veracruz 91070, México

GEMMA, J.N.

Department of Biological Sciences, Ranger Hall, University of Rhode Island, Kingston, Rhode Island 02881, USA

GROOTJANS, A.P.

Department of Plant Biology, University of Groningen, P.O. Box 14, 9750 AA Haren, The Netherlands

HEREDIA, A.

Sciences School. Baja California University, Km 106 Carr. Tijuana Ensenada. 22800 Ensenada, B.C., México

HESLENFELD, P.

EUCC – The Coastal Union, P.O. Box 11232,
NL-2301 EE Leiden, The Netherlands

HESP, P.A.

Geography and Anthropology, Louisiana State University, 227 Howe/Russell, Geoscience Complex, Baton Rouge, Louisiana 70803, USA

HOUSTON, J.A.

Sefton Coast Life Project. Formby Council Offices. Freshfield Road,
Formby L37 3PG, UK
Current address: Ecosystems Ltd., 4 Three Tuns Lane, Formby,
Merseyside, L37 4AJ, UK

JUNGERIUS, P.D.

University of Amsterdam, c/o Oude Bennekomseweg 31,
NL-6717 LN Ede, The Netherlands

KLIJN, J.A.

Alterra, P.O. Box 125, 6700 AC Wageningen, The Netherlands

KNOWLES, T.

Conservation Ecology Research Unit, Department of Zoology and Entomology, University of Pretoria, Pretoria 0002, South Africa

Kooijman, A.M.

Institute for Biodiversity and Ecosystem Dynamics – Physical Geography, University of Amsterdam, Nieuwe Achtergracht 166, 1018 WV, Amsterdam, The Netherlands

Koske, R.E.

Department of Biological Sciences, Ranger Hall, University of Rhode Island, Kingston, Rhode Island 02881, USA

Lammerts, E.J.

State Forestry Service, P.O. Box 1726, 8901 CA, Leeuwarden, The Netherlands

Lubke, R.A.

Botany Department, Rhodes University, P.O. Box 94, Grahamstown, 6140, South Africa

Martínez M.L.

Departamento de Ecología Vegetal, Instituto de Ecología, A.C., km 2.5 Antigua Carretera a Coatepec 351, Xalapa, Veracruz 91070, México

Maun, M.A.

Department of Plant Sciences, University of Western Ontario, London, Ontario N6A 5B7, Canada

Meulen, F. van der

Coastal Zone Management Centre, National Institute for Coastal and Marine Management, Ministry of Transport, Public Works and Water Management, P.O. Box 20907, 2500 EX, The Hague, The Netherlands

Moreno-Casasola, P.

Instituto de Ecología, A.C., Departamento de Ecología Vegetal, km 2.5 Antigua Carretera a Coatepec, Xalapa, Veracruz 91070, México

Niemand, L.

Conservation Ecology Research Unit, Department of Zoology and Entomology, University of Pretoria, Pretoria 0002, South Africa

Oliveira, P.S.

Departamento de Zoología, Universidade Estadual de Campinas, C.P. 6109, Campinas SP, 13083-970, Brazil

Pammenter, N.W.

School of Life and Environmental Sciences, George Campbell Building, University of Natal, Durban, 4041 South Africa

Parra-Tabla, V.

Departamento de Ecología, F.M.V.Z., Universidad Autónoma de Yucatán, Apdo. 4–116, Mérida (Itzimná), 97000, México

Pickart, A.J.

Humboldt Bay National Wildlife Refuge, 6800 Lanphere Road, Arcata, California 95521, USA

Psuty, N.P.

Institute of Marine and Coastal Sciences, Rutgers University, New Brunswick, New Jersey 08901, USA

Rico-Gray, V.

Departamento de Ecología Vegetal, Instituto de Ecología, A.C., Apdo. 63, Xalapa, Veracruz 91070, México

Rincón, E.

Instituto de Ecología, Universidad Nacional Autónoma de México, Apartado Postal 70–275, D.F: 04510, México

Ripley, B.S.

Botany Department, Rhodes University, P.O. Box 94, Grahamstown, 6140, South Africa

Sigüenza, C.

Department of Botany and Plant Sciences, University of California, Riverside, California 92521, USA

Vázquez, G.

Departamento de Ecología Vegetal, Instituto de Ecología, A.C., km 2.5 Antigua Carretera a Coatepec #301, Xalapa, Ver. 91070, México,

Wassenaar, T.D.

Conservation Ecology Research Unit, Department of Zoology and Entomology, University of Pretoria, Pretoria 0002, South Africa

Wiedemann, A.M.

The Evergreen State College, Olympia, Washington 98505, USA

Zunzunegui, M.

Department of Plant Biology and Ecology, The University of Seville, Ap. 1095, 41080 Sevilla, Spain

I What Are Sand Dunes?

1 A Perspective on Coastal Dunes

M.L. Martínez, N.P. Psuty, and R.A. Lubke

1.1 Coastal Dunes and Their Occurrence

Coastal dunes are eolian landforms that develop in coastal situations where an ample supply of loose, sand-sized sediment is available to be transported inland by the ambient winds. They are part of unique ecosystems which are at the spatial transition between continental/terrestrial and marine/aqueous environments. Coastal dunes are part of the sand-sharing system composed of the highly mobile beach and the more stable dune. A large variety of coastal dune forms are found inland of and above the storm-water level of sandy beaches and occur on ocean, lake, and estuary shorelines. They are distributed worldwide in association with sandy beaches, producing a wide range of coastal dune forms and dimensions related to spatial and temporal variations in sediment input and wind regime (Gimingham et al. 1989; Nordstrom et al. 1990; Carter et al. 1992; Pye 1993; Hesp 2000). They tend to exist wherever barrier islands or wave-dominated depositional coastal landforms occur (Fig. 1.1).

1.2 Relevance of Coastal Dunes

Because they are found almost in all latitudes, the climate and biomes developing on coastal dunes are very diverse, covering ecological habitats which range from polar to tropical latitudes, and from deserts to tropical rain forests (Snead 1972; van der Maarel 1993a, b; Kelletat 1995) Thus, one of the most outstanding features in these ecosystems is their broad distribution and ecological diversity (in terms of geomorphological dimensions, environmental heterogeneity, and species variability). Yet, despite their seeming abundance on the global level, many coastal dune ecosystems have been severely degraded as a result of an excessive exploitation of natural resources, chaotic demographic expansion, and industrial growth. For a long time, coastal dunes

Ecological Studies, Vol. 171
M.L. Martínez, N.P. Psuty (Eds.)
Coastal Dunes, Ecology and Conservation

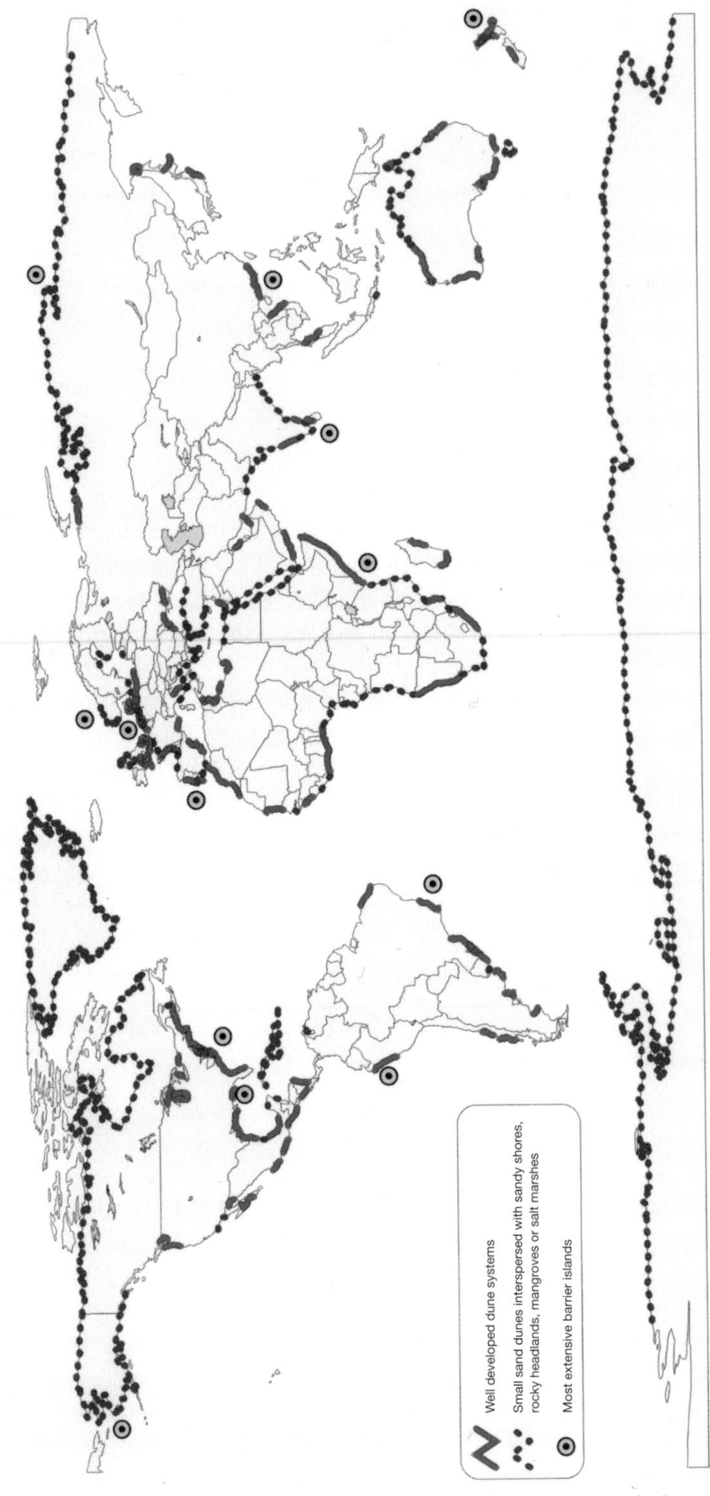

Fig. 1.1. Widespread distribution of coastal dunes is coincident with the worldwide occurrence of wave-dominated sandy beaches and with coastal barrier systems. Further, coastal dunes exist more extensively where areas of persistent coastal sand supply and dominant onshore winds favor episodes of inland transport to create broad dune fields. Active as well as stabilized coastal dunes of various scales and morphologies are recognized as products of coastal dynamics. (The map was generated from descriptions in van der Maarel 1993a, b)

have been used for many different purposes: coastal defense, water catchments, agriculture, mining, housing and tourism (Carter 1991). In addition, as part of the coastal landscape, dune areas serve as locations of groundwater recharge and assist in the retention of freshwater as a buffer against saltwater intrusion. Furthermore, specialized vegetable and fruit crops are grown in interdune depressions (van der Maarel 1993b). All these activities result in economic benefits to the human populations.

Besides their economic relevance, coastal dunes have intrinsic value related to their spatial and temporal dynamics at the sharp boundary between land and sea. For instance, they have been sites of important ecological research projects since the 19th century. Early studies conducted in these environments generated some of the first ecological theories that help to understand how ecological systems function. For example, during the early days of ecology as a science, Henry Chandler Cowles (1899) studied the spatial and temporal associations of the vegetation of Lake Michigan sand dunes. Cowles assumed that vegetational changes in space paralleled vegetation changes in time and, based on this assumption, his studies yielded the first evidence of succession. Furthermore, he was the first to develop a dynamic perspective of the interaction between vegetation and geological formations. The studies by Cowles were seminal for Frederic Clements who, almost 20 years later (1928), further developed the successional theory of plant communities.

Coastal dunes also represent an important cultural value. For example, in New Zealand the earliest human settlements occurred in coastal dunes (Hesp 2000). Thus, many of their dune areas contain archeological evidence of their cultural Maori heritage. In Peru, the early hydraulic civilizations extended down valley into the migrating coastal dune fields (Parsons 1968). In The Netherlands, the Dutch dunes have been portrayed by many painters, and the dunes also figure in some patriotic Dutch folk songs.

1.3 Current Conservation Status

The influence of humans on the coastal environment is large and has occurred for a long time. In particular, the widespread marketing of coastal recreation has increased drastically in the last 50 to 80 years, which has led to the deterioration of many previously scenic coasts and well-preserved coastal ecosystems. Currently, a large proportion of the worldwide human population lives within 10 km of the coastline. An example of the high degree of modification to coastal dunes occurs in New Zealand, where more than 115,000 ha of drifting dunes have been converted to forestry and agricultural activities during the last 80 years (Hesp 2000). In The Netherlands, large parts of the inner old dunes have been excavated, and the sand was used for the expansion of towns and cities into the low areas inland of the dunes (Carter 1991). In the USA, 70 % of the population visit beaches when they go on vacation. In Australia, 83 % of the population lives near the coast, 25 % within 3 km, resulting in increasing pressure on the coast (Hesp 2000). These proportions probably hold true for many countries in the world. As a result, many coastal dune systems of the world are in advanced stages of degradation and in many cases native and endemic species have been eliminated and replaced by introduced exotics (Grootjans et al. 1997). Other coastal dunes have been completely removed in the process of providing living space for the encroaching human population. A consequence of this reduction and removal of coastal dune topography is that the potential for storm surge damage has increased noticeably in the coastal zone. Further, the rate of change associated with construction in the coastal zone (and the loss of irreplaceable ecosystems) is occurring several times faster (two to three times as fast in the US) than the changes occurring inland. Additional impacts on coastal dunes associated with human activities are water extraction, trampling, invasive species, grass encroachment, sea-level rise, and climate change. The result of this worldwide intensive and consumptive use of coastal dunes is that many dune systems are already irreversibly altered and lost. Fortunately, there remain impressive stretches of the world's coast that still preserve pristine or minimally-disturbed ecosystems, incorporating a wide variety of coastal dune settings and ecosystems. Because it is likely that human development and activities on the coast will continue, these minimally-disturbed areas are in urgent need for appropriate management and conservation policies to ensure that they will retain their characteristics and will be available to future generations.

1.4 Aims and Scope of the Book

The continuation of scientific investigations into the understanding of dune processes and the functioning of this portion of the coastal ecosystem will contribute important data necessary for the enlightened stewardship of these dynamic and naturally evolving coastal morphologies, and it will promote their conservation, protection, enhancement and wise utilization as appropriate. Because of their relative economic importance, sand dunes have been studied for a long time. The oldest known study on the vegetation of coastal dunes was performed in 1835 by Steinheil (van der Maarel 1993a). There are also a number of more recent books which either focus on coastal dunes (van der Meulen et al. 1991; Carter et al. 1992; van der Maarel 1993a, b; García-Novo et al. 1997; Grootjans et al. 1997; Packham and Willis 1997; Wiedemann et al. 1999; Hesp 2000) or mention them briefly (Seeliger 1992). However, an examination of the literature indicates that there is a strong emphasis on the mid-latitude dunal systems and a lack of attention given to low latitude situations, those areas where much of the modern exploitation and coastal touristic development is occurring.

Until now, there have been no compilations of coastal dune system studies in which the geomorphology, community dynamics, ecophysiology, biotic interactions, environmental problems, and conservation were addressed, especially incorporating both tropical and temperate latitudes. The modest communication and collaboration among scientists studying coastal dunes in tropical and temperate latitudes are factors that generate scientific isolation and limit the potential for comparing data, performing interdisciplinary studies, and coordinating joint research programs. This book aims at narrowing the gap. The goal is to gather information on the state-of-the-art studies on coastal dunes, covering a range of topics from dune geomorphology and community dynamics to ecophysiology and the environmental problems and management strategies and policies that are necessary for their conservation. The basic idea is to bring this information to the attention of an international forum interested in coastal dunes. This volume does not pose the final answer. More likely, the diverse array of contributions contained herein will stimulate further research that will lead to a better understanding of these ecosystems and to the generation of improved conservation and management strategies.

This book is directed mainly to graduate students who are interested in biological and environmental sciences. It may be part of a reading list for undergraduates, but the discussion and insightful analysis will occupy the graduates rather than undergrads. The book will also be useful to those with an interest in conservation biology and coastal management that seek information on various topics, ranging from coastal sand dune distribution in the world, to plants and animals, biotic interactions, environmental problems, and different management tools. Protection and wise management of coastal sand dune systems can be achieved only if the ecosystem dynamics are better understood.

We have invited experts from throughout the world to contribute to the different sections of the book. Their enthusiastic response reveals the uniqueness of a book where researchers focused on temperate and tropical dunes are gathered for the first time.

The book is divided into six Sections: I. A general description of coastal dunes; II. The flora and fauna; III. Living in a stressful environment; IV. Biotic interactions; V. Environmental problems and conservation, and finally, VI. The coastal dune paradox: conservation vs. exploitation?

Section I begins by defining and describing in detail coastal dunes, which are always changing in shape and location because of the dynamics of the coastal system. Their complex and changing topography generates a high environmental heterogeneity. Norb Psuty describes the geomorphology of coastal dunes and illustrates the fundamental control of sediment supply in the system. Patrick Hesp, in turn, focuses on the distribution of coastal dunes in the mid-latitudes and tropics and revisits if there really is a paucity of dunes in the tropics.

The high environmental heterogeneity of the dune environment provides specialized habitats for many organisms. There are many adapted plant and animal species (native or even endemic) that occur in these environments. Many of these species are currently rare or endangered. Dune vegetation is very diverse: lichens and bryophytes abound in Europe and the polar regions, whereas plants with seeds (especially palms) are more diverse in the tropics. Arthropods, mollusks, amphibians, reptiles, birds and mammals are the animal groups best represented. In each community, recognizable sets of species appear and disappear along typical marine gradients, making zonation a common feature.

Section II addresses dune vegetation as a characteristic and a formational agent. Al Wiedemann and Andrea Pickart integrate the information from temperate latitudes, and present a case study from North America; Roy Lubke focuses on the vegetation dynamics in the tropical dunes and the role of invasive vs. non-invasive species, and Ab Grootjans and colleagues analyze the vegetation dynamics and processes of the most humid places within the dunes: the slacks. The information on dune vegetation has proven to be highly relevant for conservation programming. In contrast with the relatively well studied dune vegetation, the paucity of data on dune animals is evidence that the fauna of dune environments have been less well studied than the vegetation. In spite of this limitation, innovative data are presented by Rudy van Aarde and collaborators, who relate habitat rehabilitation to the conservation of vertebrates inhabiting these ecosystems.

Flora and fauna found in coastal dunes are greatly affected by substrate mobility, extremely high temperatures, drought, flooding, salinity, and a scarcity of nutrients. They show morphological, physiological, and behavioral responses to these limiting conditions. The different mechanisms required to survive in such stressful environments are addressed in Section III, which, because of the lack of information for animals, focuses only on plants. Anwar Maun contributes a thorough review of the different responses of plants to burial, whereas Brad Ripley and Norm Pammenter pursue the understanding of plant responses to the restricted budgets of water and nutrients, especially for dune pioneer species. Francisco García-Novo and his colleagues bring together a set of plant responses to the dune environment by using plant functional types in relation to environmental constraints.

In addition to the important role of the abiotic environment, interspecific interactions between the organisms that live on these ecosystems play a key role in community dynamics. Biotic interactions, explored in Section IV, cover a wide variety of plant and animal relations, ranging from arbuscular mycorrhizae (Rick Koske et al.), to algae and phanerogams (Gabriela Vázquez), plant–plant interactions (M. Luisa Martínez and José García-Franco) and ant-animal interactions (Víctor Rico-Gray et al.).

The need of management and conservation policies for coastal dune and barrier island ecosystems becomes evident from the studies above and has led to the implementation of different strategies and technologies throughout the world. Environmental problems and their potential solutions are diverse, and they are the purview of Section V. Annemieke Kooijman discusses the effectiveness of nature management options in the dune environment, especially when grass encroachment depletes biodiversity. Frank van der Meulen and co- authors, in addition, suggest a more flexible approach for coastal conservation, allowing geomorphological processes to occur, and they evaluate the costs and benefits of this decision. Gert Baeyens and M. Luisa Martínez, in turn, cover the issue of exploitation and protection of introduced animal life on sand dunes. Ileana Espejel et al. propose the usage of different indicators (environmental, functional, and structural) as feasible decision-making tools. Patricia Moreno-Casasola analyzes the current conservation status of coastal dunes in the tropics and presents a case study of coastal zone management in which local inhabitants are actively involved.

Finally, Section VI presents a discussion of the conflict common to all natural systems: the balancing of conservation with exploitation. The spatial continuum of the dune environment establishes the basis of the energetics framework of coastal ecosystems and is responsible, to varying degrees, for primary production, habitat formation, and shoreline evolution. Prior to human interference, coastal dune habitats were evolving and they will continue to evolve even with human husbandry. However, what is important in coastal conservation is to allow the natural systems to evolve, to change, and to do so in a natural pace. When humans interfere, the natural dynamics of the system are altered. Any change in the system may lead to an upending of the balance and to an exacerbation of the rate of change of the coastal dune system as a whole as well as in its components.

Therefore, the challenge is to manage the parts as well as the entirety of the system better. Each community is vital to the ecological integrity and functioning of the coastal dune complex. An improved understanding of this interaction is vital because modern coastal area management and planning strategies have increasingly adopted the systems approach as a basis for habitat preservation and enhancement. Based on the above, Heslenfeld and coauthors present general principles regarding evaluation criteria on an international level. Lastly, the final chapter is a discussion on the theme of dune conservation as a viable option given the current status and trend of the coastal environment.

The vast information generated in this book covers a wide variety of themes. Our aim and scope are to bring forth concepts on the physical basis for coastal dune development and the trend of the spatial and temporal evolution of the foredune and the inland dune continuums. We also intend to show that only through an integrated approach of the physical system in which geomorphology and inventories of the flora and fauna are complemented by information on the ecophysiological responses and the role of biotic interactions can we strive to attain an integrated coastal zone management. Certainly, a flexible and multidisciplinary approach is fundamental for coastal conservation, and the costs and benefits of conservation should be considered. Further, coastal management must integrate the needs and the participation of all of the local and regional stakeholders in the development of appropriate strategies and polices.

References

Carter RWG (1991) Coastal environments. Academic, London

Carter RWG, Curtis TGF, Sheehy-Skeffington MJ (eds) (1992) Coastal dunes. Geomorphology, ecology and management for conservation. Proc 3rd Eur Dune Congr. Galway. Ireland 17–21 June 1992. Balkema, Rotterdam

Clements FE (1936) Nature and structure of the climax. J Ecol 24:252–284

Cowles HC (1899) The ecological relations of the vegetation on the sand dunes of Lake Michigan. Bot Gaz 27:95–117

García-Novo F, Crawford RMM, Díaz-Barradas MC (eds) (1997) The ecology and conservation of European dunes. Univ de Sevilla

Gimingham CH, Ritchie W, Willetts BB, Willis AJ (eds) (1989) Coastal sand dunes. Proc R Soc Edinb B96

Grootjans AP, Jones P, van der Meulen F, Paskoff R (eds) (1997) Ecology and restoration perspectives of soft coastal ecosystems. J Coastal Conserv Special Feature 3:1–102

Hesp PA (2000) Coastal sand dunes. Form and function. Massey University. Rotorua Printers, New Zealand

Kelletat D (1995) Atlas of coastal geomorphology and zonality. J Coastal Res Spec Issue 13

Nordstrom K, Psuty N, Carter B (1990) Coastal dunes. Form and process. Wiley, Chichester

Packham JR, Willis AJ (1997) Ecology of dunes, salt marsh and shingles. Chapman & Hall, Cambridge

Parsons JR (1968) The archeological significance of mahamaes cultivation on the coast of Peru. Am Antiquity 33:80–85

Pye K (ed) (1993) The dynamics and environmental context of aeolian sedimentary systems. Geol Soc Spec Publ No 72

Seeliger U (1992) Coastal plant communities of Latin America. Academic Press, New York

van der Maarel E (1993a) Dry coastal ecosystems: polar regions and Europe. Elsevier, Amsterdam

van der Maarel E (1993b) Dry coastal ecosystems: Africa, America, Asia and Oceania. Elsevier, Amsterdam

van der Meulen F, Witter JV, Ritchie W (eds) (1991) Impact of climatic change on coastal dune landscapes of Europe, Special edn. Landscape Ecology, vol 6no 1/2). SPB Academic Publishing, The Hague, pp 5–113

Snead R E (1972) Atlas of world physical features. Wiley, New York

Wiedemann AM, Dennis R, Smith F (1999) Plants of the Oregon coastal dunes. Oregon State Univ, Eugene

2 The Coastal Foredune: A Morphological Basis for Regional Coastal Dune Development

N.P. Psuty

2.1 Conceptual Setting

Coastal dunes are ubiquitous elements of the dune-beach system that exist along the shores of many water bodies in the world where waves and currents interact with available sediment and local vegetation to create combinations of form and habitat at the water-land interface. They occur in a variety of dimensions from minor hummocks of 0.5 m to huge ridges measuring more than 100 m in elevation; from a single, shore-parallel, linear ridge with a width of a few tens of meters to a complex of dune forms that extend inland tens of kilometers. From a geomorphological perspective, the commonality associated with this myriad of forms and situations is the amassing of sand to create a depositional landform proximal to the shoreline.

The fundamental concept in coastal geomorphology is that processes of wind, waves, and currents act upon the sediments to produce a set of landforms that are causally related. This relationship is described as a process-response model and it is the conceptual foundation for all geomorphological inquiry. In coastal areas of adequate sediment supply, the coastal process-response model is the beach profile: the accumulation of sand that extends from the offshore bar, through the dry beach, and into the adjacent coastal foredune where vegetation stabilization is a further active element of the morphological process (Fig. 2.1). This dune-beach profile is the basic sand-sharing system whose components respond to variations in energy level and to mobilization of sand from one portion to another. Each component episodically stores and releases sand in an exchange of sediment, a classic closed system.

The coastal foredune is the uppermost and inlandmost component of the sand-sharing system. It has accumulated sand in association with a range of pioneer vegetation types to create a positive landform perched above the dry sand beach. It is the most conservative portion of the profile, undergoing

Ecological Studies, Vol. 171
M.L. Martínez, N.P. Psuty (Eds.)
Coastal Dunes, Ecology and Conservation

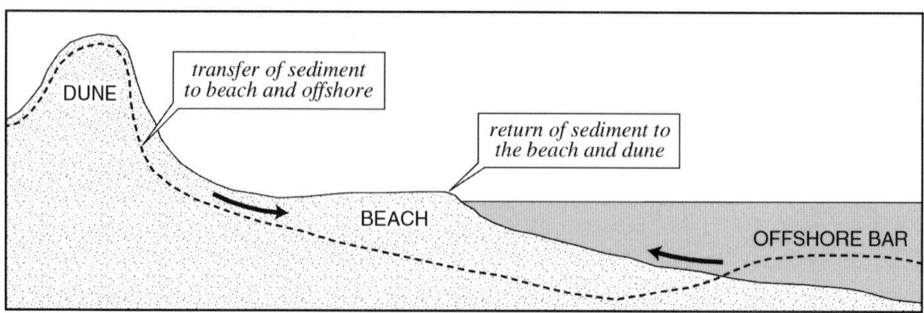

Fig. 2.1. Exchange of sand within the sand-sharing system

dimensional and temporal changes of far less magnitude and frequency than the sand beach or the offshore zone. In this simple profile, the coastal foredune exists at the boundary between the coastal processes to its seaward and continental processes landward. However, many coastal zones are not so simple as this profile and there are many instances of variable dune configurations and areas immediately inland of the dune-beach profile that appear to be morphodynamically related to the processes active in the sand-sharing system. Within this complexity lie some of the basic issues regarding the geomorphology of coastal dunes. Such as, are there constraints to the configuration of the foredune relative to some aspect of the delivery system? Is there a developmental sequence and what are the variables that affect the morphological characteristics of the foredune? How is the dunal topography inland of the dune-beach profile integrated into the regional dynamics? Is all coastal dunal topography site specific?

2.2 The Dichotomies of Inquiry

The inquiry into the formational processes and the configuration of coastal dunes has produced a spate of conferences and papers (Gimingham et al. 1989; Nordstrom et al. 1990; Carter et al. 1992; Pye 1993; Favennec 1997). A review of these contributions reveals a number of geomorphological dichotomies in coastal dunes that cannot be completely bridged. Sherman (1995) describes a division in coastal dune studies that is largely the product of different scales of study. He identifies research on sediment transport processes, on the dune form itself, and of regional dune systems. The most elemental inquiry focuses on the instantaneous or event-driven transfers of sand in the beach profile that are narrowly confined in both space and time (microscale). The early research products of Bagnold (1941) and Belly (1964) laid a foundation for the theory and concepts of eolian transport in ideal situations. However, the beach is not an ideal planar surface and these early eolian transport equations do not provide satisfactory solutions to the quantities of sand moved about in the beach/dune environment. More recent research has focused on non-linear beach characteristics such as variable slope, topography, and vegetation type and density that affect sand transport in this milieu (Hotta 1985; Arens 1994; Dijk et al. 1999). Namikas (2002) has described the nature of non steady-state flows that spatially restrict transport across beach zones. Reviews of the influences of these "coastal variables" on rates and mechanics of winds transport are represented in McEwan and Willetts (1993), Trenhaile (1997), Sherman et al. (1998), and Bauer and Sherman (1999).

Mesoscale inquiry proceeds from the very spatially and temporally limited sediment transport studies to episodic accumulation of sand and the creation of foredune morphology. This is also a conceptual and methodological jump because the emphasis shifts to the response end of the geomorphological model. As noted by Sherman (1995) and Livingstone and Warren (1996), there is a temporal discontinuity between short-term sediment transport and the multi-year time span of morphological foredune development. Inquiry into the latter topic tends to focus on net dimensional topographical changes. Questions about rates and vectors of change relate the quantity and distribution of sand accumulation in the foredune in distinct temporal spans to the effectiveness of sediment delivery into various portions of the foredune cross section (Gares 1992; Arens 1997), or to the effectiveness of vegetation to trap and collect sand (Chap. 4: Wiedemann and Pickart; Chap. 5: Lubke; Chap. 8: Maun). Others track the mobility of the foredune as a morpho-/sedimentological unit in response to major storm events (Ritchie and Penland 1988) or to multi-year displacements of foredune crestlines both alongshore and cross-shore (Psuty and Allen 1993). Still others try to tighten the process-response association by relating vectors of transport to changes in foredune configuration (Hesp and Hyde 1996). Kurz (1942) and Hesp (1989) associate vegetation types with the development of foredune configuration from scattered hummocks to a coherent ridge, whereas Bate and Ferguson (1996), Garcia Novo et al (Chap. 10), and Martinez and Garcia-Franco (Chap. 13) relate stages in inland transport to interaction with vegetation types. This mesoscale inquiry at the multi-year to decadal level emphasizes development of the foredune and subsequent transfers of sediment inland of the dune-beach profile.

Macroscale inquiry involves the investigations into major dune complexes that have been evolving over periods of centuries to millennia (Sherman 1995). The treatises by Cooper (1958, 1967) as well as the papers by Olsen (1958) and Inman et al. (1966) describe dune systems that emanate from a coastal location but become regional geomorphologies. These sites of massive dune systems record a complex heritage of sediment availability and dune phenomena that extend tens of kilometers inland, often incorporating

Fig. 2.2. Sand dunes and sand sheets transgressing inland in coastal Peru. Inland transfer of migrating dune forms, episodically stabilized, associated with sites of river discharge

Holocene, Pleistocene, and older formations (Tinley 1985; Illenberger and Rust 1988).

Livingstone and Warren (1996) indicate a dilemma in classifying coastal dunes because some of the descriptive morphologies could easily be far inland and totally independent of the coast. Some dunes may be located near the coast because of the proximity to a source of sand (Fig. 2.2), whereas others are morphodynamically related to the ambient coastal processes (Fig. 2.3). Dune forms that are in active interchange of sediment with the beach are causally positioned to occupy the upper portion of the beach profile. Other dune forms that derive their sand from a beach source but are inland of the beach profile and essentially function independent of the beach exchange processes may be coastal dunes because of geography but not a direct product of coastal dynamics. Further, dunes located inland of the beach profile may have been stranded because of coastal progradation and thus may retain all of their characteristic morphologies, but are no longer in active exchange of sediment with the beach.

Some classifications of coastal dunes attempt to address the spatial dichotomy by describing the dune form in a sequential relationship (Pye 1983; Goldsmith 1985) or stages in stability and inland transport (Short and Hesp 1982; Short 1988). A simple distinction between the foredune (primary dune) which is in active exchange with the beach and those dune forms which are

Fig. 2.3. Coastal foredune accumulating and releasing sand in exchange with the beach. Some sand is building the seaward face of the foredune whereas transport inland is broadening the foredune crest and the leeward slopes, Island Beach State Park, New Jersey

located inland of the beach exchange (secondary dune) is a foundation for creating a morphodynamic classification (Table 2.1), which can be further segmented by specific forms and dimensions, to distinguish between coastal dunes and dunes at the coast (Davies 1980; Psuty 1989; Paskoff 1997). This classification incorporates dune morphologies developing in the foredune location as well as inland dunes separated from beach interaction, such as actively migrating parabolic dunes or other transgressive morphologies. The former is the primary dune whereas the latter are secondary dunes or features evolving from the initial establishment of primary dunes.

2.3 Dune Morphology Related to Sediment Supply and Dune-Beach Exchange

Whereas there are many situations that combine to describe site-specific coastal dune configurations, on a conceptual level, there is a foredune developmental sequence related to varying sediment availability. An appreciation that the coastal dune comprises a part of the dune-beach profile is important because it brings recognition of the holistic system within which factors of time and space affect the character of the beach and accompanying dune

Table 2.1. Coastal dune classification based on relationship to foredune sand exchange, morphology, and sequence in development, modified from Psuty (1989)

Primary dune	The foredune in the dune-beach profile. There is active exchange of sediment between these components of the profile. This is an area that has been accumulating sand and it represents a net positive sediment budget at this site. However, at various temporal scales, it could be gaining, losing, or have no net change in sediment budget. The foredune is usually a coherent, linear ridge. It may be transgressing inland as part of a shift of the entire dune-beach profile, or it could be stable in its geographical location, or shifting seaward. The foredune is dynamic and it is the only dune form that is totally dependent on a coastal location
Secondary dune	Active – Created by modification of the primary dune or by transfers of sand inland from the position of the primary dune. The general characteristic is active migration of sand represented by deflation hollows and parabolic or crescentic morphologies. The transgressive ridge form is increasingly crenulate as deflation processes modify the ridge. The secondary dune may apply to the dissected remnants of a primary dune or it may describe actively transgressing dunes inland of the primary dune. In either case, it represents a condition in which sand is being transferred inland and lost to the dune-beach sand-sharing system Stable – Dune forms that are no longer in the active foredune but are not transgressing either. These dunes may have been stranded because of coastal progradation or they may have become stabilized by vegetation during their passage inland. They may have the configuration of the linear foredune ridge if abandoned by accretion or may have any of the transgressive forms associated with previous mobility. This dune form is a paleo-feature that retains the morphology of the dunes but is not being maintained by dune formational processes
Sand sheet/washover	Areas of very active inland transfers of sand. There is no foredune form or function. These large bare sand areas may be at sites of high rates of erosion and constitute a continuous transfer of sand to inland positions, as in the case of washover fans. They could also be at locations where strong onshore winds propel sand inland to overwhelm pre-existing topography without any definitive dunal forms. Sand sheets and washovers represent a unidirectional movement of sand from its beach origin

form (Psuty 1988; Sherman and Bauer 1993; Hesp 1999). An early study by Carter (1977) drew attention to the exchange of sediment between these two components of the sand-sharing system, and a collection of papers on the concept of dune-beach interaction (Psuty 1988) repeatedly stressed that foredunes are intricately related to beach dynamics and to the sediment available to drive beach changes. In part, the emphasis on dune-beach interaction added wave processes to the dynamics of foredune development. Essentially, those processes that mold the dimensions of the beach also affect foredune dimensions directly by scarping and sediment removal, or they provide sources of sediment to be transferred into the foredune topography. Vellinga (1982) and Kriebel (1986) describe storm events that mobilize some of the foredune mass, promote sand exchange, and produce a morphologic response. Psuty (1989) and Davidson-Arnott and Law (1996) applied the concepts of changing sediment budget in the beach to changing sediment budget in the adjoining foredune along several kilometers of beach and identified spatial and temporal associations of transfers into the foredune.

Beach–dune interaction is a key ingredient in the morphodynamic classification of beaches proposed by Short and Hesp (1982) and Short (1988) from their studies in Australia. In an original organization of hydrodynamic energies interplaying with sufficient sediment, this classification describes a continuum of morphological response to ambient processes that range from very dissipative conditions (mobilization of offshore sediment supply) to very reflective conditions (no sediment mobilization). The modal conditions that give rise to sand transfer to the beach also support inland storage of sand in primary and secondary dunes.

Psuty (1988), Arens (1994), and Hesp (1999) favor the concepts of a continuum of morphological responses to ambient conditions and identify sediment availability as the dominant variable that drives development of foredune characteristics. Part of the rationale is that the foredune is an accumulation form and there must be a positive sediment balance at some time in the creation of the foredune for it to exist. In addition, inherent in either of the approaches is the overriding concept of a spatial and temporal continuity that draws together the different combinations of foredune types in a sequential pattern. And, importantly, these approaches support a developmental sequence that passes through stages in a continuum with diagnostic morphodynamics to describe position and potential shift along the continuum. Hesp (1999) further incorporates vegetation into the mix of variables that affect stability and mobility relative to foredune dynamics and sediment supply to create a broadly-based foredune model. Subsequent inland transfers of sand interact with inland dunal habitats in a wealth of transgressive morphologies (Chapman 1964; Willis 1989).

2.4 Continuum Scenario

There are a number of scenarios that demonstrate conditions of sediment balance and foredune morphologies in the context of a continuum. The conditions of sediment discharge at a river mouth and the downdrift sequence of associations offers an excellent spatial continuum, as does an elongating spit or a barrier island that is being displaced alongshore. Sequences of secondary dune development that foster inland systems can be incorporated within the coastal scenarios as temporal extensions or episodes of transgression.

2.4.1 River Mouth Discharge

This sequence relates the concept of a point source of sediment input and the relationship of dune morphology to the quantity of sand available at distances away from that source (Fig. 2.4). The general scenario incorporates discharge with adequate sand to cause shoreline accretion and seaward displacement of the shoreline at the river mouth but with slower accretion at increasing distance from the mouth. Near the river mouth, the coastal topography would incorporate many low abandoned foredune ridges (Fig. 2.5). Foredunes at this site would be small because beach accretion restricts time to transfer sand from the active beach to the adjacent foredune. Thus, rapid beach accretion and seaward displacement of the total beach profile would lead to a new foredune location and strand the previously-developed foredune morphology.

At increasing distances from the river mouth, the quantity of sand delivery would decrease and the rate of beach accretion would slow, increasing the duration of sediment transport into the foredune site. This combination of tradeoffs in sediment delivery is the basis for the morphodynamic continuum that is fundamental to the pattern of foredune development. Conceptually, the number of foredune ridges should decrease away from the river mouth discharge site until there is only one large foredune ridge in active exchange of sediment between the beach and dune. The presence of a large single fore-

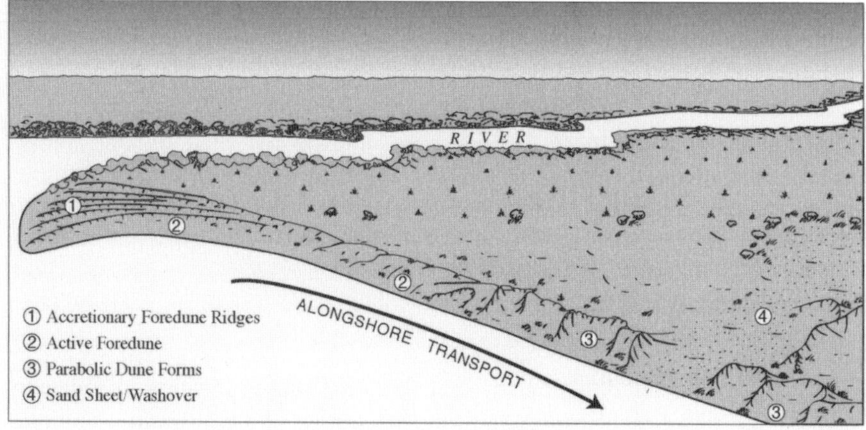

Fig. 2.4. Coastal foredune continuum related to a river mouth discharge site on a wave-dominated shoreline

Fig. 2.5. Curving lineations of abandoned foredune ridges, ca. 1.0–1.5 m relief, in an area of rapid shoreline accretion, Grijalva River, Tabasco, Mexico

dune ridge may exist beyond the point of shoreline stability. It may exist where there is slow shoreline erosion and the opportunity for transgression inland while the foredune is being maintained. Farther along the continuum, the negative beach sediment budget will be combined with a negative sediment budget in the foredune and the form will lose volume, become dissected, and incorporate morphologies that represent reduction of the dune-beach interaction such as blowouts and parabolic dune forms (Fig. 2.6). Still farther along the continuum, the dune forms will continue to diminish in dimension and extent. Eventually, the continuum proceeds to the situation where the foredune is essentially nonexistent and the profile migrates inland with washover fans or sand sheets penetrating into and across the interior morphologies.

The portion of the foredune continuum involving inland transgression of form and sediment can also lead to development of secondary dune landforms that are separated from coastal processes. This is an important element of the continuum because it is a spatial/temporal situation wherein inland transfer creates and maintains transgressive dune fields and sand sheets, thereby enabling the dichotomy of a primary coastal dune system that is dependent upon coastal processes and secondary dune systems that exist in the coastal area dependent upon mobilized sand.

Fig. 2.6. Downdrift of the Grijalva River mouth, the coastal foredune is 5–8 m in height, with blowouts and evidence of dissection. The foredune is actively transgressing inland over older coastal topography, near Tupilco, Tabasco, Mexico

2.4.2 Scenario Complexity

Wave-dominated shorelines at river mouths are excellent sites to demonstrate the application of the foredune continuum associated with the sequence of sediment supply and spatial arrangement of morphologies.

A site from the western coastal portion of the large Mezcalapa deltaic plain in the state of Tabasco, Mexico, records the interplay of variable coastal sediment supply, foredune development, and coastal dune habitats (Fig. 2.7). At times in its geomorphological development, the western Mezcalapa delta distributaries were sites of primary discharge and buildout. At other times, these distributaries were abandoned or were channels with very little discharge. The regional coastal geomorphology near the current Rio Tonalá river mouth provides evidence for a variety of forms and processes associated with discharge shifts. Initially, there is a general separation of geomorphological features that are produced by the fluvial processes of stream flow and channel development in the delta versus another group of features that are the product of wave, current, and wind processes at a shoreline. The contact of the two forms is largely shore parallel and consists of a transgressive dunal ridge of 10–15 m at the inner margin of these coastal forms that terminate or truncate the adjacent low-lying fluvial topography. Seaward of this dunal ridge is a topography of low abandoned foredune ridges, 1–2 m in local relief, whose

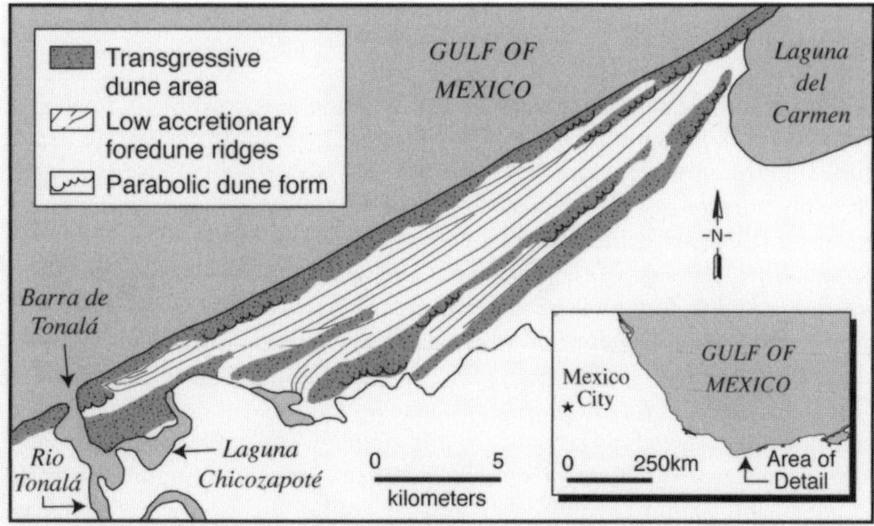

Fig. 2.7. Geomorphologic association of foredune types associated with changing sediment supply at the Rio Tonalá, Tabasco, Mexico

pattern displays a general arcing seaward in the vicinity of the discharge point of the Rio Tonalá. The arcing pattern is interrupted subsequently by a large foredune ridge, 5–10 m, that has transgressed inland and truncated the trends of the smaller foredunes. This large dunal ridge is quasi-shore parallel and is succeeded seaward by another series of low ridges that once again arc toward the Rio Tonalá mouth. This sequence is repeated several times to produce a variety of coastal dune habitats within short distances. The modern beach at this location is the site of an actively inland transgressing foredune.

The transgressive/regressive dune sequence and its ecological niches have been created because of the variation of sediment supply discharged through distributaries along the western margin of the delta interacting with hurricanes and strong frontal storms. The low ridges are abandoned foredunes associated with rapid coastal progradation and seaward displacement. The larger transgressive dune ridges are primary and secondary coastal dunes that have developed during times of shoreline erosion but with sufficient sand transferred inland to maintain a net positive sediment budget. This was the situation that created the transition at the junction of the coastal and fluvial topographies, and the alternation of sediment supply has been repeated several times to create the larger foredune ridges cutting through the low curving abandoned foredunes. Some of the inland ridges remained actively-transgressing inland with blowouts and parabolic dune forms despite being separated from the sand source at the shoreline, thereby becoming a complex of secondary dune forms.

2.5 General Model

Although the coastal foredune is part of the dune-beach profile in form and function, it is possible to separate it conceptually from the beach. Both the beach and the foredune are forms of sediment accumulation and they each develop morphological responses to their short term and long term sediment balance. The beach widens and narrows in response to changes in sediment supply. The dune gains and loses height and width relative to a gain or loss of sand. A stylized depiction of the relationship between the two parts of the profile relates the sequence of morphological expression in the foredune as driven by sediment supply (Fig. 2.8). The essential gradient in the model is from conditions of very high to very low sediment input and a differentiation of beach versus foredune response to this gradient. The output of the model is the sequence of foredune morphology, either spatial or temporal or both, along a continuum of evolutionary development of the foredune. The depiction is dimensionless. The association is relative sediment supply and it is subject to leakage from the sand-sharing system, so that inland transfers to sustain dune development landward of the foredune could be affecting sediment balance. Indeed, there is a position in the continuum that maximizes the potential for a transgressive foredune as well as inland transfer and secondary dune system development. Such a situation may accompany dissection of the foredune through development of parabolic forms that transgress inland while remaining part of the foredune (Fig. 2.8).

The essential element of the model is that with a positive sediment supply to the beach, the dimensions of the foredune are inversely related to the rate of beach accretion. High rates of shoreline progradation do not permit much time for sand transfers into the foredune position, and thus the foredune accumulation is never very great. As beach progradation slows, the opportunity to transfer sand to the foredune increases and thus there is an increase in the dimensions of the foredune related to the association of tradeoffs described above. The most problematic component of the model is the situation when the shoreline is stable or with a minor negative sediment budget. Evidence from the field (Psuty and Allen 1993) and numerous examples of well-developed foredunes on eroding coasts suggest that conditions do exist where the transfer of sand into and to the lee of foredune is similar to the losses on the seaward side of the foredune during erosion episodes. The example of large foredune forms at the shoreline near the mouth of the Rio Tonalá at present and at several instances in the past suggests that foredunes can persist during times of regional negative budget affecting beach position. The model uses this suite of information to depict maximum foredune growth during the period of minor regional shoreline erosion and transgression. Obviously, the foredune can also recover sediment volume and increase in dimension during episodes of beach recovery.

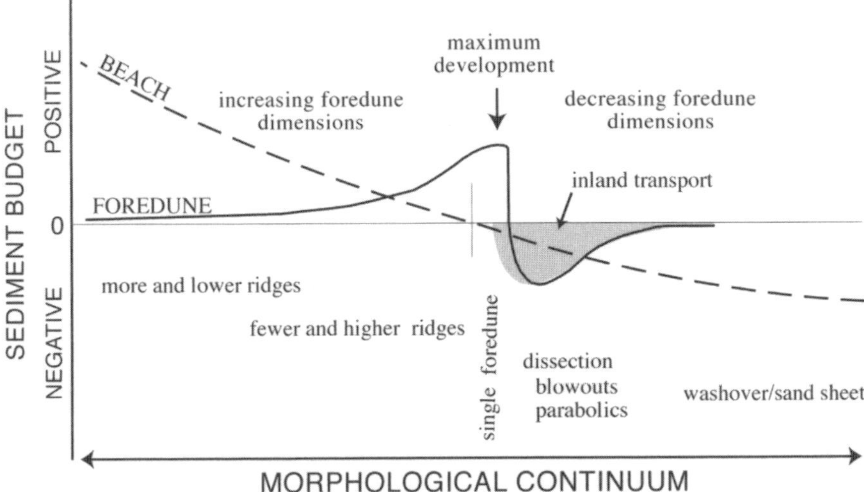

Fig. 2.8. Conceptual model of relationship of sediment budget of beach to sediment budget of foredune, inland transfer of sand, and resulting topographies in a sand-sharing system. Opportunities for maximizing foredune development (*dark line*) and maximizing inland sediment transport (*gray shading*) to support secondary parabolic dunes are closely-positioned and may overlap within the spatial/temporal continuum. With continued beach erosion, there will be no sand accumulation in the foredune to support inland transfer and thus washover and sand sheet processes will dominate the extreme negative sediment budget stage of the continuum

The transgressive portion of the continuum may also be the time of dune morphology decoupling from the interaction with coastal processes and to the establishment of parabolic or secondary dunes. However, at some point in the negative budget scenario, the foredune must be losing more sediment than is being transferred to it leading to dissection of the foredune ridge and reduction of its morphological identity.

Variation in sediment budget can drive considerable complexity in coastal foredune development, in ecological niches and sheltered areas, and in subsequent transfers of sediment and morphology inland. Indeed, secondary dune topography in the coastal zone may exist after the beach and foredune are completely eroded, such as cliff-top dunes and inland dune systems. Importantly, the foredune model establishes that there is a place in the continuum that favors inland transfer and is supportive of regional dune development. The conceptual basis for secondary coastal dune formation may assist in understanding the evolutionary establishment of coastal dune forms and habitats as an episodic inland transfer of sand that supports dune form and function away from its coastal origin. Further, a recognition of the dichotomies attendant to the scales of foredune research and concepts should prove helpful in interrelating morphology and vegetation types following the efforts of Godfrey et al. (1979), Hesp (1989, 1999), Sarre (1989) Wiedemann and Pickart (Chap. 4), and Maun (Chap. 8).

2.6 Humans as a Variable

Whereas there is natural variation in sediment availability and transfers in the coastal zone that result in a sequence of primary and secondary dune morphologies, there are likewise human manipulations of the processes, the sediments, and the landforms. It is difficult to escape the imprint of humans because the effects can range from direct manipulation to indirect influence of processes and responses (Davis 1956; Walker 1990; McLachlan and Burns 1992; Nordstrom 2000). Human impacts do not render the conceptual coastal dune continuum model invalid, but they do impart another variable because humans can alter the sediment budget, mold and destroy dune morphologies, and displace shorelines. In essence, human activities can overcome the slow migrations in morphological response along the continuum to produce large jumps from one point to another, or a stepped response to an alteration of form or sediment supply, as described when beach fill provides a superabundant source of sand for eolian transport into the foredune system (van der Wal 2000; Marques et al. 2001). Or conversely, human actions can support a steady state scenario and continuously balance the sediment delivery to allow a desired mode in the sequence to be maintained (Chap. 15: Kooijman; Chap. 16: van der Meulen et al.; Chap. 16: Moreno-Casasola; Chap. 20: Heslenfeld et al.). Human impacts are part of the system and it is folly to ignore the role of human agents in manipulating processes, sediment availability, as well as morphological/ecological responses.

2.7 Conclusions

Dune-beach interaction and sand-sharing are key elements in understanding development of the foredune under a variety of sediment budget scenarios. Conceptually, the foredune stores and releases sediment as it waxes and wanes in concert with the erosional or accretional trends of the adjacent beach. Spatially, the foredune may pass through a sequential morphological development associated with alongshore distance from a sediment source. Temporally, the same sequential development may occur as the sediment supply varies. At stages in the model, inland transfers of sediment from the foredune may support development of transgressive coastal dune forms that broaden the areal extent of the dunal features. Further, episodic oscillations of shoreline position may also strand foredune morphologies inland of the active shoreline. Human activities are additional factors affecting coastal dune characteristics in any stage of the process-response scenarios.

References

Arens SM (1994) Aeolian processes in the Dutch foredunes. PhD Diss, Univ of Amsterdam

Arens SM (1997) Transport rates and volume changes in a foredune on a Dutch Wadden island. J Coast Cons 3:49–56

Bagnold RA (1941) The physics of blown sands and desert dunes. Methuen, London

Bate G, Ferguson M (1996) Blowouts in coastal foredunes. Landscape Urban Plann 34:215–224

Bauer BO, Sherman DJ (1999) Coastal dune dynamics: problems and prospects. In: Goudie AS, Livingstone I, Stokes S (eds) Aeolian environments, sediments and landforms. Wiley, New York, pp 71–104

Belly PY (1964) Sand movement by wind. Tech Memo No 1, US Army Corps of Engrs, Coast Eng Res Ctr, Fort Belvoir, Virginia

Carter RWG (1977) The rate and pattern of sediment interchange between beach and dune. In Tanner WF (ed) Coastal sedimentology. Florida St Univ, Tallahassee, pp 3–34

Carter RWG, Curtis TFG, Sheehy-Skeffinton M (eds) (1992) Coastal dunes: geomorphology, ecology and management for conservation. Balkema, Rotterdam

Chapman VJ (1964) Coastal vegetation. Macmillan, New York

Cooper WS (1958) Coastal sand dunes of Oregon and Washington. Geol Soc Am Mem 72

Cooper WS (1967) Coastal sand dunes of California. Geol Soc Am Mem 101

Davidson-Arnott RGD, Law MN (1996) Measurement and prediction of long-term sediment supply to coastal foredunes. J Coast Res 12:654–663

Davies JL (1980) Geographical variation in coastal development, 2nd edn. Longmans, New York

Davis JH (1956) Influence of man upon coast lines. In: Thomas WH Jr (ed) Man's role in changing the face of the earth. University of Chicago Press, Chicago, pp 504–521

Dijk PM van, Arens SM, Boxel JH van (1999) Aeolian processes across transverse dunes II: modelling the sediment transport and profile development. Earth Surf Proc Land 24:319–333

Favennec J, Barrère P (eds) (1997) Biodiversité et protection dunaire, Lavoisier, Paris

Gares PA (1992) Topographic changes associated with coastal dune blowouts at Island Beach State Park, NJ. Earth Surf Proc Landforms 17:589–604

Gimingham CH, Ritchie W, Willetts BB, Willis AJ (eds) (1989) Coastal sand dunes. Proc R Soc Edinb B96

Godfrey PJ, Leatherman SP, Zaremba R (1979) A geobotanical approach to classification of barrier beach systems. In: Leatherman S (ed) Barrier Islands. Academic, New York, pp 99–126

Goldsmith V (1985) Coastal dunes. In Davis RA (ed) Coastal sedimentary environments. Springer, Berlin Heidelberg New York, pp 171–236

Hesp PA (1989) A review of biological and geomorphological processes involved in the initiation and development of incipient foredunes. Proc R Soc Edinb B96:181–201

Hesp PA (1999) The beach backshore and beyond. In: Short AD (ed) Handbook of beach and shoreface morphodynamics. Wiley,, New York, pp 145–169

Hesp PA, Hyde R (1996) Flow dynamics and geomorphology of a trough blowout. Sedimentology 43:505–525

Hotta S (1985) Wind blown sand on beaches. PhD Diss, Univ of Tokyo
Inman DL, Ewing GC, Corliss JB (1966) Coastal sand dunes of Guerrero Negro, Baja California, Mexico. Bull Geol Soc Am 77:787–802
Illenberger W, Rust I (1988) A sand budget for the Alexandria coastal dune field, South Africa. Sedimentology 35:513–521
Kriebel DL (1986) Verification study of a dune erosion model. Shore Beach 54:13–20
Kurz H (1942) Florida dunes and scrub, vegetation and geology. Geol Bull No 23. Florida Dept of Conservation, Tallahassee
Livingstone I, Warren A (1996) Aeolian geomorphology: an introduction. Addison-Wesley/Longman, London
Marques MA, Psuty NP, Rodriguez R (2001) Neglected effects of eolian dynamics on artificial beach nourishment: the case of Riells, Spain. J Coast Res 17:694–704
McEwan IK, Willetts BB (1993) Sand transport by wind: a review of the current conceptual model. In: Pye K (ed) The dynamics and environmental context of aeolian sedimentary systems. Geol Soc, London, pp 7–16
McLachlan A, Burns M (1992) Headland bypass dunes on the South African coast: 100 years of (mis)management. In: Carter RWG, Curtis TFG, Sheehy-Skeffinton M (eds) Coastal dunes: geomorphology, ecology and management for conservation. Balkema, Rotterdam, pp 71–79
Namikas S (2002) Field evaluation of two traps for high-resolution aeolian transport measurements. J Coast Res 18:136–148
Nordstrom KF (2000) Beaches and dunes of developed coasts. Cambridge University Press, Cambridge
Nordstrom KF, Psuty NP, Carter RWG (eds) (1990) Coastal dunes: processes and morphology. Wiley, Chichester
Olsen JS (1958) Lake Michigan dune development, vols II and III. J Geol 56:345–351, 56:413–483
Paskoff R (1997) Typologie géomorphologique des milieux dunaires européens. In: Favennec J, Barrère P (eds) Biodiversité et protection dunaire, Lavoisier, Paris, pp 198–219
Psuty NP (ed) (1988) Dune/beach interaction. J Coastal Res Special Issue No 3
Psuty NP (1989) An application of science to management problems in dunes along the Atlantic coast of the USA. Proc R Soc Edinb B96:289–307
Psuty NP, Allen JR (1993) Foredune migration and large scale nearshore processes. In: List JH (ed) Large scale coastal behavior '93. USGS open file report 93-381, pp 165–168
Pye K (1983) Coastal dunes. Prog Phys Geog 7:531–557
Pye K (ed) (1993) The dynamics and environmental context of aeolian sedimentary systems. Geol Soc Spec Pub No 72
Ritchie W, Penland S (1988) Rapid dune changes associated with overwash processes on the deltaic coast of south Louisiana. Mar Geol 81:97–112
Sarre R (1989) The morphological significance of vegetation and relief on coastal foredune processes. Z Geom (Suppl) 73:17–31
Sherman DJ (1995) Problems in the modeling and interpretation of coastal dunes. Mar Geol 124:339–349
Sherman DJ, Bauer BO (1993) Dynamics of beach-dune systems. Prog Phys Geog 17:413–447
Sherman DJ. Jackson DWT, Namikas SL, Wang J (1998) Wind-blown sand on beaches: an evaluation of models. Geomorphology 22:113–133
Short AD (1988) Holocene coastal dune formation in southern Australia: a case study. Sediment Geol 55:121–142

Short AD, Hesp PA (1982) Waves, dune and beach interactions in southeastern Australia. Mar Geol 48:259–284

Tinley KL (1985) Coastal dunes of South Africa. South Africa Nat Sci Prog, Rep No 109

Trenhaile AS (1997) Coastal dynamics and landforms. Oxford University Press, New York

Vellinga P (1982) Beach and dune erosion during storm surges. Coast Eng 6:361–387

van der Wal D (2000) Grain-size-selective aeolian sand transport on a nourished beach. J Coastal Res 16:896–908

Walker HJ (1990) The coastal zone. In: Turner B (ed) The earth as transformed by human action. Cambridge University Press, Cambridge, pp 271–294

Willis AJ (1989) Coastal sand dunes as biological systems. Proc R Soc Edinb B96:17–36

3 Coastal Dunes in the Tropics and Temperate Regions: Location, Formation, Morphology and Vegetation Processes

P.A. Hesp

3.1 Introduction

The following is an attempt to examine the differences between coastal dunes occurring in the tropics and those mid-latitude (mostly temperate) areas outside the tropics but principally within 50°N and 50°S. The tropics lie between 23.5°S and 23.5°N of the equator bounded by the Tropics of Cancer and Capricorn.

3.2 Climatic Conditions in the Tropics

Climatic conditions can act as major controls on whether aeolian dunes are able to form, the types of coastal dunes that form and dune-field stabilsation processes, as well as the prevailing vegetation biomes, vegetation growth rates and the structural types present. Given this, the following briefly outlines the climatic regions and the associated terrestrial biomes which occur in the tropics.

Within the latitudes of 23.5°N and S of the equator, and near the coast, there are two principal climatic regions (following the Köppen-Geiger classification system), namely, tropical climates and dry, arid/semiarid climates (Fig. 3.1). Mesothermal climates (particularly humid subtropical climates) are also present but occur to a much lesser extent (Christopherson 2000). Within the tropical climatic regions, tropical rain forest and tropical monsoon climate types principally occur on the east coasts of South America, Central America, on the west coast of Africa between around 4°S and 10°N of the equator, and much of SE Asia and in South Asia, the west coast of India and the coast of Bangladesh. These regions are dominated by the equatorial and tropical rain forest terres-

Ecological Studies, Vol. 171
M.L. Martínez, N.P. Psuty (Eds.)
Coastal Dunes, Ecology and Conservation

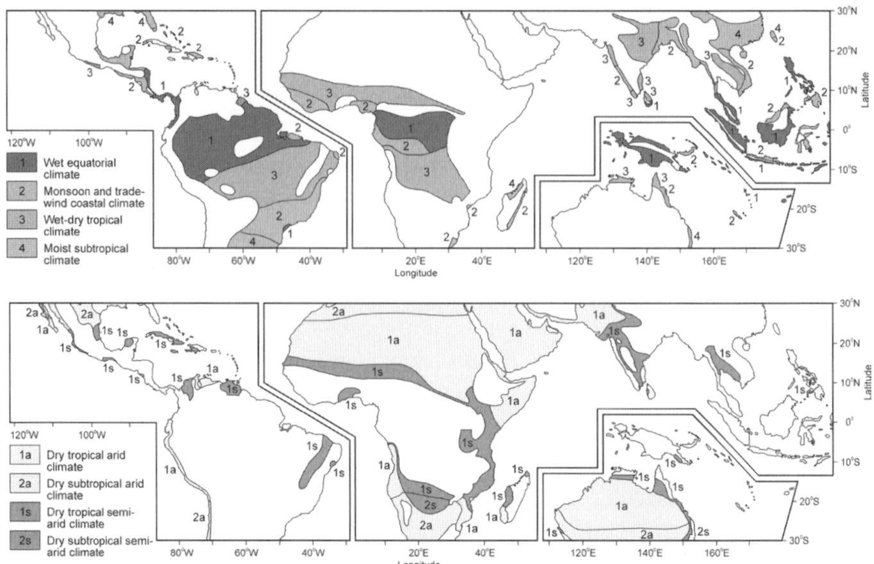

Fig. 3.1. World maps indicating the main regions of wet to moist and monsoon and trade wind tropical and subtropical climatic regions (1a), and dry arid to semi-arid tropical and subtropical climatic regions. (Modified from Strahler and Strahler 1997)

trial biome. Tropical savannah climate regions principally occur in central Brasil (along with seasonal forest and scrub), the west coast of Central America and Mexico, the east coast of Africa below the equator, and the east coast of India and northern Australia. Most of these regions are dominated by tropical savannah biome except for most of the west coast of Central America and Mexico (predominantly warm desert and semi-desert biome) and northern Australia (tropical seasonal forest and scrub biome). Dry arid and semiarid climates dominate the climate of west coast South America, the African west coast (desert and tropical savannah biome, excluding the above ~4–10° west coast region) and east coast north of the equator, the Arabian Peninsula, and the west coast of Australia. Humid subtropical climate types occur between around 18°S and the Tropic of Capricorn in eastern Australia (tropical rain forest biome) and to a minor extent near the Tropic's in Brazil (equatorial and tropical rain forest biome) and China (broadleaf and mixed forest biome) (Christopherson 2000).

3.3 The Location of Coastal Dunes in the Tropics

In general, there are very few dune fields in SE Asia. Verstappen (1957) describes a small transgressive dune field at Parangtritis, Java, Indonesia. It is the largest dune field in Indonesia with dunes reaching up to 15 m . It is fronted by a high energy intermediate to dissipative surf zone (Hesp, pers. observ.) and while Verstappen (1957) and Bird (1985) argue that it was initiated by human disturbance, the local conditions (coastline orientation to southeasterly dry monsoon winds, a winter dry season, high energy surf zone, and a sand supply) indicate that it may be a natural occurrence modified by human pressures (Hardjosuwarno and Hadisumarno 1993). Two other significant dune fields with dunes up to 20 m in height occur at Pasirbesi and Puger in Java (Hardjosuwarno and Hadisumarno 1993).

Very few sand dunes are present in Thailand (Pitman 1985). In some areas beach ridges may have small aeolian caps. There are no dunes in Singapore, and, as noted below, there is very limited dune development in Malaysia (Swan 1971; Teh 1985, 1992).

There is only one area of significant dune-field development in the Philippines (Fig. 3.2). A large parabolic dune field (the LaPaz dune field) occurs on the northwest coast above 17° N (Alex Pataray, National Institute of Geological Sciences, University of the Philippines, pers. comm.). Parts of the coast of Vietnam from the Bay of Along (~19° N) to Cape Vung Tau (also called Cape Saint-Jacques; 10° N) in the south also has significant transgressive dune

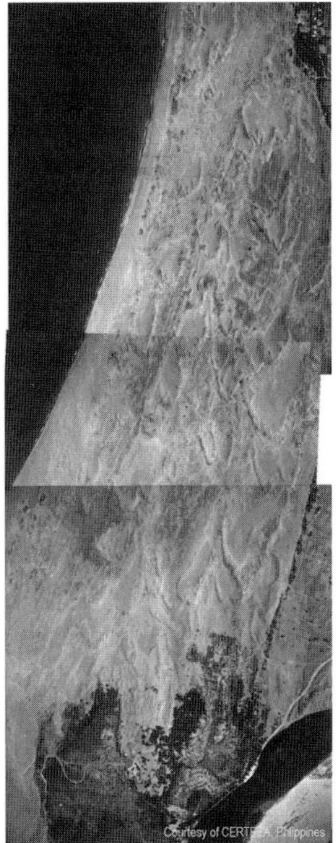

Fig. 3.2. Photo-mosaic of the parabolic dune field on the western coast of Ilococ Norte, northern Luzon, Philippines, located at 18°13′–18°15′N, 120°31′–120°34′E. The approximate scale of the central photograph is 1.5 km across the centre of the photograph from the shoreline to the edge of the photograph. (Photo courtesy of Alex Pataray)

fields (Lam Cong Dinh 1998), although local information suggests that these dunes were formerly vegetated (and most probably a different dune type) prior to defoliation during the Vietnam War.

Extensive parabolic and transgressive dune fields occur in NE Queensland and the Northern Territory in both tropical monsoon and savannah climates (Pye 1983a–c; Lees et al. 1990; Shulmeister and Lees 1992; Shulmeister et al. 1993; Swan 1979b). Massive transgressive dune fields and very large foredune plains have developed all along the tropical coast of Brazil (De Lacerda et al. 1993; Dillenburg et al. 2000) and, in fact, some of the largest dunes and dune fields are within approximately 2–3° of the equator (Maia et al. 1999; Fig. 3.3).

Small dune fields occur on the tropical West African coast (Lee 1993), and Cuba (Borhidi 1993). The Galapagos Islands are typified by a single foredune in most cases (van der Werff and Adsersen 1993). Small foredunes and relict foredunes up to 10 m in height occur in parts of the West Indies (Stoffers 1993; Gooding 1947; Davidson-Arnott, pers. comm.), and small foredunes and relict foredunes occur at Cox's Bazaar in Bangladesh (Alam et al. 1999). Dunes are also found in Hawaii (Stearns 1970; Richmond and Mueller-Dombois 1972), the Seychelles (Piggott 1968), Ghana (Talbot 1981), west and east coast India (Kunte 1995; Sanjeevi 1996) and Christmas Island (Valencia 1977).

Coastal dunes of various types (foredunes, foredune plains, transgressive dune fields) occur along 300 km of the Sri Lanka coast lying within 10° of the equator (Swan 1979a). Swan notes that the best dune development occurs in areas of strong, persistent onshore winds and long dry season.

One dune field comprising very large parabolic dunes occurs in Fiji at the mouth of the Sigatoka River where the river effluent and energy do not allow the coral reef to form. Two small areas of dunes also occur on Yasawa and Vatulele (Fiji) (Dickinson 1968; Nunn 1990).

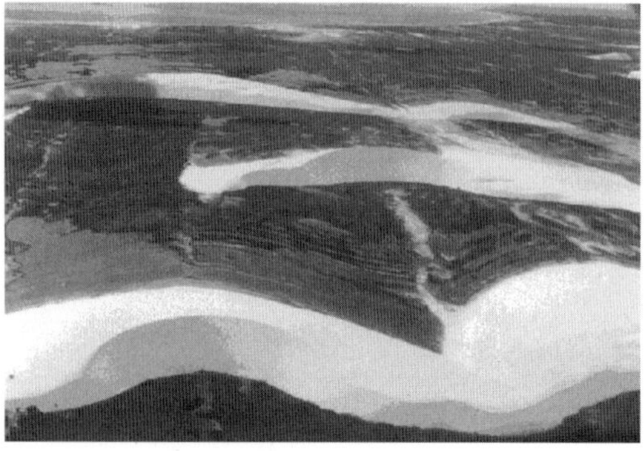

Fig. 3.3. Large (up to 50 m high) barchanoidal transverse dunes near Jericoacoara, Ceará, NE Brasil (3°S). Interdune deflations plains are rainwater flooded in the short wet season and 'cuspate vegetation marks' are formed by vegetation growth along the upwind margins of the dunes. Trailing ridges are formed by vegetation growth on the outside margins of the dune wings and horns (Photograph L.P. Maia)

Extensive vegetated and active transgressive dune fields occur on the Gulf coast of Mexico in latitudes 19–21°N. The dune fields comprise several phases and the active dunes are characterised by a variety of dune types including barchanoidal, transverse, aklé and large star dunes (see below).

There are many coastal dune fields in the arid and semi-arid tropics, including extensive ones in Namibia (Penrith 1993; Lancaster 1989; Hesp and Hastings 1998; Boucher and Le Roux 1993), the west coast Mexico extending from north of the Tropic to just below it (particularly Baja California, Mexico; Fryberger et al. 1990; Johnson 1993; Murillo de Nava and Gorsline 2000), Peru (Finkel 1969), Chile (Araya-Vergara 1986), NW Western Australia from the Tropic of Capricorn to approximately Cape Baskerville in Dampier Land (Hesp and Chape 1984). Small coastal dune fields (often just small foredunes, nebkha and relict foredunes) occur on the Asian Red Sea coast (Zahran 1993), in Somalia (Pignatti et al. 1993), the UAE, Saudi Arabia (Gheith and Abou Ouf 1996), Kuwait (Zahran 1993), Oman (Ghazanfar 1999), western Sahara, Mauritania (Hemminga and Nieuwenhuize 1990; Kocurek et al. 1991) and Senegal. Some of these dune fields lie along the edge of desert dune fields and in some cases are either barely distinguishable from the desert dunes (e.g. Namibia; Chile–Peru; Baja California, Mexico; Oman) or the desert extends to the coast from inland areas (e.g. Mauratania; the Rub' al Khali – Jafura sand sea in Saudi Arabia extending from within the tropics to the adjacent coast outside the tropics (Anton and Vincent 1986).

The presence of dunes and dune fields as listed above largely debunks the notion that sand dunes were either very poorly developed, or largely absent in the "humid tropics" (e.g. Jennings 1964; Pye 1983 c).

3.4 Are There Differences Between Tropical and Temperate Coastal Dunes Types and Processes?

The following sections address the question of whether there are, in fact, any real differences between dunes that develop in the tropics versus dunes that evolve outside the tropics. Since few comparative studies have been carried out, the discussion remains tentative. It should also be borne in mind that coastal dunes existing and developing across the actual Tropics of Capricorn and Cancer (i.e. around 20–25°N/S) are unlikely to be different from those dunes at 23.5° (N/S), and therefore probably have little regard for the tropic lines drawn on world maps!

3.5 Foredunes

Foredunes may be classified into two types, incipient and established foredunes. Incipient foredunes are new, or developing foredunes forming within pioneer plant communities. They may be formed by sand deposition within discrete or relatively discrete clumps of vegetation, or individual plants, or driftwood, flotsam etc. (types 1a and 1b of Hesp 1989), forming shadow dunes, vegetation mounds and nebkha. These may form at various locations ranging from the immediate backshore to back-barrier flats (Carter et al. 1992). In toto, such development often eventually comprises an incipient foredune zone. Such foredunes may be seasonal if formed in annual plants, and require invasion by perennial plants in order to survive. Plant species type is important in determining morphological development; species such as the tall, dense *Ammophila* tend to produce higher, more hummocky peaked dune forms than lower, more spreading, rhizomatous plants such as *Spinifex* or *Ipomoea* which produce lower, less hummocky dune forms (Hesp 2002).

Incipient foredunes may also form on the backshore by relatively laterally continuous alongshore growth of pioneer plant seedlings in the wrack line or spring high tide region, and/or by rhyzome growth onto the backshore region (types 2a and 2b of Hesp 1989). Morphological development principally depends on plant density, distribution, height and cover, wind velocity and rates of sand transport. Plant growth, density and distribution can also vary seasonally, and therefore seasonal growth rates (low [or even absent in high latitudes] in winter, high in spring) strongly influence patterns of sand transport and deposition on incipient (and established) foredunes (Davidson-Arnott and Law 1990; Law and Davidson-Arnott 1990). Secondary factors such as the rate of occurrence of swash inundation, storm wave erosion, overwash incidence, and wind direction can also be important in determining subsequent dune evolution (Hesp 2002).

3.5.1 Flow Dynamics in Vegetation

The flow dynamics within and over individual plants and continuous plant canopies varies considerably. Relatively continuous plant canopies variously impact the wind/sand flow depending on plant density, shape or morphology, distribution and height (e.g. Buckley 1987; Aylor et al. 1993; Raupach 1992; van Dijk et al. 1999). Van Dijk et al. (1999) demonstrate in their modelling that as plant height increases, dune height increases and dune length decreases. Such work verifies field observations (Hesp 1989; Arens et al. 2001). High, dense canopies (e.g., grasses such as *Ammophila* sp., shrubs and trees such as *Atriplex* sp. in Western Australia and poplars on the Great Lakes in Canada) act to reduce flow velocities very rapidly (Hesp 1989; Niedoroda et al. 1991;

Jacobs et al. 1995). Sand transport (saltation and traction) is markedly reduced from the leading edge. Incipient foredunes tend to be asymmetric with the short slope to seawards. Lower plant canopies (e.g. *Spinifex* sp., *Uniola* sp., *Ipomoea* sp.), act to reduce the flow and transport more slowly so that there is often a gradual downwind reduction in transport and asymmetric dunes are formed with the short slope on the downwind (lee slope) side (Hesp 2002).

3.5.2 Tropical Versus Temperate Foredune Trends and Morphologies

While it may be seen from the brief discussion above that a significant number of factors influence incipient (and subsequently) established foredune morphology, plant morphology and density and dominant mode of growth are critical factors, since these strongly influence sediment transport and deposition within plant canopies.

While temperate coastal zones can have a number of pioneer plant species, there is commonly one or two dominant species responsible for forming foredunes. Such species include grasses and sedges such as *Ammophila arenaria, A. littoralis, Carex* sp. *Festuca* sp. *Elymus farctus, Spinifex* sp., *Panicum* sp., *Spartina* sp., *Sporobolus* sp. and other species such as *Eryngium*. In contrast, low, creepers or trailing plants are more common in the tropics (but with varying numbers of grass, sedge and herb species).

There is commonly a gradient in coastal dune vegetation species with latitude, such that an individual or group of species, particularly pioneer species, dominate one latitudinal region and slowly give way to another species or group with an increase or decrease in latitude (Johnson 1982; Cordazzo and Seeliger 1988; Hesp 1991; Moreno-Casasola 1988, 1990, 1993; de Lacerda et al. 1993; Pfadenhauer 1993). As one trends from temperate climates towards the tropics in both eastern and western Australia, Brazil and South Africa (e.g., Doing 1981; Tinley 1985; Frazier 1993; Weisser and Cooper 1993; Müller 1980; pers. observ.; de Lacerda et al. 1993) there is a trend from a predominance of grasses, herbs and subshrubs to creepers, and a particular predominance of creepers in the tropics (e.g. Whitmore 1975; Moreno-Casasola 1988). The creepers are dominated by *Ipomoea* and *Canavalia*, both of which are low, prostrate, rhizomatous species which can very rapidly grow across the backshore under accretionary conditions. This is the so-called '*Pes-caprae* formation' of Schimper (1903) or '*Ipomoea pes-caprae* – *Canavalia* associes' (Richards 1964). Their rapid seawards growth potential, plus low creeping or trailing habit tends to lead to the development of low, terrace type incipient and, in some cases established foredunes (Davies 1980, his Fig. 115; Lee 1993, his Fig. 6.4). Of course, a slow rate of sediment supply and low winds probably aid this morphological development and may account for the limited foredune development seen in parts of the humid tropics. Thus, all other factors

Fig. 3.4. **a** *Spinifex sericeus* incipient foredune ridge, Mahia Penninsula, New Zealand. **b** *Ipomoea* incipient foredune terrace at Tioman Island, Malaysia

being equal (and there are rather a lot of factors!), one should see a general tendency, at least with incipient foredunes, for foredune *ridges* to be formed in the distal parts of the tropics (away from the equator) and within temperate zones where pioneer grasses tend to dominate (e.g. Hesp 1991; Polunin and Walters 1985). Foredune *terraces* should be more common nearer the equator (Fig. 3.4a, b).

3.6 Gross Dune-Field Morphology

If one considers the gross, large-scale morphology of entire transgressive and parabolic dune fields (i.e. ignore the dune types present on them), one can distinguish at least two major forms, namely tabular and buttress types (Tinley 1985; Hesp et al. 1989; Hesp and Thom 1990). Tabular dune fields tend to be broad plateau-type sand bodies (Hesp and Thom 1990; see Figs. 23.3 and 23.4 of Wiedemann 1993), while buttress dune fields are triangular, landward ascending ramps (similar to the buttresses of rain forest trees; Tinley 1985).

In very general terms, buttress dune fields are more common in the tropics and adjacent humid subtropics than in temperate areas (with some exceptions below). Buttress dune fields occur where dune fields are migrating obliquely onshore or perpendicularly onshore and where sand is migrating or advancing into tall forest margins. Tropical forest may be a more capable barrier to dune migration than temperate forest due to greater growth rates (see below), higher species richness and the height tropical forest trees may grow to. Thus, dune fields would build up against such forest becoming more buttress-like compared to migration over temperate and Mediterranean grasslands, heathlands or forest. For example, if one compares the development of parabolic and transgressive dune fields on the Australian west coast above the Tropic of Capricorn with the east coast above the Tropic, there are no buttress-type dune fields on the west coast where dunes are typically migrating into grassland or low heath (Hesp and Chape 1984; Hesp and Curry 1985; Hesp and Morrissey 1984). Pioneer tropical species may be more capable of colonising and stabilising dune slip faces and precipitation ridges, thereby increasing the chances of creating high, marginal, landward ascending precipitation ridges.

Examples of such buttress dune fields are common along the east coast of South Africa extending from below the tropic and up into Mozambique and Kenya (Tinley 1985; Weisser and Cooper 1993; Weisser and Marques 1979; Frazier 1993). Some of the parabolic and transgressive dune fields of central and NE Queensland are commonly buttress-type dune fields, although there is also a strong Pleistocene dune inheritance in many cases. The large sand islands extending north from around Stradbroke Island (28°S) and many of

Fig. 3.5. a A buttress dune field at Florianopolis, Brazil. High, active precipitation ridges are climbing up and over relict precipitation and trailing ridges. **b** The same dune field illustrating the gross buttress dune field form. A wide, very wet deflation plain lies seaward of the active dunes

the dune fields north to Cape York (around 10°45′S) can be included in this group (Thompson 1983; Pye 1983a, b; Batianoff and McDonald 1980).

In the Torres area of Brazil (around 29°S), tropical forest species are locally present due to the closeness of the mountain range and locally higher rainfall. In this area, buttress-type transgressive dune fields occur with high marginal precipitation ridges (Fig. 3.5a, b). To the south and north of the Torres region, the tropical elements disappear, and the dune fields are wide, low and tabular with remarkably low, small precipitation ridges.

Exceptions occur due to factors such as Pleistocene inheritance and accommodation space. In some cases, Holocene dunes are a relatively thin veneer of sand overlying extensive, multiple Pleistocene dune phases and these may provide the base, underlying ramp morphology (Pleistocene inheritance) upon which the Holocene dunes have climbed landwards to form buttress dune fields. In other cases, a buttress-type gross morphology may be the only form possible if the accommodation space for dune building is small or steep bedrock occurs adjacent to the beach.

Note that there are many sections of coastline within the tropics that are characterised by foredune plains rather than parabolic and transgressive dune fields (e.g. Dominguez et al. 1987; Dillenburg et al. 2000). The writer is presently not aware of any major gross morphological differences between foredune plains in the tropics and temperate regions. Swales in the humid tropics may be much wetter, have more permanent and seasonal wetlands and contain mangrove swamps in some areas (e.g., the massive foredune plains of Eighty Mile Beach, NW Western Australia (Hesp, pers. observ.).

3.7 Rate of Dune-Field Vegetation Colonisation and Re-Vegetation Processes

Vegetation processes, particularly growth rates, are, on average, greater in the tropics than in temperate regions (Schimper 1903; Walter 1973; Collinson 1977; Kellman and Tackaberry 1997). Very rapidly growing plants are quite common in the tropics, especially in those tropical areas that experience more rainfall, and less seasonal drought or dry period.

Thus, one would anticipate that vegetative colonisation of, and re-vegetation processes in dune fields should occur at a faster rate in the tropics that in temperate regions. This should mean that there would be a greater chance of preservation of original dune form following vegetative colonisation and stabilisation in the tropics than in temperate regions. Figure 3.6 illustrates a transverse dune slipface being colonised and stabilised by *Croton punctatus* and *Chamaechrista chamaecristoides* species in El Quixote, Mexico. Comparison of aerial photographs over several years indicates that this process occurs very rapidly, and that preservation of original to near-original dune form can

Fig. 3.6. A transverse dune slip face in the El Quixote transgressive dune field in Mexico being rapidly stabilised by *Croton punctatus* and *Chamaechrista chamaecristoides* species

be quite high. Such near-perfect preservation of original dune form does not appear to be as common in temperate regions (Thom et al. 1992).

3.8 Types of Dune-Field Vegetation Colonisation and Dune Morphologies

In the Veracruz region of Mexico (centred around 19–21°N) on the east coast (mean annual temperature 23 °C, rainfall 1600 mm; Moreno-Casasola 1993), there are a number of both active and vegetated transgressive dune fields. The vegetated dune fields are far more extensive than the active ones and are characterised by multiple, long, narrow, relatively straight, sometimes sinuous ridges. These ridges are formed in two principal ways; either by the vegetative colonisation of the margins of transverse and aklé dune ridges, or by precipitation dune ridge formation on the edge of the active dune field (Fig. 3.7).

As a transverse dune migrates downwind, and in most cases in this region, alongshore, it may be colonised along the seaward and landward margins by vegetation. This colonisation may take place in at least two ways as follows:

1. Pioneer (backshore/foredune) vegetation such as *Ipomoea*, typically growing initially on the backshore, colonises the base of the dune side slope,

Fig. 3.7. Vertical aerial photograph of the Dona Juana transgressive dune field in Mexico. Note the multiple, subparallel trailing ridges present in the vegetated portion of the dune field

Fig. 3.8. Vertical aerial photograph of the Ibiraquera transgressive dune field, Brazil showing transverse dunes (*1*), trailing ridges formed from the margins of transverse dunes (*2*), and gegenwalle ridges formed within the deflation basins (*3*). Ground distance across the photograph (*left* to *right*) is 4 km

grows up the slope and partially stabilises the dune ridge margin and crest. Other pioneer and intermediate species propagate within the *Ipomoea* and assist in the stabilisation process.

2. Pioneer (transgressive dune field) vegetation such as *Croton punctatus* and *Chamaecrista chamaecristoides* in Mexico (Martínez and Moreno-Casasola 1998) colonise the outside slope margin (and eventually the crest) of the transverse dunes. The plants may relatively uniformly colonise the whole slope and eventually the crest, or may form individual hummocks (or nebkha) along the dune slope crest trapping the sediment and forming a ridge as the dune migrates away. Ridge formation takes place in similarity to parabolic dune trailing ridge formation – the outside slope and crest of the ridge is relatively stable and vegetated while the main part of the dune migrates away. Because the ridge is initially formed at the highest point on the dune (the slip face crest side), as the dune moves away, the upwind dune slope edge is held by the vegetation and the "inside" part is eroded forming a ridge. As it does, it leaves the vegetated crest edge behind and forms trailing ridges (Fig. 3.8). Two trailing ridges can be formed at the same time on each side of a transverse dune. If there are several transverse dunes in the dune field, as is common, multiple trailing ridges can be formed at one time.

Fig. 3.9. **a** Ibiraquera dune field, Brazil. *Spartina ciliata* and *Andropogon* species colonising the margins of transverse dunes as the dunes migrate downwind. **b** Ibiraquera dune field, Brazil. Looking north and upwind 180° around from the position of **a** showing where the transverse dune once was, the inside erosion of the stabilised edge of the dune and the formation of the trailing ridge

Similar trailing ridges form behind barchans, mega-barchans, and barchaoidal transverse dunes in NE Brazil near the equator (Maia et al. 1999; Jimenez et al. 1999), and are common in transgressive dune fields in Brazil down to 27–28°S (Figs. 3.3, 3.9a, b).

In the Mexican dune fields mentioned above, and in Baja California Mexico dune fields (mostly lying up to a few degrees north of the Tropics), various species (*Croton punctatus* and *Chamaecrista chamaecristoides* in the Veracruz region) commonly form en echelon chains of nebkha hummocks forming quite long, straight, hummocky ridges also. These occur following the establishment of one shadow dune behind a plant. Seeds are presumably trapped in the low velocity, leeward protected shadow zone and establish a new plant. The process continues, leading to the development of en echelon shadow dunes and nebkha aligned downwind in the direction of the dominant wind. These are formed almost anywhere in the dune fields from deflation plains to upper dune crests and higher dune margins (Fig. 3.10).

Observations in other temperate transgressive dune fields (e.g. Cooper 1958, 1967; Goldsmith 1978; Hesp and Thom 1990; Hunter et al. 1983; McLachlan et al. 1987; Hesp et al. 1989; Borowka 1990; Orme 1990) indicate that such within-dune field trailing ridge development and re-vegetation processes is uncommon to absent (although quite common in parabolic dune fields). The

Fig. 3.10. En echelon nebkha and shadow dunes forming ridge lines at Guerrero Negro, Baja California, Mexico. This area is outside the tropics at 28°N, but illustrates one typical pattern in semi-arid to arid coastal dune fields seen both within and adjacent to the tropics

answer again may lie in factors such as species presence, the adaptive strategies of tropical – subtropical dune-field specialists, greater rates of plant growth, or other factors. Clearly more research is required.

3.9 Conclusions

Tentative conclusions are as follows:

1. Taller grasses, sedges and a variety of pioneer species dominate the incipient foredune zone in temperate regions while low, creepers dominate in the tropics. One should see a general tendency, at least with incipient foredunes, for foredune *ridges* to be formed in the distal parts of the tropics (away from the equator) and within temperate zones where pioneer grasses tend to dominate, and foredune *terraces* should be more common nearer the equator.
2. In very general terms, buttress dune fields are more common in the tropics and adjacent humid subtropics where the dune fields are migrating obliquely onshore to onshore (not alongshore) than in temperate areas where tabular dune fields are more common (with some exceptions). On average, precipitation ridges should be higher in the tropics compared to temperate regions.
3. The vegetative colonisation of, and re-vegetation processes in dune fields should occur at a faster rate in the tropics than in temperate regions. This should mean that there would be a greater chance of preservation of original dune form following vegetative colonisation and stabilisation in the tropics than in temperate regions.
4. The modes of vegetation colonisation within tropical dune fields appear to be greater than in temperate dune fields. The result is that within-dune-field trailing ridge development is common, and a variety of re-vegetation processes and locations occur in the tropical dune fields compared to temperate dune fields.

Acknowledgements. My thanks to Marisa Martinez for her superb hospitality, for making me write this paper and for showing me the Mexican dune fields; to Caroline (Thais) Martinho for her assistance in the literature search; Tim O'Dea for his unstinting support; Sergio Dillenburg, Lauro Calliari, Luiz Tomazelli, Thais, Paulo Giannini, Luciana Esteves, Nelson Gruber, and Parente Maia for their fantastic hospitality, support and visits to Brazilian dune fields.

References

Alam MS, Huq NE, Rashid MS (1999) Morphology and sediments of Cox's Bazar coastal plain, SE Bangladesh. J Coastal Res 15(4)902–908

Anton D, Vincent P (1986) Parabolic dunes of the Jafurah Desert, Eastern Province, Saudi Arabia. J Arid Environ 11:187–198

Araya-Vergara JF (1986) The evolution of modern coastal dune systems in central Chile. In: Gardiner V(ed) Proc. 1st Intl Conf on Geomorphology, Part II, pp 1231–1243

Arens SM, Baas ACW, Van Boxel JH, Kalkman C (2001) Influence of reed stem density on foredune development. Earth Surf Process Landforms 26:1161–1176

Aylor DE, Wang Y, Miller D (1993) Intermittent wind close to the ground within a grass canopy. Boundary-Layer Meteorol 66:427–448

Batianoff GN, McDonald TJ (1980) Capricorn Coast sand dune and headland vegetation. Tech Bull No 6, Queensland Dept Primary Ind, Brisbane

Bird ECF (1985) Indonesia. In: Bird ECF, Schwartz ML (eds) The World's coastline. Van Nostrand Reinhold, New York, pp 879–888

Borowka RK (1990) The Holocene development and present morphology of the Leba Dunes, Baltic coast of Poland. In: Nordstrom KF, Psuty NP, Carter RWG (eds) Coastal dunes: form and process. Wiley, London, pp 289–314

Borhidi A (1993) Dry coastal ecosystems of Cuba. In: van der Maarel E (ed) Dry coastal ecosystems: Africa, America, Asia and Oceania. Ecosystems of the world 2B. Elsevier, Amsterdam, pp 423–452

Boucher C, Le Roux A (1993) Dry coastal ecosystems of the South African west coast. In: van der Maarel E (ed) Dry coastal ecosystems: Africa, America, Asia and Oceania. Ecosystems of the world 2B. Elsevier, Amsterdam, pp 75–88

Buckley R (1987) The effect of sparse vegetation on the transport of dune sand by wind. Nature 325:426–428

Carter RWG, Bauer BO, Sherman DJ, Davidson-Arnott RGD, Gares PA, Nordstrom KF, Orford JD (1992) Dune development in the aftermath of stream outlet closure: Examples from Ireland and California. In: Carter RWG, Curtis TGF, Sheehy-Skeffington MJ (eds) Coastal dunes: geomorphology, ecology and management for conservation. Proc 3. European Dunes Congress, pp 57–69

Christopherson RW (2000) Geosystems. An introduction to physical geography. 4th edn. Prentice Hall, Englewood Cliffs, 626 pp

Collinson AS (1977) Introduction to world vegetation. Allen and Unwin, London, 201 pp

Cooper WS (1958) Coastal sand dunes of Oregon and Washington. Geol Soc Am Mem 72:169 pp

Cooper WS (1967) Coastal sand dunes of California. Geol Soc Am Mem 104:131 pp

Cordazzo CV, Seeliger U (1988) Phenological and biogeographical aspects of coastal dune plant communities in southern Brazil. Vegetatio 75:169–173

Davidson-Arnott RGD, Law MN (1990) Seasonal patterns and controls on sediment supply to coastal foredunes, Long Point, Lake Erie. In: Nordstrom KF, Psuty NP, Carter RWG (eds) Coastal dunes: form and process, pp 177–200

Davies JL (1980) Geographical variation in coastal development. Longman, London

De Lacerda LD, De Araujo DSD, Maciel NC (1993). Dry coastal ecosystems of the tropical Brazilian coast. In: van der Maarel E (ed) Dry coastal ecosystems: Africa, America, Asia and Oceania. Ecosystems of the world 2B. Elsevier, Amsterdam, pp 477–493

Dickinson WR (1968) Singatoka dune sands, Viti Levu (Fiji). Sediment Geol 2:115–124

Dillenburg SR, Roy PS, Cowell PJ, Tomazelli LJ (2000) Influence of antecedent topography on coastal evolution as tested by the shoreface translation model (STM). J Coastal Res 16(1):71–81

Doing H (1981) Phytogeography of the Australian floristic kingdom. In: Groves RH (ed) Australian vegetation. Cambridge University Press, Cambridge, pp 3–25

Dominguez JML, Martin L, Bittencourt ACSP (1987) Sea level history and Quaternary evolution of river-mouth associated beach ridge plains along the ESE Brasilian coast: A summary. In: Nummendal D, Pilkey OH, Howard JD (eds) Sea level fluctuation and coastal evolution. SEPM Spec Publ No 41, Tulsa, OK, pp 115–127

Finkel HJ (1969) The barchans of southern Peru. J Geol 67:614–647

Frazier JG (1993) Dry coastal ecosystems of Kenya and Tanzania. In: van der Maarel E (ed) Dry coastal ecosystems: Africa, America, Asia and Oceania. Ecosystems of the world 2B. Elsevier, Amsterdam, pp 129–150

Fryberger SG, Krystinik LF Schenk CJ (1990) Tidally flooded back-barrier dunefield, Guerrero Negro area, Baja California, Mexico. Sedimentology 37(1):1–23

Ghazanfar SA (1999) Coastal vegetation of Oman. Estuarine Coastal Shelf Sci 49:21–27

Gheith AM, Abou Ouf M (1996) Geomorphological features and sedimentological aspects of some coastal and inland sand dunes, Jeddah Region, Saudi Arabia. Arab Gulf J Sci Res 14(3):569–593

Goldsmith V (1978) Coastal dunes. In: Davis RA (ed) Coastal sedimentary environments. Springer, Berlin Heidelberg New York, pp 171–235

Gooding EGB (1947) Observations of sand dunes of Barbados, British West Indies. J Ecol 34:111–125

Hardjosuwarno S, Hadisumarno S (1993) Dry coastal ecosystems of the southern coast of Java. In: van der Maarel E (ed) Dry coastal ecosystems: Africa, America, Asia and Oceania. Ecosystems of the world 2B. Elsevier, Amsterdam, pp 189–196

Hemminga MA, Nieuwenhuize J (1990) Seagrass wrack-induced dune formation on a tropical coast (Banc d'Arguin, Mauritania). Estuarine, Coastal Shelf Sci 31:499–502

Hesp PA (1989) A review of biological and geomorphological processes involved in the initiation and development of incipient foredunes, In: Gimmingham CH, Ritchie W, Willetts SS, Willis AJ (eds) Coastal sand dunes. Proc R Soc Edinb 96B:181–202

Hesp PA (1991) Ecological processes and plant adaptations on coastal dunes. J Arid Environ 21:165–191

Hesp PA (2002) Foredunes and blowouts: initiation, geomorphology and dynamics. Geomorphology 48:245–268

Hesp PA, Chape S (1984) A 1:3 million map of the coastal environment of Western Australia. Central Map Agency, WA Dept of Lands and Survey, Perth, WA

Hesp PA, Curry P (1985) A land resource survey of the Fall Point Coastline, Broome, WA WA Dept Agric Tech Rep 38

Hesp PA, Hastings K (1998) Width, height and slope relationships and aerodynamic maintenance of barchans. Geomorphology 22:193–204

Hesp PA, Morrissey J (1984) A Resource Survey of the Coastal Lands from Vlamingh Head to Tantabiddi Well, West Cape Region, WA Dept Agric Tech Rep No 24

Hesp PA, Thom BG (1990) Geomorphology and evolution of active transgressive dunefields. In: Nordstrom KF, Psuty NP, Carter RWG (eds) Coastal dunes: form and process. Wiley, London, pp 253–288

Hesp PA, Illenberger W, Rust I, McLachlan A, Hyde R (1989) Some aspects of transgressive dunefield and transverse dune geomorphology and dynamics, south coast, South Africa. Z Geomorph Suppl 73:111–123

Hunter RE, Richmond BR, Alpha TR (1983) Storm-controlled oblique dunes of the Oregon coast. Bull Geol Soc Am 94:1450–1465

Jacobs AFG, van Boxel JH, El-Kilani RMM (1995) Vertical and horizontal distribution of wind speed and air temperature in a dense vegetation canopy. J Hydrol 166:313–326

Jennings JN (1964) The question of coastal dunes in tropical humid climates. Z Geomorphol 8:150–154

Jimenez JA, Maia LP, Serra J, Morais J (1999) Aeolian dune migration along the Ceará coast, northeastern Brasil. Sedimentology 46:689–701

Johnson AF (1982) Dune vegetation along the eastern shore of the Gulf of California. J Biogeogr 9:317–330

Johnson AF (1993) Dry coastal ecosystems of northwestern Mexico. In: van der Maarel E (ed) Dry coastal ecosystems: Africa, America, Asia and Oceania. Ecosystems of the world 2B. Elsevier, Amsterdam, pp 365–374

Kellman M, Tackaberry R (1997) Tropical environments. The functioning and management of tropical ecosystems, 380 pp

Kocurek G, Havholm KG, Deynoux M, Blakey RC (1991) Amalgamated accumulations resulting from climatic and eustatic changes, Akchar Erg, Mauritania. Sedimentology 38:751–772

Kunte PD (1995) On some aspects of barrier islands of the west coast, India. J Coastal Res 11(2):508–515

Lam Cong Dinh (1998) Fixation des dunes vives par *Casuarina equisetifolia* au Vietnam. Bois For Trop 256(2):35–41

Lancaster N (1989) The Namib sand sea – dune forms, processes and sediments. Balkema, Rotterdam, 178 pp

Law MN, Davidson-Arnott RGD (1990) Seasonal controls on aeolian processes on the beach and foredune. Canadian Symp on Coastal Sand Dunes. pp 49–67

Lee JA (1993) Dry coastal ecosystems of West Africa. In: van der Maarel E (ed) Dry coastal ecosystems: Africa, America, Asia and Oceania. Ecosystems of the world 2B. Elsevier, Amsterdam, pp 59–70

Lees BG, Lu Y, Head J (1990) Reconnaissance thermoluminescence dating of northern Australian coastal dune systems. Quat Res 34:169–185

Maia LP, Jimenez JA, Freire GSS, Morais JO (1999) Dune migration and Aeolian transport along Ceará northeastern Brasil, downscaling and upscaling Aeolian induced processes. In: Kraus NC, McDougal WG (eds) Coastal sediments. Proc 4th Intl Symp Coastal Eng and Science of Coastal Sediment Processes, 99, pp 1220–1232

Martinez ML, Moreno-Casasola P (1998) The biological flora of coastal dunes and wetlands: *Chamaecrista chamaecristoides* (Colladon) I. & B. J Coastal Res 14(1):162–174

McLachlan A, Ascaray C, du Toit P (1987) Sand movement, vegetation succession and biomass spectrum in a coastal dune slack in Algoa Bay, South Africa. J Arid Environ 12:9–25

Moreno-Casasola P (1988) Patterns of plant species distribution on coastal dunes along the Gulf of Mexico. J Biogeogr 15:787–806

Moreno-Casasola P (1990) Sand dune studies on the eastern coast of Mexico. Proc Canadian Symp on Coastal Sand Dunes, pp 215–230

Moreno-Casasola P (1993) Dry coastal ecosystems of the Atlantic coasts of Mexico and Central America. In: van der Maarel E (ed) Dry coastal ecosystems: Africa, America, Asia and Oceania. Ecosystems of the world 2B. Elsevier, Amsterdam, pp 390–405

Müller P (1980) Biogeography. Harper and Row, New York, 377 pp

Murillo de Nava JM, Gorsline DS (2000) Holocene and modern dune morphology for the Magdalena coastal plain and islands, Baja California Sur, Mexico. J Coastal Res 16(3):915–925

Niedoroda AW, Sheppard DM, Devereaux AB (1991) The Effect of Beach Vegetation on Aeolian Sand Transport. In: Kraus NC, Gingerich KJ, Kriebel DL (eds) Coastal sediments. Am Soc Civil Engineers, New York. pp 246–260

Nunn PD (1990) Coastal processes and landforms of Fiji: their bearing on Holocene sea-level changes in the south and west Pacific. J Coastal Res 6(2):279–310

Orme AR (1990) The instability of Holocene coastal dunes: the case of the Morro dunes, California. In: Nordstrom KF, Psuty NP, Carter RWG (eds) Coastal dunes: form and process. Wiley, London, pp 315–336

Penrith ML (1993) Dry coastal ecosystems of Namibia. In: van der Maarel E (ed) Dry coastal ecosystems: Africa, America, Asia and Oceania. Ecosystems of the world 2B. Elsevier, Amsterdam, pp 71–74

Pfadenhauer J (1993) Dry coastal ecosystems of the temperate Atlantic South America. In: van der Maarel E (ed) Dry coastal ecosystems: Africa, America, Asia and Oceania. Ecosystems of the world 2B. Elsevier, Amsterdam, pp 495–500

Piggott CJ (1968) A soil survey of the Seychelles. Tech Bull No 2, Land Resources Division, Dir of Overseas Survey, 89 pp

Pignatti S, Moggi G, Raimondo FM (1993). Dry coastal ecosystems of Somalia. In: van der Maarel E (ed) Dry coastal ecosystems: Africa, America, Asia and Oceania. Ecosystems of the world 2B. Elsevier, Amsterdam, pp 31–36

Pitman JI (1985) Thailand. In: Bird ECF, Schwartz ML (eds) The World's coastline. Van Nostrand Reinhold, New York, Chap 105, pp 771–787

Polunin O, Walters M (1985) A guide to the vegetation of Britain and Europe. Oxford University Press, Oxford, 238 pp

Pye K (1983a) Formation and history of Queensland coastal dunes. Z. Geomorphol Suppl 45:175–204

Pye K (1983b) The coastal dune formations of northern Cape York Peninsula, Queensland. Proc R Soc Qld 94:37–42

Pye K (1983 c) Coastal Dunes. Prog Phys Geogr 7(4):531–557

Raupach MR (1992) Drag and drag partition on rough surfaces. Boundary-Layer Meteorol 60:375–395

Richards PW (1964) The Tropical rain forest. An ecological study. Cambridge University Press, Cambridge, 450 pp

Richmond T de A, Mueller-Dombois D (1972) Coastline ecosystems on Oahu, Hawaii. Vegetatio 25(5–6):367–400

Sanjeevi S (1996) Morphology of dunes of the Coromandel coast of Tamil Nadu: A satellite data based approach for coastal landuse planning. Landscape Urban Plann 34:189–195

Schimper AFW (1903) Plant-geography upon a physiological basis. Clarendon Press, Oxford, 839 pp

Shulmeister J, Lees BG (1992) Morphology and chronostratigraphy of a coastal dunefield; Groote Eylandt, northern Australia. Geomorphology 5:521–534

Shulmeister J, Short SA, Price DM, Murray AS (1993) Pedogenic uranium/thorium and thermoluminescence chronologies and evolutionary history of a coastal dunefield, Groote Eylandt, northern Australia. Geomorphology 8:47–64

Stearns HT (1970) Ages of dunes on Hawaii. BP Bishop Mus Occ Pap 24:49–72

Stoffers AL (1993) Dry coastal ecosystems of the West Indies. In: van der Maarel E (ed) Dry coastal ecosystems: Africa, America, Asia and Oceania. Ecosystems of the world 2B. Elsevier, Amsterdam, pp 407–421

Strahler A, Strahler A (1997) Physical geography. Wiley, New York

Swan B (1971) Coastal geomorphology in a humid tropical low energy environment: The islands of Singapore. J Trop Geog 3343–61

Swan B (1979a) Sand dunes in the humid tropics: Sri Lanka. Z Geomorphol NF 23(2)152–171

Swan B (1979b) The presence of sand dunes in a tropical low energy zone, Friday Island, Torres Strait (Australia). Rev Geomorphol Dyn 28:61–72

Talbot MR (1981) Holocene changes in tropical wind intensity and rainfall: evidence from southeast Ghana. Quat Res 16:202–220

Teh TS (1985) Peninsular Malaysia/Indonesia In: Bird ECF, Schwartz ML (eds) The World's coastline. Van Nostrand Reinhold, New York, Chap 106, pp 789–795
Teh TS (1992) The permatang system of Peninsular Malaysia: An overview. In: Tjia HD Sharifah MS Abdullah (eds) The coastal zone of Peninsular Malaysia, pp 42–62
Thom BG, Shepherd MJ, Ly C, Roy P, Bowman GM, Hesp PA (1992) Coastal geomorphology and Quaternary geology of the Port Stephens- Myall Lakes Area. Dept of Biogeography and Geomorphology, ANU Monograph No 6. ANUTech PL Canberra
Thompson CH (1983) Development and weathering of large parabolic dune systems along the subtropical coast of eastern Australia. Z Geomorphol Suppl-Bd 45:205–225
Tinley K (1985) Coastal dunes of South Africa. S Afr Nat Sci Prog Report No 109, 300 pp
Valencia MJ (1977) Christmas Island (Pacific Ocean): reconnaissance geological observations. Atoll Res Bull. 197
Van der Werff HH, Adsersen H (1993) Dry coastal ecosystems of the Galapagos Islands. In: van der Maarel E (ed) Dry coastal ecosystems: Africa, America, Asia and Oceania. Ecosystems of the world 2B. Elsevier, Amsterdam, pp 459–475
Van Dijk PM, Arens SM, van Boxel JH (1999) Aeolian processes across transverse dunes II: Modelling the sediment transport and profile development. Earth Surf Process Landforms 24:319–333
Verstappen HTh (1957) Short note on the dunes near Parangtritis (Java). Tijd Kon Nederl Aard Gen 74:1–6
Walter H (1973) Vegetation of the Earth. The English University Press, London, 237 pp
Weisser PJ, Cooper KH (1993) Dry coastal ecosystems of the South African east coast. In: van der Maarel E (ed) Dry coastal ecosystems: Africa, America, Asia and Oceania. Ecosystems of the world 2B. Elsevier, Amsterdam, pp 109–128
Weisser PJ, Marques (1979) Gross vegetation changes in the dune area between Richards Bay and the Mfolozi River, 1937–1974. Bothalia 12(4):711–721
Whitmore TC (1975) Tropical rain forests of the Far East. Clarendon Press, Oxford, 282 pp
Wiedemann AM (1993) Dry coastal ecosystems of northwestern North America. In: van der Maarel E (ed) Dry coastal ecosystems: Africa, America, Asia and Oceania. Ecosystems of the world 2B. Elsevier, Amsterdam, pp 341–358
Zahran MA (1993) Dry coastal ecosystems of the Asian Red Sea coast. In: van der Maarel E (ed) Dry coastal ecosystems: Africa, America, Asia and Oceania. Ecosystems of the world 2B. Elsevier, Amsterdam, pp 17–30

II The Flora and Fauna of Sand Dunes

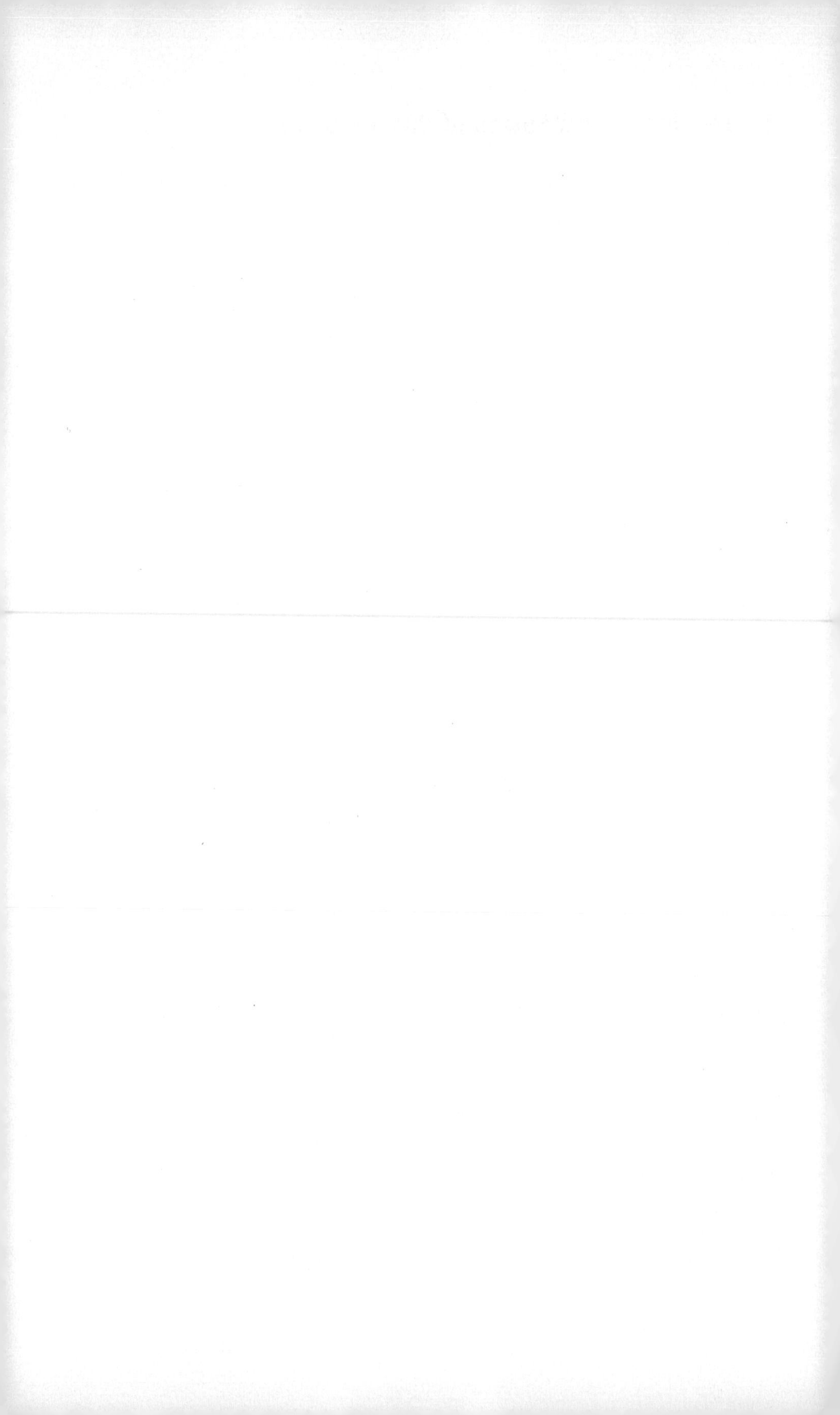

4 Temperate Zone Coastal Dunes

A.M. WIEDEMANN and A.J. PICKART

4.1 Coastal Temperate Zone Climates

Almost half of the world's coastal areas are included in the temperate climatic zone, and of these areas, many support coastal dune systems. The objective of this chapter is to characterize the vegetation of these dune areas and to show how their plants can be sensitive indicators of small climatic shifts as well as the transitions to adjoining polar and Mediterranean/tropical climates. The use of the word "temperate" as in "temperate zone climate" implies a physical habitat that experiences neither wide extremes of temperature nor very low levels of precipitation. In its broadest sense, it refers to the areas of the Earth lying between 23.5° and 66.5° latitudes on both sides of the equator. These boundaries, however, are determined by astronomical factors that have very little to do with the physical and biological factors governing the growth and distribution of plants. For the purposes of this chapter, the classification of Bailey (1958), used by Van der Maarel (1993) in the *Dry Coastal Ecosystem* volumes, will be the basis for defining "temperate zones" based on physical data. Bailey uses a number of parameters in the classification: maximum mean temperature of the warmest month, minimum mean temperature of the coldest month, annual precipitation, and percent precipitation in summer. There are two types of temperate zones: "wet winter temperate" (WWT) and "wet summer temperate" (WST). Most temperate coastal zones fall into the wet summer group. Only a few, chiefly along the west coast of North America and in New Zealand, are of the wet winter type. All of the coastal zones of Europe, from the Baltic Sea to approximately the north shores of the Mediterranean Sea (about 60°N to 45°N) are of the wet summer type. In both Hemispheres temperate zones are bordered poleward by the subpolar (SPO) and toward the equator by the summer-dry subtropical (SDS) (Mediterranean) climatic zones.

The dune systems of these temperate coastal zones are characterized by a variety of distinct plant habitats: upper beach, foredune, backdune or sand plain, deflation plain, slack, and old stabilized dune, this last defining the inner margin of the Holocene dune field. Distinctive plant communities

Ecological Studies, Vol. 171
M.L. Martínez, N.P. Psuty (Eds.)
Coastal Dunes, Ecology and Conservation

occupy these habitats. The upper beach and foredune are occupied by plants tolerant of salt spray, strong winds, and sand burial. These communities tend to be permanent because of the special adaptations required to grow and thrive. On the lee side of the foredune and the broad sand plain frequently found farther inland begins a succession with a set of plants that initially stabilize the sand. These sand-tolerant grasses and forbs gradually cover the sand surface with a dense mat-like layer of vegetation. Once this dune mat is fully developed, shrub species (and sometimes tree species concurrently) become established. The "shrub stage" may persist a long time or only briefly, depending on microclimatic conditions (such as the presence of ground water, the distance from shoreline, and salt spray effects, as well as a source of tree seeds). On the "most temperate" coastal areas, a forest eventually develops that contains many tree and shrub species found in adjoining habitats. Where there is a gradual shift to Mediterranean climates, scrub or chaparral develops and remains more or less permanently.

Table 4.1 is a floristic characterization of selected temperate zone coastal dune areas based on regions delineated by Van der Maarel (1993). Easily seen is the almost universal occurrence of similar species and genera on the upper beach and the active sands of the foredune. Behind the foredune, where stabilization is taking place, there is less similarity. At the shrub stage of vegetation development, communities are more heterogenous depending upon variation in the microclimates of the site. Finally, at the arborescent stage (if there is one) there is identity with the regional tree flora. This has been noted and discussed in many ways (see, e.g., Doing 1985).

The climatic zones, no matter what the basis for their delineation, do not have sharp boundaries, but change gradually. In general, plants signal the transition from one zone to another quite clearly, either as single species or as small groups of species. Because of the ameliorating effect of the sea on local climate, beach species tend to be almost azonal in their occurrence, having worldwide distributions at the species level (e.g., *Honkenya peploides, Cakile maritima*). On the foredunes identity is more at the generic level, especially in the Northern Hemisphere. In fact there seems to be a continuity in species and growth form around the Northern Hemisphere. Consider, e.g., *Carex macrocephala* and *Carex kobomugi* linking Asia and North America or *Leymus mollis* and *Leymus arenaria* linking northern Europe and North America. The connections in the Southern Hemisphere are not so clear; the small number of temperate climatic zones and the long isolation may be factors. However, the physiognomy and vegetation zonation are similar everywhere. There is, however, one universal link. The obligate psammophyte *Ammophila arenaria* (along with its subspecies and its congener, *Ammophila breviligulata*), has a distribution, both natural and human-induced, that lies roughly between 32° and 60° on both sides of the equator. Because of the effectiveness of these species in stabilizing active dunes, they (chiefly *Ammophila arenaria*) have been introduced into almost every part

Table 4.1. Floristic characterization of selected temperate zone (wet summer and wet winter) coastal dune areas. Areas were selected on the basis of their floristic and physiognomic representation of the major temperate zone dune systems. Location codes in parentheses refer to relevant chapters in Van der Maarel (1993), the letter A or B refers to volume, the number to the chapter in the volume. Species listed are typical of the locality, but not peculiar to it

Location	Beach and foredune	Dune mat[a]	Stabilized shrub	Old dune forest
S Norway	*Cakile maritima/Honkenya peploides*	*Cladonia* spp.	*Empetrum nigrum*	*Quercus robor*
58°N (A9)	*Ammophila arenaria/Elymus farctus*	*Sedum acre*	*Calluna vulgaris*	
Netherlands	*Cakile maritima/Elymus farctus*	*Festuca rubra*	*Hippophae rhamnoides*	*Quercus robor*
52°N (Al7)	*Ammophila arenaria*	*Oenothera ammophila*	*Salix arenaria*	*Betula pendula*
Japan	*Glehnia littoralis/Lathyrus japonicus*	*Calystegia soldanella*	*Rosa rugosa*	*Quercus dentata*
44°N (B13)	*Carex macrocephala/Leymus mollis*	*Carex kobomugi*	*Celastrus orbiculata*	
New Zealand	*Cakile edentula/Ammophila arenaria*	*Euphorbia glauca*	*Lupinus arboreus*	*Kunzea ericoides*
44°S (B15)	*Desmoschoenus spiralis*	*Calystegia soldanella*	*Sambucus nigra*	*Podocarpus totara*
New Hampshire	*Honkenya peploides*	*Hudsonia tomentosa*	*Myrica pensylvanica*	*Pinus rigida*
USA – 42°N (B22)	*Ammophila breviligulata*	*Lechea marítima*	*Prunus marítima*	*Acer rubrum*
S. Carolina	*Uniola paniculata*	*Iva imbricata*	*Ilex vomitoria*	*Quercus virginiana*
USA – 32°N (B22)	*Panicum amarum*	*Spartina patens*	*Myrica cerifera*	
Oregon	*Ambrosia chamissonis*	*Festuca rubra*	*Vaccinium ovatum*	*Pinus contorta*
USA – 47°N (B23)	*Ammophila arenaria/Leymus mollis*	*Solidago spathulata*	*Gaultheria shallon*	*Tsuga heterophylla*
Pt. Reyes	*Cakile marítima*	*Eriogonum latifolium*	*Ericameria ericoides*	Coastal scrub[b]
USA – 38°N (B24)	*Leymus mollis/Ambrosia chamissonis*	*Erigeron glaucus*	*Lupinus chamissonis*	
Argentina	*Cakile maritime*	*Panicum urvilleanum*	*Baccharis genistifolia*	*Acacia cavenia*
39°S (B33)	*Spartina coarctata*		*Celtis spinosa*	

[a]Dense ground cover of herbaceous species on back dunes and sand plains
[b]Coastal scrub and chaparral represent the transition to a Mediterranean climate

of the world. In Europe, with its long history of planting the species widely for "coastal defence" *Ammophila arenaria* is found on all coastal dune areas of the European temperate zone (Huiskes 1979). Its northern-most occurrence seems to be southern Finland, at 60°N (Hellemaa 1998). It was introduced into South Africa in the 1870s and widely planted. It did not naturalize, but maintained populations without spreading (noninvasive) (Hertling 1997). In Australia (Wiedemann, pers. observ.), it has been planted in the Perth and Sydney areas, but did not spread here either. It flowers poorly, if at all, and does not compete well with the native *Spinifex* species. Its poor growth at the latitudes of these three places (32°S) indicates the species is probably at the equatorial limit of its range. A little farther south, on the southwest coast of Western Australia (35°S), it grows well but still does not spread to any significant extent. At Wilson's Promontory in Victoria, southeast Australia (40°S), it grows and reproduces vigorously. It also grows well in Tasmania (41°–44°S) and is well established in New Zealand (40°–45°S). In 1927 it was introduced to the Falkland Islands (55°S) and today is a major component of the littoral vegetation (Moore 1968).

On the east coast of North America, the native *Ammophila breviligulata* has a natural distribution from eastern Canada (50°N) to South Carolina (32°S), where it is replaced by the ecologically equivalent *Uniola paniculata* (Stalter 1993). *A. breviligulata* is found also on the shores of the Great Lakes (Maun 1993). It occurs today in scattered small patches on the Pacific coast, but was introduced in stabilization plantings. From the moment in 1869, when *Ammophila arenaria* was first planted on the active dunes that today underlie the city of San Francisco, the species flourished. Within 75 years, both through natural spread and large-scale planting, *A. arenaria* spread along the entire west coast of N. America, from about 34°N at Los Angeles to Vancouver Island (49°N), then to 54°N on the Queen Charlotte Islands (Breckon and Barbour 1974). This distribution on the west coast of North America is of interest in two respects.

First, it is continuous between its latitudinal limits, from the subpolar in the north to full Mediterranean in the south; only a few upper beach species with global distributions share this characteristic. Secondly, it is on the west coast of North America that the first alarms were raised with respect to the effect of the grass on native plant communities and on the very morphology of the dunes themselves.

4.2 Coastal Dunes of Western North America

This very long, continuous temperate zone coastline, the very active "dune restoration" activities along its length, and the intimate knowledge the authors have of its history and characteristics, make this area suitable to present as a case study. Coastal dunes are a common feature along the 2091 km shoreline from Cape Flattery at the northwestern-most tip of Washington State to San Diego in California (48° to 32°N). Dunes occur on 610 km of this shoreline.

There are small dune systems on the west coast of Vancouver Island (49°N) and the northeast corner of Graham Island of the Queen Charlottes (54°N). The overall directional trend of the coast is west of south along the north half; at Cape Mendocino, about half way down, it rounds to east of south to the Mexican border. The annual wind regime is fairly consistent along the entire coast, with local deviations and a somewhat weakening of the general pattern southward (Cooper 1958, 1967). In summer (June to August) onshore winds from the sector N-NW greatly predominate. They result both from off-shore high pressure centers and sea-land winds. These winds have a high average velocity. In winter (December to March) onshore winds from the sector S-SW related to seasonal low pressure centers predominate. The centers produce frontal storms that bring heavy rains and strong south to southwesterly winds.

These wind regimes, along with abundant sand supply, an extensive receptive shore, and variations in coastal trend, produce a variety of dune forms (Wiedemann 1984). North and south of the Columbia River, accreting shorelines have resulted in a series of beach ridges parallel to the shore. South of the river at least nine ridges can be seen, all presently stabilized. Farther south on the Oregon coast very large parabola dunes (many over 1 km in length) have been produced, mostly by the winter winds. At one location "nested" parabolas, over 4 km in length have resulted from cycles of active sand movement and subsequent stabilization by plants. In California most of the dune systems are made up of parabola dunes (Fig. 4.1). On the central Oregon coast are extensive active dune fields (Fig. 4.2) extending several km inland with massive winter transverse dunes (crests 1000 m long and 50 m high) moving northward, formed and driven by the winter winds. Smaller transverse dunes are formed and driven southeasterly by the summer winds.

Neither of these two kinds of dunes is vegetated. The foredune has a significant history. Formerly present in southern California as a shoreline zone of large hummocks (or mounds) and as an upper beach ridge in northern California, early accounts and aerial photographs indicate it was absent along the Oregon coast (Cooper 1958), except for the prograding shoreline at the mouth of the Columbia River in the north. The present day foredune along the west coast is entirely the result of the introduction and spread of *Ammophila arenaria,* a high, broad foredune developed in less than 100 years. The native flora of the foredune has been almost entirely replaced by this aggressive species (Wiedemann 1998; Wiedemann and Pickart 1996).

Table 4.2 demonstrates the significant shift in climate along the Pacific coast as reflected in the distribution of selected dune species. It is distinctly wet winter from Neah Bay at the northern tip of the Washington coast to

Fig. 4.1. Parabola dunes at Humboldt Bay, northern California. View is to the south. The distance from shore to parabola tip exceeds 1000 m. (Photograph by A. Wiedemann, June 1983)

Fig. 4.2. The winter transverse dunes of the central Oregon coast. View is to the south, the dunes moving toward the observer (note slip faces). Average crest length is about 1000 m and height above dune base can exceed 50 m (Cooper 1958) (Photograph by Oregon Dept. of Transportation, May 1972)

Table 4.2. There is a pronounced climatic shift at about 37°N, reflected in precipitation, temperature, and species distribution. Santa Cruz is about halfway down the coast between Neah Bay and San Diego. Only one species, *A. chamissonis*, continues into the true Mediterranean climate. The temperate zone species begin dropping out at Newport, the remainder are out by Santa Barbara

Locality Latitude	Neah Bay 48°N	Ilwaco 46°N	Newport 44°N	Coos Bay 43°N	Arcata 41°N	Santa Cruz 37°N	Santa Barbara 34°N	San Diego 32°N
Mean maximum temperature (°C), warmest month	15	19	17	19	16	24	23	25
Mean minimum temperature (°C), coldest month	3	2	3	3	5	3	4	8
Mean annual precipitation (mm)	1960	2060	1680	1610	1060	760	410	250
Precipitation in summer (%)*	23	22	27	18	14	8	6	9
Ambrosia chamissonis	- - -	- - -	- - -	- - -	- - -	- - -	- - -	- - -
Ammophila arenaria	- - -	- - -	- - -	- - -	- - -	- - -	- - -	
Fragaria chiloensis	- - -	- - -	- - -	- - -	- - -	- - -	- - -	
Abronia latifolia	- - -	- - -	- - -	- - -	- - -	- - -	- - -	
Leymus mollis	- - -	- - -	- - -	- - -	- - -			
Lathyrus japonicus	- - -	- - -	- - -	- - -	- - -			
Carex macrocephalus	- - -	- - -	- - -					
Phacelia argentea				- - -	- - -			
Eriogonum latifolium				- - -	- - -	- - -		
Abronia maritima						- - -	- - -	- - -

about Coos Bay on the southern Oregon coast. From there to Arcata, on the north coast of California, the gradual shift to a Mediterranean climate is reflected in the greatly decreased precipitation. New species appear in the transition zone. From Santa Cruz southward the climate is distinctly Mediterranean.

4.3 Conservation and Management

The introduction and spread of *Ammophila* has resulted in major changes in the dune landscape. A highly successful competitor, *A. arenaria* has "taken over" or created most of the existing foredunes throughout its current range. In most places it has virtually eliminated the native dune-forming species and the distinctive low, open, rounded and high diversity foredunes created by them (Fig. 4.3). The native dunegrass, *Leymus mollis,* was once prevalent on beaches and foredunes of the western U.S coast north of 38°N (Barbour and Johnson 1988). Significant stands of this grass are now restricted to dune systems at Point Reyes and Humboldt Bay, California. In addition, *A. arenaria* has invaded many back dune ridges and completely stabilized some formerly active dunes. At Bodega Bay, California, where *A. arenaria* was extensively planted (Cooper 1967), the entire parabola dune system has been virtually frozen under a dense blanket of *A. arenaria.* Along the central Oregon coast the massive high foredune created by *A. arenaria* (Fig. 4.4) has cut off all sand supply from the beaches to the back dunes. It is not known how dependent the high winter transverse dunes are on a continual supply of sand from the beach, but the lack of sand has resulted in the wind eroding an extensive backdune sand plain to the water table – a "deflation plain". Vegetation develops quickly and follows the progressive development of the deflation plain toward the base of the high dunes (Fig. 4.5). The perceived "danger" (according to "dune managers") is that this scenic dune landscape will soon vanish under a dense mat of vegetation.

Despite the widespread invasion of *A. arenaria* worldwide, management (control and eradication) of this species has until recently been confined to the west coast of North America. In his seminal work on the coastal dunes of Oregon and Washington, Cooper (1958) noted the topographic and vegetation changes brought about by this species: higher, more sharply ridged, densely vegetated foredunes, and the colonization of open dune fields and dune mat leading to a hummocky topography dominated by *Ammophila* and overall to a repressive effect on the native dune flora. His observations were soon followed by alarm over the impact of the species on coastal dune biodiversity (Breckon and Barbour 1974; Barbour et al. 1976). By the early 1980s, experimental trials in controlling the species had begun at what is now the Humboldt Bay National Wildlife Refuge (Lanphere Dunes Unit) by The Nature

Fig. 4.3. Lower, more rounded profile of foredune at Humboldt Bay, California, after restoration (removal of *A. ammophila*). Vegetation is dune mat. (Photograph by A. Pickart)

Conservancy (Van Hook 1985). The first successful eradication program, using manual techniques, was completed there between 1992–1996 (Pickart and Sawyer 1998). Similarly, recent research on the impacts (primarily loss of native plant habitat) of the species in Australia and New Zealand (Duncan 2001; Heyligers 1985; Humphries 1996) has led to the initiation of control efforts: attempts to stop its spread and, if possible to eradicate it.

There are currently numerous *A. arenaria* control programs being implemented along the US west coast that are intended to restore natural dune processes. Methods of control include one or a combination of techniques: manual removal, excavation/burial with heavy equipment, burning, and application of the herbicide glyphosate (Pickart and Sawyer 1998). Projects relying on manual removal enjoy the advantage of unaided native plant recovery through the dispersal of relict native plants and the ameliorating conditions (relative stability, fertility, and moisture) created by decaying rhizomes of *A. arenaria* left in place. However, this method is extremely labor-intensive and mostly suitable for areas in which *A . arenaria* has not become extensively established.

The original project at Humboldt Bay resulted in eradication of 10 acres (4 ha) of dense *A. arenaria* growing on the foredune and adjacent dune ridges.

Fig. 4.4. Typical steep profile foredune vegetated with *Ammophila arenaria* prior to restoration. Humboldt Bay, California. (Photograph by A. Pickart)

This project was carried out over a 4-year period (1992–1996) at a total cost of US$ 350,000. By the final year of treatment native cover was 36 % of that measured in comparable, uninvaded areas; by 2002 it had reached 100 % (unpubl. data). Removal of *A. arenaria* returns the dunes to an early stage of vegetation development, adding back in the important agent of instability. Species diversity is predictably lower (by approximately 10 %) than in comparable native areas. Restored conditions favor those species that rely on disturbance and openings in the vegetation, including the federally listed endangered annual species *Layia carnosa*, which occurred at increased densities in the year following restoration.

Although the first restoration efforts targeted only the earliest and most obvious plant invaders, advances in awareness and understanding of the dune ecosystem has led both to earlier detection of problems and to more systems-based approaches to restoration. Restoration efforts now focus on re-establishing dune processes, and involve management of multiple taxa in an integrated fashion. These include, in addition to *A. arenaria, Lupinus arboreus* (native south of 30°N, but introduced and invasive on dune systems to the north), *Carpobrotus edulis,* and a suite of annual grasses (including *Bromus diandrus, Vulpia bromoides, Briza maxima,* and *Aira* spp.). Vegetated dunes south of 38°N are also susceptible to invasion by *Ehrharta calycina,* which has led to the conversion of large areas of native dune scrub to non-native grass-

Fig. 4.5. Dune mat vegetation on the central Oregon coast with shrub (*Arctostaphylos*) and tree *(Pinus)* vegetation becoming established. (Photograph by A. Wiedemann)

land (Pickart 2000), and *Conocosia pugioniformis* which is in a relatively early stage of invasion, but has the potential to impact foredune and dune scrub communities (Albert and D'Antonio 2000).

The type and severity of these invasions vary. Some invoke complex ecosystem changes. For example, the intrusion of *Lupinus arboreus* into the native, herbaceous dune mat of northern California results in soil enrichment that triggers invasions of other plant species (Pickart et al. 1998). Together and individually, however, these invasions share the important consequence of greatly reducing or eliminating sand movement. Whereas plant succession in dunes moves the system naturally towards stability, non-native invasions greatly accelerate the process (Wiedemann and Pickart 1996). One result is rapid local extinction rates, with long-term consequences for the re-establishment of early successional species after the infrequent, large-scale tectonic events that periodically rejuvenate dunes in the Pacific Northwest (Clark and Carver 1992; Leroy 1999).

The full extent of the impacts of plant invasions on dune systems worldwide is unknown. There are only a few places in which any assessment has been attempted. At the Humboldt Bay dunes, a geographic information system (GIS) was employed to map and classify dune vegetation (Aria 1999). The results indicated that 52 % of total dune vegetation, or 82 % of non-forested vegetation, was dominated by introduced species. A rough estimate in 1997 of total area occupied by *A. arenaria* along the west coast of the US south of Florence, Oregon, exceeded 6000 ha (Pickart 1997). Clearly, "the *Ammophila* problem," and that of other invasive dune species, is of a magnitude far exceeding the resources available to address it. Careful prioritization of conservation efforts is essential, divided appropriately among preservation, restoration, and management, including steps taken to slow or prevent further invasions.

References

Albert M, D'Antonio C (2000) *Conicosia pugioniformis*. In: Bossard CC, Randall JM, Hoshovsky MC (eds) Invasive plants of California's wildlands. University of California Press, Berkeley, pp 116–119

Aria KT (1999) Using aerial photographs rectified with a geographic information system to map coastal dune vegetation and land use in Humboldt County, California. MS Thesis, Humboldt State Univ, Arcata, California

Bailey HP (1958) An analysis of coastal climates, with particular reference to humid mid-latitudes. In: RJ Russell (ed) Proc 2nd Coastal Geography Conf, Washington, DC, pp 23–56

Barbour MG, Johnson AF (1988) Beach and dune. In: Barbour MG, Major J (eds) Terrestrial vegetation of California. California Native Plant Soc Spec Publ 9, Sacramento, pp 223–261

Barbour MG, DeJong TM, Johnson AF (1976) Synecology of beach vegetation along the Pacific Coast of the United States of America: a first approximation. J Biogeogr 3:55–69

Breckon GJ, Barbour MG (1974) Review of North American Pacific coast beach vegetation. Madrono 22:333–60

Clarke SH, Carver GA (1992) Late Holocene tectonics and paleoseismicity, southern Cascadia subduction zone. Science 255:188–92

Cooper WS (1958) Coastal sand dunes of Oregon and Washington. Geol Soc Am Mem 72, Boulder, CO

Cooper WS (1967) Coastal dunes of California. Geol Soc Am Mem 104, Boulder, CO

Doing H (1985) Coastal foredune zonation and succession in various parts of the world. Vegetatio 61:65–75

Duncan MC (2001) The impact of *Ammophila arenaria* on indigenous dune plant communities at Mason Bay, Stewart Island, New Zealand. MS Thesis, Univ of Otago, New Zealand

Hertling UM (1997) *Ammophila arenaria* (L.)Link (Marram Grass) in South Africa and its potential invasiveness. PhD Diss, Rhodes Univ, S Africa

Heyligers PC (1985) The impact of introduced plants on foredune formation in southeast Australia. In: Dodson JR, Westoby W (eds) Are Australian ecosystems different? Proc Ecol Soc Aust 14:23–42

Hellemaa P (1998) The development of coastal dunes and their vegetation in Finland. Fennia 176–1:111–221

Huiskes AHL (1979) *Ammophila arenaria* (L.) Link – biological flora of the British Isles. J Ecol 67:363–382

Humphries SE (1996) Australian national weeds strategy: what are the lessons? In: Lovich J, Randall J, Kelly M (eds) Proc California Exotic Plant Pest Council Symp, vol 2, pp 21–29

Leroy TH (1999) Holocene sand dune stratigraphy and paleoseismicity of the North and South Spits of Humboldt Bay, northern California. MS Thesis, Humboldt State Univ, Arcata, California

Maun M (1993) Dry coastal ecosystems along the Great Lakes of North America. In:Van der Maarel E (ed) Dry coastal ecossytems, vol 2B. Elsevier, Amsterdam, pp 299–316

Moore DM (1968) The vascular flora of the Falkland Islands, Sci Rep No 60, British Antarctic Survey. Natural Env Res Council, London

Pickart AJ (1997) Control of European beachgrass (*Ammophila arenaria*) on the West Coast of North America. In; Kelly M, Wagner E, Warner P (eds) Proc California Exotic Pest Plant Council Symp, vol 3, pp 82–90

Pickart AJ (2000) *Ehrharta calycina, Ehrharta erecta*, and *Ehrharta longiflora*. In: Bossard, CC, Randall JM, Hoshovsky MC (eds) Invasive plants of California's wildlands. University of California Press, Berkeley, pp 164–170

Pickart AJ, Sawyer JO (1998) Ecology and restoration of northern California coastal dunes. California Native Plant Soc, Sacramento

Pickart AJ, Miller LM, Duebendorfer TE (1998) Yellow bush lupine invasion in northern California coastal dunes: I. Ecological impacts and manual restoration. Restoration Ecol 6:59–68

Stalter R (1993) Dry coastal ecosystems of the eastern United States of America. In: Van der Maarel E (ed) Dry coastal ecosystems, vol 2A. Elsevier, Amsterdam, pp 317–340

Van der Maarel E (1993) (ed) Ecosystems of the world. Dry coastal ecosystems 2A,B. Elsevier, Amsterdam

Van Hook SS (1985) European beachgrass. Fremontia 12:19–20

Wiedemann AM (1984) The ecology of Pacific Northwest coastal sand dunes: a community profile. US Fish and Wildlife Service FWS/OBS-84/04

Wiedemann AM (1998) Coastal foredune development, Oregon USA. J Coastal Res SI(26):45–51

Wiedemann AM, Pickart A (1996) The *Ammophila* problem on the northwest coast of North America. Landscape Urban Plann 34:287–99

5 Vegetation Dynamics and Succession on Sand Dunes of the Eastern Coasts of Africa

R.A. LUBKE

5.1 Introduction

Studies on dunes systems on the eastern coast of Africa (Fig. 5.1) have not been a priority but recent studies of the dynamics of dunes have been undertaken (Tinley 1985; Lubke et al. 1997), of particular interest being the large prograding dune systems on the KwaZulu Natal coast (Moll 1969; Pammenter 1983; Avis 1992) and the impressive transgressive dune fields of the Eastern Cape coast (Lubke 1983; Lubke and Avis 1988; Talbot and Bate 1991).

Some coastal systems have been disturbed by the invasion of alien species and their introduction (Shaughnessy 1980); their effects on the dune systems have been documented by Hertling (1997). The control of aliens (e.g. *Acacia cyclops* from Australia) is often essential as they disrupt ecosystems, but some species, (e.g. *Ammophila arenaria*), are potentially good stabilisers, not invasive and do not displace indigenous species (Hertling 1997; Lubke and Hertling 2001).

This chapter is a critical analysis of natural prograding and eroding dunes and mobile dune fields providing a baseline for successional change in dune systems, and the effects of aliens and predictions made regarding the changes on the dynamics of the dune systems are described. On the basis of this research suggestions are made regarding management of dune systems.

5.2 Successional Change Along the South-Eastern African Coast

Studies along this coastline (Fig. 5.1) have elucidated the successional changes and processes under different initial conditions and the effects of indigenous pioneer species as opposed to introduced alien plant invasion in the dune ecosystem.

Ecological Studies, Vol. 171
M.L. Martínez, N.P. Psuty (Eds.)
Coastal Dunes, Ecology and Conservation

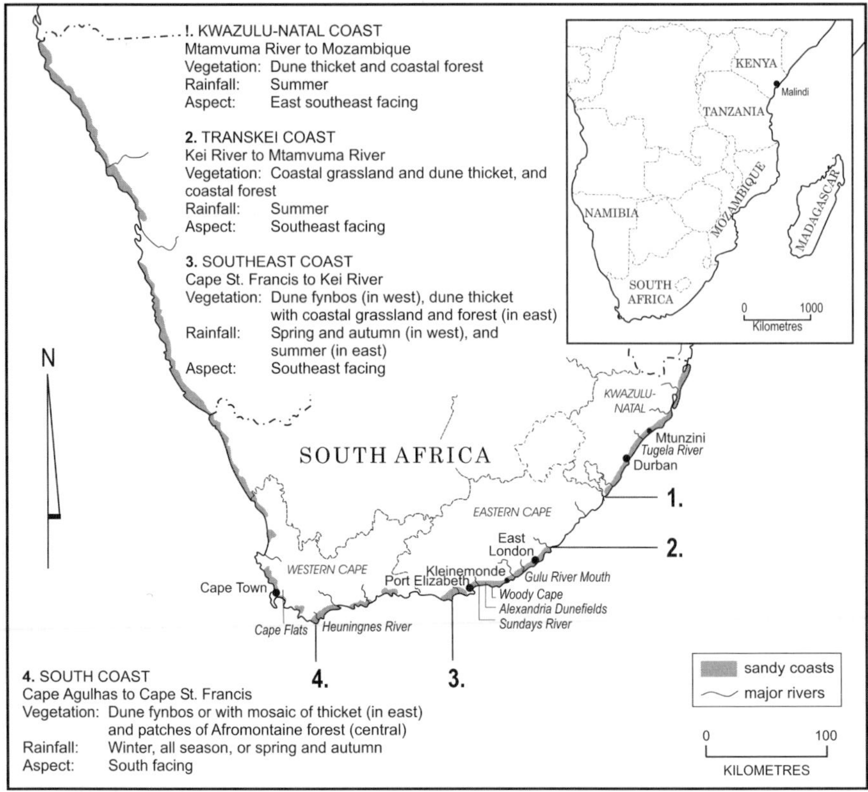

Fig. 5.1. Map of the southeast African coastal regions

The coastal dune communities (Fig. 5.2) are often quite distinct consisting of monospecific stands of pioneer plants. In other cases there may be a gradation from the pioneers into thickets so that the different communities are not that distinct. Piecing together the dynamics of successional change has taken place by analysis of the spatial relationships between the various communities and also through temporal studies of fixed plots over an extended time period.

5.2.1 Studies on Prograding Dune Fields

Along prograding coastlines there is a continuous supply of sand and new habitats are available for colonisation by pioneer species. The dunes at the Mlalazi Nature Reserve, at Mtunzini on the Kwa-Zulu/Natal coast (Fig. 5.1), provide an excellent example of succession on a prograding coastline (Fig. 5.2A). High summer rainfalls and supply of sand due to soil erosion from sugarcane fields into the Tugela River ensures a continuous deposition of

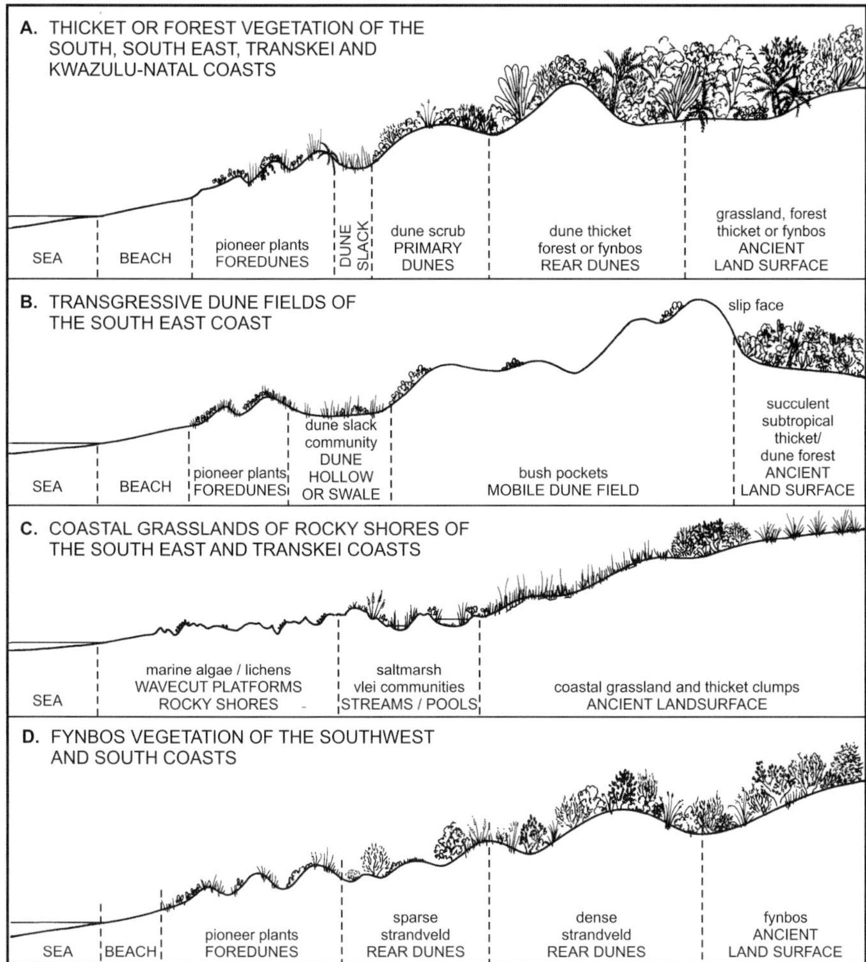

Fig. 5.2. Types of coastline and the coastal vegetation. **A** Thicket and/or forest are the climax vegetation on many prograding dune fields. **B** Community change on transgressive dune fields, which are often eroding, is not that obvious. **C** Coastal grasslands are climax vegetation types on rocky shores. **D** Coastal fynbos is climax vegetation on the south coast dunes, where alien species, e.g. *Acacia cyclops* and *Ammophila arenaria* are often present

sand and silt at the river mouth and by longshore drift northwards to the Mlalazi dunes. First well documented by Moll (1954), Weisser et al. (1982) related successional changes to the chronology of dune development using aerial photographs to date the parallel dune ridges of the region. They showed that dune advancement occurred in pulses with dune advancement from 1.2 m year^{-1} (1937–1957) to 3.56 m year^{-1} (1957–1965) to 5.67 m year^{-1} (1965–1977) and in 1982/1983 a dune ridge would be formed about every five

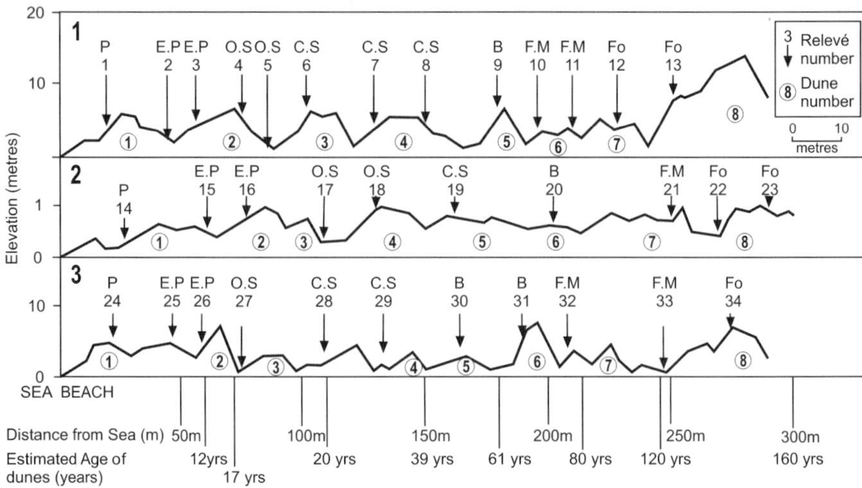

Fig. 5.3. Profile diagrams of the three transects sampled across the Mtunzini dune field, showing position of sample sites in relation to topographical variation and community type recognised in the field, and the position of the eight dune ridges. The *x*-axis is both distance from the sea and age (inferred from Table 5.1). *P* Pioneer; *EP* enriched pioneer; *OS* open dune scrub; *CS* from Avis 1992)

years. The age of the dune ridges were estimated in this way for a study carried out in 1987 (Table 5.1). A quantitative study of this dune system was carried out by sampling 34 random relevés along three transects across the dune ridges (Fig. 5.3).

Seedlings of *Scaevola plumieri* have been observed to establish on the drift line as a result of seeds rolling down from the first foredune covered with mature *Scaevola plumieri* plants. Knevel (2001) monitored seed production and establishment of *S. plumieri* over a 3-year period on different sites on similar dunes at Kleinemonde). Under high rainfall conditions seedlings rapidly establish and form hummock dunes, which coalesce to form parallel linear dune ridges. Other pioneers, such as *Ipomoea pes-caprae,* also occur but the continuous stem elongation and root production of *S. plumieri* are the important factors in sand stabilisation and foredune development. On the second dune ridge, there is an abundance of other herbaceous species, where grasses, herbs and some shrubs become established. *S. plumieri* is an accomplished pioneer when subjected to sand deposition and salt spray, but does not thrive on the second dune ridges (Pammenter 1983).

Eight communities were identified along a gradient of increasing distance from the sea and increasing age of the dunes: pioneer, enriched pioneer, open dune scrub, closed dune scrub, bush clumps, bush clump/forest margin transition, forest margin and forest (Fig. 5.4). Communities showed an increase in species richness, cover, stature and biomass.

Table 5.1. Rate of sand movement (from Weisser and Backer 1983), mean distance between dune ridges and approximate dune ages, when the study was carried out in 1987 on the Mtunzini dune field. (Avis 1992)

Dune ridge	Mean rate of sand Movement (m year^{-1})	Mean distance from previous ridge (m)	Age (years) from previous ridge	Approximate accumulative age (years)
1	5.67	5	5	5
2	5.67	39.3±5.6	39.5/5.67=6.9	12
3	5.67	25.3±5.5	25.3/5.67=4.5	17
4	4.5	41.3±5.8	41.3/4.5=9.2	26
5	2.4	30.7±3.7	30.7/2.4=12.8	39
6	1.25	27.3±1.76	27.3/1.25=27.8	61
7	1.25	34.7±4.4	34.7/1.25=27.8	89
8	1.25	39.3 ± 1.8	39.3/1.25=31.4	120

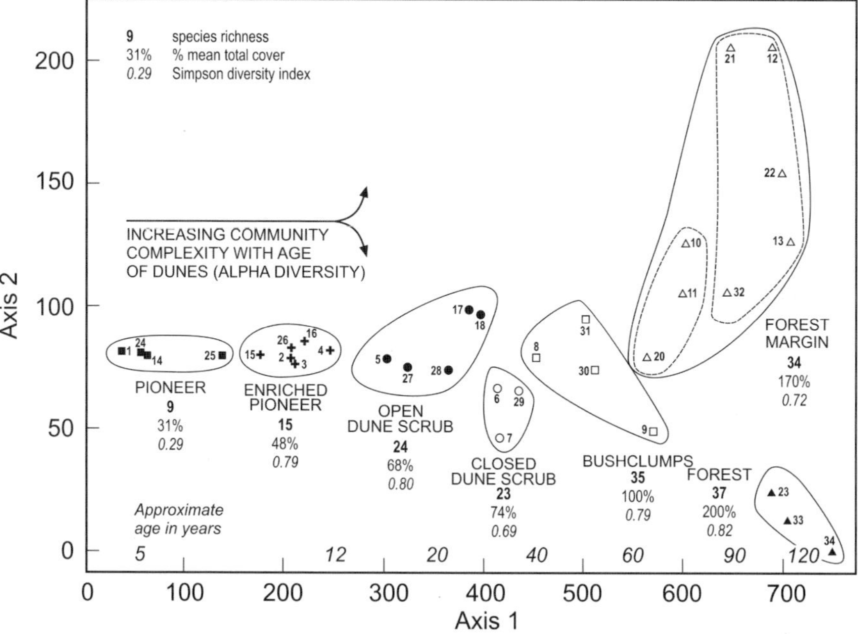

Fig. 5.4. Ordination of relevés of the Mtunzini dune field, showing the distribution of plant communities identified by TWINSPAN age of the communities estimated from aerial photographs. (Avis 1992)

As succession progresses, species diversity increases, but sometimes reaches a maximum value prior to the climax community, whereupon it decreases (Margalef 1968). Simpson Diversity Index (from density data) shows a sharp increase from pioneer to enriched pioneer communities, and remains fairly constant except for decreases in closed dune scrub and forest margin communities. This variability in the alpha diversity along the chronosequence supports Drury and Nisbet's (1973) statement that diversity is a result of microtopographic and other influences, and that it is not uniformly expressed in all parts of the community.

One would expect beta diversity to be highest and the coefficient of community (CC) lowest, for communities furthest apart, since they do not have many shared species (Table 5.2). The CC value decreases steadily as communities become more disjunct. The pioneer community has no similarity with forest margin and forest communities, with no shared species. Likewise, open dune scrub has a lower CC value with forest margins and forest than closed dune scrub, and values for the latter are lower than for bush clumps. The number of shared species decreases steadily from pioneer to forest communities suggesting that species turnover is very rapid along this 300-m successional gradient.

The edaphic factors also showed changes along the chronosequence, which are usually associated with successional changes in vegetation. Total organic

Table 5.2. Matrix showing the number of species shared by the eight communities (*italic*), the species richness in each community (**bold**) and the Sorenson's co-efficient of community (CC), in a study of succession on the Mtunizini dune field

Community	Pioneer	Enriched pioneer	Open dune scrub	Closed dune scrub	Bush clump	Bush clump forest margin	Forest margin	Forest
Pioneer	**8**	0.695	0.516	0.139	0.093	0.045	0	0
Enriched pioneer	*8*	**15**	0.666	0.526	0.320	0.235	0.122	0.076
Open dune scrub	*7*	*13*	**24**	0.765	0.508	0.333	0.103	0.065
Closed dune scrub	*3*	*10*	*18*	**23**	0.620	0.406	0.175	0.100
Bush clump	*2*	*8*	*15*	*18*	**35**	0.732	0.579	0.500
B/Forest margin	*1*	*6*	*10*	*12*	*26*	**36**	0.685	0.547
Forest margin	*0*	*3*	*3*	*5*	*20*	*24*	**34**	0.845
Forest	*0*	*2*	*2*	*3*	*18*	*21*	*30*	**37**

matter increased (0.5–6.1 %) with community complexity and pH decreased (8.5–7.3). An increase in the exchangeable bases was also noted, and the substratum became more stable with increasing distance from the sea. Changes in the soil properties can be related to changes (increases) in community complexity and also represent an increase in total biomass of the more complex communities. Salisbury (1925) first showed that organic content of dune soils increased along progressive dune ridges, and that soil changes could be correlated with changes in the vegetation. He also noted a leaching of carbonates with increasing age, and a change in pH from alkaline to acid. Lubke (1983) and Lubke and Avis (1982) noted similar edaphic changes with dune community succession at Kleinemonde in the Eastern Cape.

The Mtunzini area was most suitable for a comparative study on plant succession, since the age of the dunes was determined and the communities related to a chronosequence. Similar results were found by Olson (1958) and Morrison and Yarranton (1973), but in both cases dunes were much older (thousands of years). The rate of succession at Mtunzini is rapid due to the high rainfall and rapid accretion of sand along this coastline (Table 5.1). The succession is probably along a single pathway, at least in the early stages when pioneer communities are very similar, as indicated by their relative distribution along the median of axis 2 of Fig. 5.4. The position of relevés from forest margin and forest groups on the outer edges of axis 2 suggests that multiple pathways may occur at this stage. A multiple pathway succession, with three trajectories from pioneer to woodland vegetation was described by van Dorp et al. (1985) in the Netherlands over a 10–15-year period. DCA of the vegetation data supported the hypothesis that this increase in community complexity is related to an increase in the age of dunes, and species with similar environmental requirements appeared to be grouped together along the chronosequence gradient (Fig. 5.4) rather than evenly distributed along a continuum.

5.2.2 Studies on Transgressive Dune Fields and Partially Eroding Coastlines

In other regions of the coast there is no distinct chronological sequence, since the coastline is not prograding uniformly and is often eroding (Fig. 5.2B). Avis and Lubke (1996) sampled relevés ranging from pioneer communities to closed dune thicket in a number of localities at Kleinemonde, east of Port Alfred. Results were similar to those obtained at Mtunzini, except that pioneer foredune communities are dominated by a number of pioneer species (see Fig. 5.6), whereas the dune thicket was dominated by a single tree, the coastal white milkwood, *Sideroxylon inerme*. Although not linearly positioned along a transect across the dune field, the various communities sampled were located along a gradient of increasing community complexity (Avis and

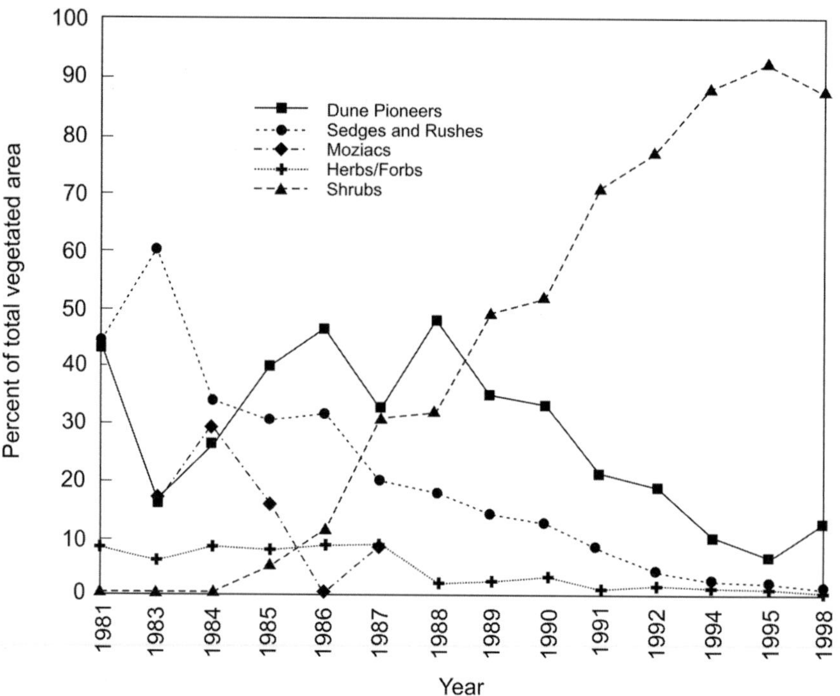

Fig. 5.5. The percentage cover of the major plant species groups over 20 years in a dune slack at Kleinemonde, Eastern Cape. (Lubke and Avis 2001)

Lubke 1996). At Kleinemonde the foredunes provide indirect facilitation by protecting the more mesic dune slacks from factors such as salt spray and sand movement, whereas this facilitation is more direct on prograding systems such as Mtunzini.

Studies on the dune slacks, which act as centres of diversity within the mobile dune field, were carried out by establishing fixed plots, which were monitored over 5, 17 (Lubke and Avis 1988, 2000) and 20 years (Fig. 5.5). The dynamic changes in the dune slack community result from the movement of sands parallel to the coastline, from the west to the east, due to predominantly southeasterly or westerly winds in the region (Fig. 5.6). As transgressive dunes advance they bury the western side of the slack, while other pioneer species are colonising the eastern margins. Vegetation can establish in these slacks due to the high water table and the protection afforded by the foredunes. Our site was covered with numerous *Scirpus nodosus* seedlings in 1978: and other herbaceous species, such as *Vellereophyton vellereum* and *Chironia decumbens*, became abundant in the early stages (Lubke and Avis 1982, 1988), but were replaced by woody species after about four years (Fig. 5.5). This study is continuing and as yet larger trees, i.e. *Brachylaena discolor, Sideroxylon inerme,* etc. have not yet been recorded. It supports concepts presented in an

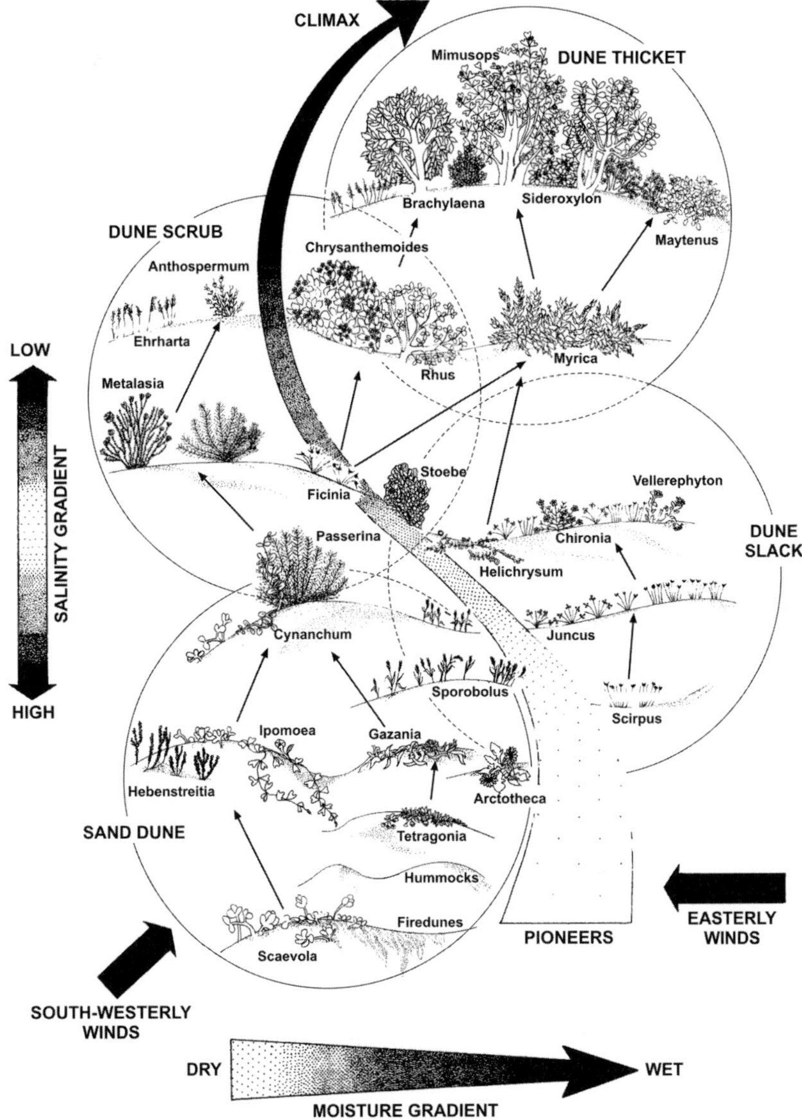

Fig. 5.6. Hypothetical model of dune succession at Kleinemonde. (Lubke and Avis 1988)

earlier model of succession for this region (Lubke and Avis 1988; Fig. 5.6). We have postulated that dune thicket communities (Fig. 5.2B) will not progress because rainfall in the eastern Cape is erratic with long drought periods. Sufficient rainfall for an extended period is crucial in the successional development of patches of thicket vegetation from dune slack vegetation into the transgressive and migratory dune field.

5.2.3 Studies on Rocky Shores and Eroding Coastlines

Along some of the southeast coast and Transkei coast (Fig. 5.1), there are rocky shores with wave-cut platforms and eroding cliffs (Fig. 5.2C). Above the rocky shores and cliffs coastal grassland is the climax vegetation. Detailed studies on these grasslands (Judd 2000) have revealed much information about the types of communities, but little about the origin of the grasslands. In some areas along these eroding shores bays and stretches of sandy dunes occur with prograding shorelines. Here, successional changes to forest or thicket are similar to that of the southeast coast. KwaZulu-Natal coastal grasslands are secondary in origin developing where thicket or forest was cleared for grazing or cultivation of crops (Weisser 1978; Lubke et al. 1991). Thus, these grasslands may also be secondary in these regions but this warrants further study.

5.3 Changes in Dune Succession Due to Invasive Aliens

Alien plant invasion has a large influence on the indigenous flora and plant communities of southern Africa (Shaughnessy 1980; Richardson et al. 1997).

5.3.1 The Effect of *Ammophila arenaria* as a Dune Pioneer on the Southern Cape Coast

Hertling and Lubke (1999a, 2000) and Lubke and Hertling (2001) studied the distribution of *Ammophila arenaria* (marram grass) communities and the dynamics of the system in various coastal dune systems, especially on the south coast (Fig. 5.1) where coastal fynbos is dominant (Fig. 5.2D). Marram grass becomes well established as a pioneer on foredunes in regions where the rainfall is high enough and consistent without lengthy drought periods. However, unlike other invader plants (see Sects. 5.3.2 and 5.3.3) it does not show traits of an outwardly aggressive behaviour in ecosystems. It forms dense stands and is able to tolerate burial like *Scaevola plumieri* as has been recorded by Maun and Lapierre(1984) in *A. breviligulata*, a similar species on the Lake Michigan dune shores. Although apparently excluding indigenous species, a quantitative assessment showed that the *A. arenaria* communities appear similar to those of indigenous dune plant communities with respect to species richness. Simpson's species diversity indices were, however, considerably lower (1.55 ± 0.08 and 2.39 ± 0.14 respectively), indicating a higher abundance of *A. arenaria* relative to other species in the marram stands. Although having a slight negative impact, *A. arenaria* does not show extreme dominance to the exclusion of other species, as it does on the North American Pacific coast (Weidemann and Pickart 1996).

Studies were carried out in a stabilisation area in the vicinity of the mouth of the Heuningnes River, at De Mond Nature Reserve, (Fig. 5.1) in the Southern Cape by Lubke and Hertling, (2001) in order to determine the succession from monospecific *Ammophila arenaria* stands. Stabilisation started in the late 1930s as the mouth was repeatedly blocked by drift sands in the winter, causing flooding of the Augulhas Plain. Between 1942 and 1958, 283 ha were stabilised and in 1996, when our study was undertaken, the stabilised area extended over 90 ha. Aerial photographs and a map showing dates of stabilisation were used to document the ages of the dune stabilisation and sampling was carried out in 20x1 m^2 quadrats in 42 stands which could be dated. Evidence for succession was detailed by Detrended Correspondence Analysis (DCA) of the stands, which were split into communities identified by TWINSPAN (Fig. 5.7). We found that stands stabilised in the 1980s now have a rich and dense dune scrub vegetation. This could be due to favourable

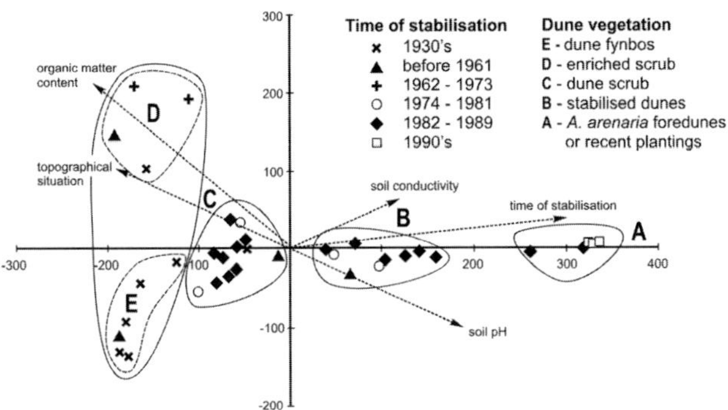

Dune fynbos / enriched dune scrub				
E. dune fynbos 22 - 60 years	**D.** *enriched scrub* 35 - 60 years	**C.** dune scrub 13 - 35 years	**B.** stabilised dunes 6 - 20 years	**A.** *A. arenaria* foredunes or recent plantings 3 - 10 years
Ischyrolepis eleocharis	*Rhus crenata*	*Myrica cordifolia*	*Psoralea repens*	*Ammophila arenaria*
Thamnochotus insignis	*Rhus laevigata*	*Metalasia muricata*	*Chironia baccifera*	*Elymus distichus*
Phylica ericoides	*Myrica quercifolia*	*Ficinia lateralis*	*Dasispermum suffruticosum*	*Arctotheca populifolia*
Euclea racemosa	*Knowltonia capensis*	*Passerina rigida*	*Ammophila arenaria*	*Didelta carnosa*
Agathosma collina	*Chasmanthe aethiopica*	*Helichysum patulum*	*Helchysum praecinctum*	*Senecio elegans*
Passerina paleacea		*Pentaschistis eriostoma*	*Trachyandra divaricata*	
Rhus glauca		*Nylandtia spinosa*	*Sutherlandia frutescens*	
Helichrysum dasyanthum		*Otholobium bracteolatum*		
Felicia zeyheri		*Ehrharta villosa*		

Fig. 5.7. Detrended correspondence analysis of De Mond stands results in the differentiation of successional stages at De Mond Nature Reserve. The important characteristic species are listed for the related communities. The ordination axes were also related indirectly to five environmental variables by canonical correspondence analysis. The four stages relate well to the stabilisation times of respective stands determined from a map of the stabilisation of the area. (Lubke and Hertling 2001)

habitat features such as a sheltered dune slack site with high organic matter and moisture. On the other hand, sites of an early stabilisation date can bear persistently vigorous *A. arenaria* populations if they are situated on exposed dune slopes characterised by greater sand movement, where *A. arenaria* profits from its superior sand burial tolerance. In sheltered dune slack locations, *A.arenaria* does not have this niche advantage and is replaced by other species, such as *Myrica cordifolia*, which was prevalent in these stands (e.g. community C, Fig. 5.7).

This study shows a significant example of succession involving *A. arenaria*, where the grass provides temporary stability of dune sands until indigenous dune plants take over. On a smaller scale, the succession from *A.arenaria* to indigenous plant species has been observed at other sites along the coast (Hertling 1997; Hertling and Lubke 1999b).

5.3.2 The Effect of Invasive Communities of *Acacia cyclops* in the Southern and Eastern Cape

Acacia cyclops and *A. saligna* from Australia, introduced to stabilise the sandy Cape Flats (Fig. 5.1), have invaded mobile dune fields along the southeastern coast (Richardson et al. 1997). These species have a high invasive potential (Hertling and Lubke 1999b) and in some cases, e.g. headland bypass systems, have resulted in stabilised mobile dune fields causing the lack of supply of sand to beaches in the bays upwind from the dune systems (Lubke 1985). Although not foredune pioneers, these woody species, being nodule-forming legumes, are extremely successful in low-nutrient sands and may fill a niche that is vacant on open dunes. The bird-dispersed seeds of *A. cyclops* have been carried inland into some grassland and thicket communities as well as northwards into the dunes along the Cape coast.

In South Africa, a Working for Water Programme (DWAF 2002) was introduced with the objective to remove alien invader species, thus increasing biodiversity of indigenous species, to supply more water to streams and reservoirs as the water-thirsty aliens are removed and to provide work and job security for the local people. Consequently, along some of the Eastern Cape Coast and at dune fields at Kleinemonde (Fig. 5.1) *Acacia cyclops* has been removed. Five 100-m^2 permanent plots have been established and 20 random 1-m^2 quadrats are sampled periodically to record the changes in plant diversity as indigenous species return following alien removal. These data have yet to be analysed but on open foredunes and on dune ridges the pioneers (e.g. *Scaevola plumieri* and *Ipomoea pes-caprae*) are rapidly returning, while in dune slacks some of the shrubs (e.g. *Passerina rigida)* have become established as seedlings and others (e.g. *Myrica cordifolia)* are expanding vegetatively into *A. cyclops* cleared areas. Seeds of *A. cyclops* were still found to occur in 63 % of the sample sites in a seed bank study of the dunes by Knevel (2001) so that a maintenance programme will be necessary to remove returning invading plants as they germinate.

5.3.3 The Introduction of *Casuarina equisetifolia* as a Dune Stabiliser

Along KwaZulu Natal, Mozambique and East African coasts, a commonly introduced tree is *Casuarina equisetifolia*, the Australian beefwood. These trees have been planted for shade on the hot tropical coasts and are able to grow in the pioneer and dune grassland zone very successfully (Lubke et al. 1991). However, in KwaZulu Natal, concern has been expressed as to their invasive potential, as they have spread in some areas where natural blowouts occur.

Avis (1992, 1995) made a study of the Department of Forestry records and sampled 17 sites where dune stabilisation had been carried out along the Eastern Cape coastline from 1965 to 1982. At this time, Forestry no longer uses the invasive woody species for dune stabilisation (Cobby 1988) but many invasive species persist.

At the Gulu River Mouth site, 20 km southwest of East London (Fig. 5.1), mobile sand from the dunes was thought to be responsible for silting up the river mouth, and to prevent further siltation it was decided to stabilise the dunes (Fig. 5.8). In 1975, over 6000 *Casuarina equisetifolia* trees were planted, due to its non-invasive and rapid growth rate. Indigenous shrubs such as *Rhus crenata* and *Passerina rigida*, together with pioneer species were either planted or seeded, and usually formed a number of distinct plant communities.

C. equisetifolia plantations are distinct from communities planted with indigenous vegetation, and have a species diversity of less than half that of the latter areas. Competitive exclusion results from their rapid growth rate, ability to rapidly utilise available nutrients, production of dense, fibrous roots covering several square metres of soil around each tree and copious leaf litter with an allelopathic effect on most other species. The only species able to survive was the herbaceous composite *Senecio litorosus*. Thus, despite being non-

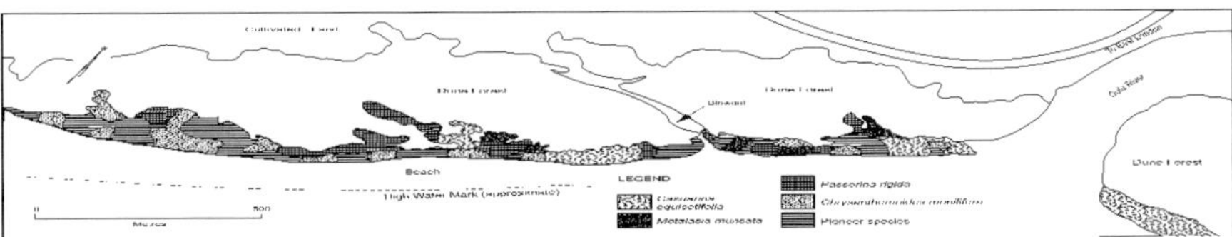

Fig. 5.8. Map showing the vegetation of the Gulu West stabilisation sites. Note the patches of *Casuarina equisetifolia* with very low species diversity, compared with other indigenous communities. (Avis 1992, 1995)

invasive this alien species is not suitable for dune stabilisation as, unlike *Ammophila arenaria,* it persists and will need to be physically removed to enable indigenous species to become established (Avis 1995).

5.4 Discussion

5.4.1 Distinguishing the Mechanism of Succession with Indigenous Pioneers

The only sequential assessment of dune communities and dune vegetation dynamics along the southern African coastline is that of Tinley (1985). Later studies have looked at areas along the east coast of Africa more exactly (e.g. Frazier 1993). Studies of the communities along gradients of increasing age as in our studies at Mtunzini and over a temporal sequence at Kleinemonde are unique on the south-east African coastline. In studies on sand dune systems at Malindi (Fig. 5.1) in Kenya, Musila et al. (2001) record similar results to Mtunzini with different plant species but without age determination of the dune system. The change in species richness which they recorded across the dune field was from 2–5 species on young dunes to 40–60 species in mature communities, and thus compares to our Mtunzini results (Table 5.2 and Avis 1992).

The distinct separation of relevés into communities (Table 5.1) along a gradient of increasing age (Figs. 5.2, 5.3 and 5.4) suggests that these communities do not integrade continuously along the environmental gradients (Whittaker 1975). They appear to form distinct, clearly separated zones more in line with the "community concept" (Clements 1916). However, this does not prove that this concept applies to the vegetation sequence at Mtunzini, since one cannot prove the one hypothesis by rejecting the other (Shipley and Keddy 1987). The situation is even more confusing when one considers that at the scale of individual species turnover (Table 5.2), there is overlap with no sharp boundaries between species, suggesting that species turnover follows the individualistic hypothesis of Gleason (1926). However, Shipley and Keddy (1987) have shown that the individualistic concept is unfalsifiable at the level of pattern analysis, allowing one to favour the "community-unit concept". The ordination diagram (Fig. 5.4) suggests an organisational structure of the communities (*vide* Clements 1916), but changes at the species level (Avis 1992) suggest that the communities integrate continuously along this gradient (Whittaker 1975) in an individualistic manner (Gleason 1926). However, ordination methods cannot provide unequivocal evidence for the continuum concept (Austin 1985), but the low level of similarity between communities less than 100 m apart (Table 5.2) suggests that species only intergrade over short distances. Possibly, as stated by Shipley and Keddy (1987), one needs to deny this dichotomy and to consider multiple working hypotheses of community structure.

Problems and shortfalls in successional theory led to the Connell & Slatyer (1977) models, and these support our Facilitation Model. There are also a number of other studies which support these models. Olson(1958) presented evidence for an autogenic succession on the Lake Michigan sand dunes, and he showed that two Connell and Slatyer models (mechanisms *vide* Picket et al. 1987) play a role in this succession.

5.4.2 Effects of Aliens and the Need for Dune Stabilisation

Many of the large-scale dune stabilisation programmes on this coastline have the primary objective of preventing sand movement which threatens human well-being. Often, therefore, the use of alien vegetation has been promoted both here and overseas (Weidemann and Pickart 1996) but potentially invasive species should not be used due to the threats they pose to adjoining ecosystems. Once the need for a stabilisation programme has been determined, the creation of functional, aesthetic ecosystems should be the primary objective of such programmes. Our studies show that indigenous species can be used successfully on our coastline (Lubke 1983) but the process is costly, time consuming and slow, often necessitating detailed studies to be undertaken prior to the initiation of a stabilisation programme and ongoing monitoring. If stabilisation is necessary under harsh, adverse conditions, *Ammophila arenaria* has been shown to be most suitable (Lubke and Hertling 2001).

5.4.3 Conservation of Biodiversity and Dune Ecosystems and Future Studies

Introduction or programmes such as the “Working for Water” clearance of aliens (DWAF 2002) have long-term goals of restoration of biodiversity and reestablishment of ecosystems. The effect of alien species on diversity in dune systems has been shown (Avis 1996) and removal of aliens at Kleinemonde has shown how diversity can increase. Our unique coastal systems (Fig. 5.2) need to be protected and fortunately many reserves are sited along coastal areas for recreational proposes. Studies on the use of indigenous species for dune stabilisation (Knevel 2001) have shown that these pioneers have the potential for dune stabilisation, but more studies on their application in the field is required for dune management.

Aspects in the dune successional process still need to be answered such as the role of the various species in facilitating the process, if in fact they do have a role. Ecophysical studies and the application of functional ecology principles need to be considered, requiring more studies on the life histories and the behaviour of the species. Finally, the explanation of the successional process of coastal grasslands along our rocky shores still needs to be studied in detail to show their relationship to the more widespread thicket climax communities on sandy shores.

Acknowledgements. I would like to thank my postgraduate students for the contribution they have made to our understanding of the dune systems through long hours in the field, laboratory and discussions on dune processes. The undergraduate students and others who also contributed in gathering data are also thanked. Finally, the Rhodes University is acknowledged for funding this research.

References

Austin MP (1985) Continuum concept, ordination methods and niche theory. Aust Rev Ecol Syst 16: 39–61

Avis AM (1992) Coastal dune ecology and management in the eastern Cape. Unpubl PhD Thesis, Rhodes Univ, Grahamstown

Avis AM (1995) An evaluation of the vegetation developed after artificially stabilizing South African coastal dunes with indigenous species. J Coastal Conserv 1:41–50

Avis AM, Lubke RA (1996) Dynamics and succession of coastal dune vegetation in the Eastern Cape, South Africa. Landscape Urban Plann 34:347–254

Cobby JE (1988) The management of diverse ecological types of the Directorate of Forestry. In: Bruton MN, Gess FW (eds). Towards an environment plan for the Eastern Cape. Rhodes Univ, Grahamstown. pp 126–163

Clements FE (1916) Plant succession. Carnegie Institute, Washington, DC, Publ 242

Connell JH, Slatyer RO (1977) Mechanisms of succession in natural communities and their role in community stability and organization. Am Nat 111:1119–1144

Drury WH, Nisbet ICT (1973) Succession. J Arnold Arboretum 54:331–368

DWAF – Department of Water Affairs and Forestry (2002) http://www.dwaf.pwv.gov.za/Projects/wfw/

Frazier JG (1993) Dry coastal ecosystems of Kenya and Tanzania. In: van der Maarel E (ed) Dry coastal ecosystems – African, America, Asia and Oceania. Ecosystems of the world 2B. Elsevier Amsterdam, pp 129–150

Gleason HA (1926) The individualistic concept of plant association. Ibid 53:7–26

Hertling UM (1997) *Ammophila arenaria* (L.) Link. (marram grass) in South Africa and its potential invasiveness. PhD Thesis, Rhodes Univ, Grahamstown, South Africa, 279 pp

Hertling UM, Lubke RA (1999a) Indigenous and *Ammophila arenaria*-dominated dune vegetation on the South African Cape coast. J Appl Veg Sci 2:157–168

Hertling UM, Lubke RA (1999b) Use of *Ammophila arenaria* for dune stabilization in South Africa and its current distribution – perceptions and problems. Environ Manage 24:467–482

Hertling UM, Lubke RA (2000) Assessing the potential for biological invasion – the case of *Ammophila arenaria* in South Africa. S Afr J Sci 96:520–527

Judd R (2000) The Coastal Grasslands of the Eastern Cape, West of the Kei River. PhD Thesis. Rhodes Univ, Grahamstown, South Africa

Knevel IC (2001) The life history of selected coastal fore dune species of Eastern Cape, South Africa. PhD Thesis, Rhodes Univ, Grahamstown, South Africa, 291 pp

Lubke RA (1983) A survey of the coastal vegetation near Port Alfred, Eastern Cape. Bothalia 14:725–738

Lubke RA (1985) Erosion of the beach at St. Francis Bay, eastern Cape. Biol Conserv 32:99–127

Lubke RA, Avis AM (1982) Factors affecting the distribution of *Scirpus nodosus* plants in a dune slack community. S Afr J Bot 1(4):97–103

Lubke RA, Avis AM (1988) Succession on the coastal dunes and dune slacks at Kleinemonde, Eastern Cape, South Africa. Monogr Syst Bot Missouri Bot Gard 25:599–622

Lubke RA, Avis AM (2000) 17 years of change in a dune slack community in the Eastern Cape. Proc IAVS Symp 2000 IAVS, Opulus Press, Uppsala, pp 35–38

Lubke RA, Hertling UM (2001) The role of European marram grass in dune stabilization and succession near Cape Agulhas, South Africa. J Coastal Conserv 7:171–182

Lubke RA, Avis AM, Phillipson PB (1991) Vegetation and floristics. Reprinted from: Environmental impact assessment, eastern shores of Lake St Lucia (Kingsa/Tojan Lease Area) Specialist Reports. Coastal and Environmental Services, Grahamstown, 741 pp

Lubke RA, Avis AM, Steinke TD, Boucher CB (1997) Coastal vegetation: In: Cowling RM, Richardson D (eds) Vegetation of South Africa. Cambridge University Press, Cape Town

Margalef R (1968) Perspectives in ecological theory. University of Chicago Press, Chicago

Maun MA, Lapierre J (1984) The effects of burial by sand on *Ammophila breviligulata*. J Ecol 72:827–839

Moll EJ (1969) A preliminary account of the dune communities at Pennington Park, Mtunzini, Natal. Bothalia 10:615–626

Morrison RG, Yarranton GA (1973) Diversity, richness and evenness during a primary sand dune succession at Grand Bend, Ontario. Can J Bot 51:2401–2411

Musila WM, Kinyamario JI, Jungerius PD (2001) Vegetation dynamics of coastal sand dunes near Malindi, Kenya. Afr J Ecol 39:170–177

Olson JS (1958) Rates of succession and soil changes on south Lake Michigan sand dunes. Bot Gaz 119:125–170

Pammenter NW (1983) Some aspects of the ecophysiology of *Scaevola thunbergii*, a subtropical coastal dune pioneer. In: McLachlan A, Erasmus T (eds) Sandy beaches as ecosystems. Developments in hydrobiology, vol 19. W Junk, The Hague, pp 675–685

Picket STA, Collins SL, Armesto JJ (1987) Models, mechanisms and pathways of succession. Bot Rev 53:335–371

Richardson DM, Macdonald IAW, Hoffmann JH, Henderson L (1997) Alien plant invasions. In: Cowling RM, Richardson DM, Pierce SM (eds) Vegetation of Southern Africa. Cambridge University Press, Cambridge, pp 535–570

Salisbury EJ (1925) Note on the edaphic succession in some dune soils with special reference to the time factor. J Ecol 13:322–328

Shaughnessy GL (1980) Historical ecology of alien woody plants in the vicinity of Cape Town, South Africa. PhD Thesis, University of Cape Town

Shipley B, Keddy PA (1987) The individualistic and community-unity-concepts as falsifiable hypotheses. Vegetatio 69:47–55

Talbot MMB, Bate GC (1991) The structure of vegetation in bush pockets of trangressive coastal dune fields. S Afr J Bot 57:156–160

Tinley KL (1985) Coastal dunes of South Africa. South African National Scientific Programmes Report, No 109. CSIR, Pretoria.

Van Dorp D, Boot R, Van der Maarel E (1985) Vegetation succession on the dunes near Oostvoorne, The Netherlands, since 1934, interpreted from air photographs and vegetation maps. Vegetatio 56:123–126

Weidemann AM, A Pickart (1996) The *Ammophila* problem on the northwest coast of North America. Landscape Urban Plann 34:287–299

Weisser PJ (1978) Changes in the area of grasslands on the dunes between Richards Bay and the Mfolozi River 1937–1974. Proc Grassland Soc S Afr 13:95–97

Weisser PJ, Backer AP (1983) Monitoring beach and dune advancement and vegetation changes 1937–1977 at the farm Twinstreams, Mtunzini, Natal, South Africa. In: Mc Lachlan A, Erasmus T (eds): Sandy beaches as ecosystems. Developments in Hydrobiology, vol 19. Junk, The Hague, pp 727–740

Weisser PJ, Garland LF, Drews BK (1982) Dune advancement 1937–1977 at the Mlalazi Nature Reserve, Natal, South Africa, and preliminary vegetation-succession chronology. Bothalia 4:127–130

Whittaker RH (1975) Communities and ecosystems, 2nd edn. MacMillan, London

6 Why Young Coastal Dune Slacks Sustain a High Biodiversity

A.P. Grootjans, E.B. Adema, R.M. Bekker and E.J. Lammerts

6.1 Introduction

Dune slacks are depressions within coastal dune areas that are flooded during the rainy season, which in Europe is during winter and spring (Boorman et al. 1997; Grootjans et al. 1998), but in the tropics during the summer (Vázquez, Chap 12). During the dry season the water table may drop far below the surface. Young dune slacks that have been formed in a natural way, by sand blowing or natural dune formation (Piotrowska 1988; Zoladeski 1991), are very poor in nutrients and at the same time very species-rich. Various life and growth forms can be present in such slacks: annuals, biennials, perennials, young shrubs and trees (Crawford and Wishart 1966; Ranwell 1972). Dune slack soils are usually calcareous, since they normally originate from recently deposited sands that contain much shell fragments. Dune slacks with acid soils occur in areas where sand has been deposited at the beach with a low initial lime content. Examples are dune areas in parts of Poland (Piotrowska 1988) and the Dutch, German and Danish Wadden Sea Islands, where initial lime contents are low (less than 2 % $CaCO_3$; Petersen 2000) and where precipitation dominates over evaporation. This leads to prominent decalcification processes in the top layer and to rapid acidification (Stuyfzand 1993). In dryer areas where evaporation dominates over precipitation, decalcification processes are less evidently expressed in the vegetation. Flooding frequencies during the wet period are decisive for the plant species composition in such dune areas (Zunzunegui et al. 1998; Munoz-Reinoso 2001).

Although many species are now restricted to the coastal area, dune slacks have very few endemic species (Van der Maarel and Van der Maarel-Versluys 1996). Many typical dune slack species can also occur in calcareous fens, fen meadows, and other types of inland wetlands. The restriction of many wetland species to the coastal area is no doubt related to the intensive land use in the mainland areas.

Ecological Studies, Vol. 171
M.L. Martínez, N.P. Psuty (Eds.)
Coastal Dunes, Ecology and Conservation

In the following, we will discuss the complex interactions between dune slack vegetation, hydrological conditions, and management in dune slacks in order to conserve or restore these ecosystems for future generations. The examples discussed will be mostly from the NW European dune areas, where dune areas have been affected very negatively by human activities, such as mass recreation, abstraction of drinking water for large cities, increased atmospheric nitrogen deposition from industrial and agricultural areas, and large-scale afforestation (Van Dijk and Grootjans 1993). This destruction of what is seen by many as the last remnants of natural ecosystems led to much societal opposition during the last decades and many restoration projects were initiated to restore dune ecosystems with a high biodiversity (Kooijman, Chap. 15).

6.2 The Dune Slack Environment

Since dune slacks are in fact temporary wetlands, most typical dune slack species have to be adapted to both wet and dry conditions. Consequently, plants that grow in dune slacks may experience severe anoxic conditions, which can be followed by sometimes very dry conditions.

6.2.1 Hydrological System

Dune slacks are not just temporary dune ponds filled with water during the wet season and evaporate water during the dry season. Although such slacks exist they are very rare. Most dune slacks are fed by various water sources. This can be precipitation water, surface water or groundwater. The latter two sources are usually calcareous while the first is acid. The hydrological situation can be more complicated, since the groundwater may come from different hydrological systems (Munoz-Reinoso 2001; Grootjans et al. 2002). In most cases the maintenance of dune slack ecosystems depends on both the amount of precipitation and groundwater discharge. Dune slacks fed by calcareous groundwater are usually situated at the low-lying periphery of the dune system, where most of the groundwater of the main hydrological system discharges. However, seepage slacks can also be found close to the top of the main hydrological system when thick clay or peat layers prevent infiltration to deeper layers and give rise to local groundwater flow towards adjoining dune slacks. Such slacks function as 'flow-through lakes' with groundwater discharge in one part of the slack and infiltration of surface water in another (Stuyfzand 1993). In a dune area with several dune slacks lying close together, slight differences in water level between the slacks may initiate groundwater flow from one slack to another (Kennoyer and Anderson 1989; Grootjans et al.

1996). Under such conditions calcareous groundwater from deeper layers can flow towards the up-gradient parts of the slack. The influx of calcareous groundwater stimulates mineralisation of organic matter and consequently the accumulation of organic matter is lower here than at the infiltration sites (Sival and Grootjans 1996).

6.2.2 Adaptations to Flooding and Low Nutrient Supply

Plant species with well-developed aerenchyma such as *Schoenus nigricans* and *Littorella uniflora* can counteract anoxia by actively leaking oxygen from the roots into the surroundings. This phenomenon is called radial oxygen loss (ROL) (Armstrong 1975). Anoxic conditions, therefore, may prevent the establishment of late-successional species without ROL capabilities. Such species are not adapted to high concentrations of reduced iron, manganese and sulphide that can occur in the rooting zone under anoxic conditions (Studer-Ehrensberger et al. 1993). Sulphide in particular can be very harmful for plant species if it remains in a reduced state (Lamers et al. 1998).

In addition to anoxia, dune slack species have to cope with nutrient-poor conditions (Schat et al. 1984). Many species, such as *Littorella uniflora, Centaurium pulchellum*, and *Radiola linoides*, have a very low nutrient demand because they are very small. Others, such as *Schoenus nigricans*, form large and long-lived tussocks. The tussock as such, is not an adaptation to nutrient poor conditions, but this particular species can very efficiently recycle nitrogen and phosphorus from senescent to juvenile shoot tissue (Ernst and Van der Ham 1988). In this way, it can capture nutrients in the tussock, making them unavailable for fast growing grasses and herbaceous species (Van Beckhoven 1995).

6.3 Succession in Dune Slacks

Natural succession in dune slacks starts with a pioneer phase in which small pioneer species establish on an almost bare soil. This wet soil is usually covered with a thin layer of green algae and laminated microbial mats (Van Gemerden 1993; Grootjans et al. 1997), and only a few phanerogamous species are adapted to the very low nutrient availability in this phase. In a later stage, where some organic matter has accumulated, pleurocarpic bryophytes and small dune slack species can establish. In this stage, the accumulation rate of organic matter increases and after 10–15 years tall grasses and shrubs appear, which eventually leads to the decline of pioneer species that require nutrient-poor and base-rich habitats. Red list species, such as *Dactylorhiza incarnata, Epipactis palustris* and *Liparis loeselii* represent such basiphilous species,

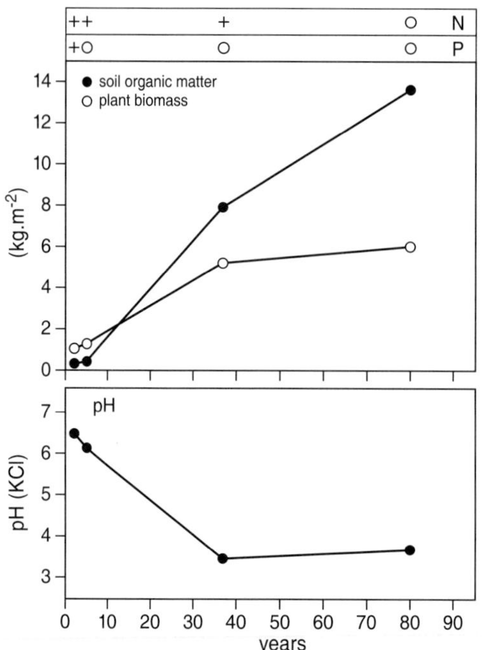

Fig. 6.1. Plant biomass, soil organic matter and soil pH measured in a chronosequence on the Dutch Wadden Sea island of Terschelling, representing various succession stages (2, 5, 37 and 80 years). Responses of the plant biomass to the addition of nitrogen and phosphorus fertilizer (quantity=16 g/m^2) are indicated above

which are most abundant in the intermediate phases, where they receive nutrients from soil mycorrhizas (Smith 1966).

The shift from pioneer stage to more mature stages usually takes place between 20–30 years (Van der Maarel et al. 1985). In some dune slacks, however, pioneer stages may last for at least 30–60 years (Petersen 2000; Adema et al. 2002). The rate of vegetation succession in dune slacks is largely controlled by the productivity of the ecosystem, the decomposition of organic matter and the recycling of nutrients within the ecosystem (Koerselman 1992; Olff et al. 1993). To monitor factors governing vegetation succession over a period of more than half a century is almost impossible to carry out. However, sod-cut experiments, in which the organic top layer is removed, are available in some dune slacks where nature managers have tried to restore pioneer stages at various time intervals. Such a spatial representation of supposed successional stages is called a chronosequence. An analysis of vegetation development in such a chronosequence (Berendse et al. 1998) showed that during the first 10 years most of the organic matter was stored in the living plants, particularly in the root system (Fig. 6.1). After about 15 years the amount of soil organic matter increased, while the pH dropped steeply, and a thick (c. 10 cm) organic layer developed. This drop in pH only occurred in dune slack which were poor in $CaCO_3$, not in calcareous soils with lime contents above 0.3 % $CaCO_3$ (Ernst et al. 1996).

6.3.1 Nutrient Limitation During Succession

Fertilisation experiments in the same chronosequence showed that the growth of the aboveground biomass was limited by both nitrogen and phosphorus in the youngest (2-year-old) sod-cut experiment (Fig. 6.1; Lammerts et al. 1999). In the 5-year-old stage and in the 37-year old stage only nitrogen was limiting, most likely because many pioneer species have a very low phosphorus demand (Willis 1963; Van Beckhoven 1995). The sedge species *Schoenus nigricans* showed no response at all to either nitrogen or phosphorus additions. This implies that, as long as phosphorus limits the growth of tall grasses, basiphilous pioneer vegetation can persist for quite some time, even when nitrogen availability increases. Buffer mechanisms that keep the soil above pH 6 appear to be crucial for maintaining low phosphorus availability.

Due to increased accumulation of organic material in later successional stages, the N-mineralisation also increases, leading to a higher availability of mineral nitrogen (Berendse et al. 1998). The phosphorus availability is also relatively high in older dune slacks on the Dutch Wadden Sea islands because soils are poor in iron here and the phosphorus is, therefore, only loosely bound to iron-organic compounds (Kooijman and Besse 2002). Tall growing, late successional grass species, such as *Calamagrostis epigejos,* rapidly increase in cover after the nitrogen and phosphorus limitations have been lifted (Ernst et al. 1996).

In the 80-year-old stage the aboveground biomass was no longer limited by nitrogen. The most likely explanation for this lack of response to nutrient addition in this old successional stage is that competition for light has become a dominant factor. Willow (*Salix repens*) and heathland species *(Empetrum nigrum, Erica tetralix*) invest in supportive, non-productive tissue and lift their photosynthetically active parts to the top of the vegetation. In this way they can in some stage overgrow the small pioneer species.

Atmospheric input of nitrogen may accelerate the accumulation of organic matter in the topsoil considerably, because the growth of most pioneer and mid-successional species is N-limited and therefore, responsive to additional supply of nitrogen.

6.3.2 Seed Banks and Succession

Seed bank research (Bekker et al. 1999) showed that most pioneer species, such as *Centaurium pulchellum*, and several *Juncus* species had long-term persistent seed banks and that many late successional species, such as *Salix repens, Eupatorium cannabinum* and *Calamagrostis epigejos* had transient seed banks (Thompson et al. 1997). The species that had long-term persistent seed banks were also the species that appeared immediately after sod cutting, even if the original pioneer vegetation had disappeared since several decades. Figure 6.2 clearly shows that the group of dune slack pioneer species as a whole had long-term persistent seed banks with seeds surviving at least 5 years but often for longer periods. The longevity index in Fig. 6.2 (after Bekker et al. 1998) was constructed from records of longevity in the database of Thompson et al. (1997). It indicates the 'average' period of seed survival in the soil. An index of 0–3 means that the seeds are transient; they cannot survive in the soil longer than 1 year. The range 4–7 indicates that seeds can survive for 1–5 years, and from 8–10 seeds survive for more than 5 years. The group of late successional species had short-lived or transient seed banks, while the seed longevity of the group of mid-successional species was intermediate. However, no data are available with respect to most typical dune slack species such as orchid species, which are critically endangered in NW-

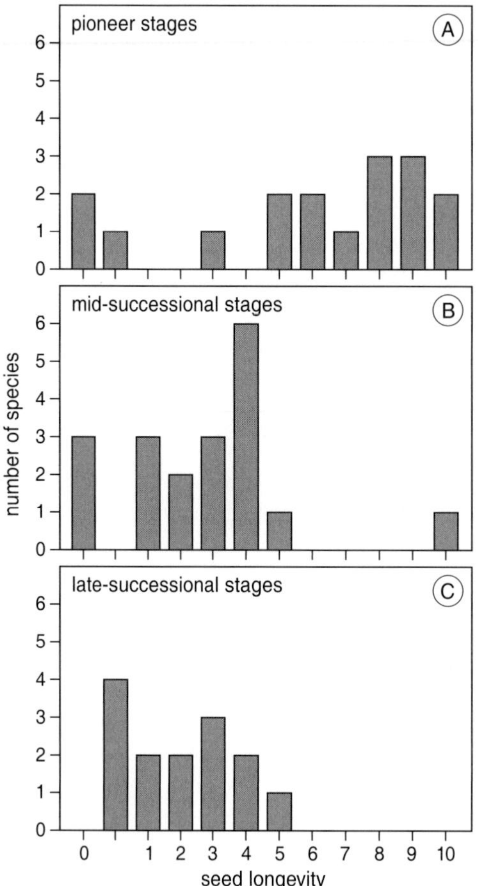

Fig. 6.2. Seed longevity of dune slack species of pioneer stages (**A**), mid-successional stages (**B**), and late-successional stages (**C**)

Europe. This means that we cannot rely on the presence of living seeds in sites where these species were found and have disappeared recently during vegetation succession.

The establishment of a new population of an endangered sedge species (*Schoenus nigricans*) can be illustrated in a recent sod-cut experiment in the dune slack Koegelwieck on the island of Terschelling. This slack is fed by calcareous groundwater that enters at the southern border. The water proceeds as surface water and becomes less calcareous due to dilution by precipitation water. When the surface water infiltrates again at the northern boundary, the water has become very calcium poor. *Schoenus nigricans* did not appear to have a long-term persistent seed bank, but had to establish new populations

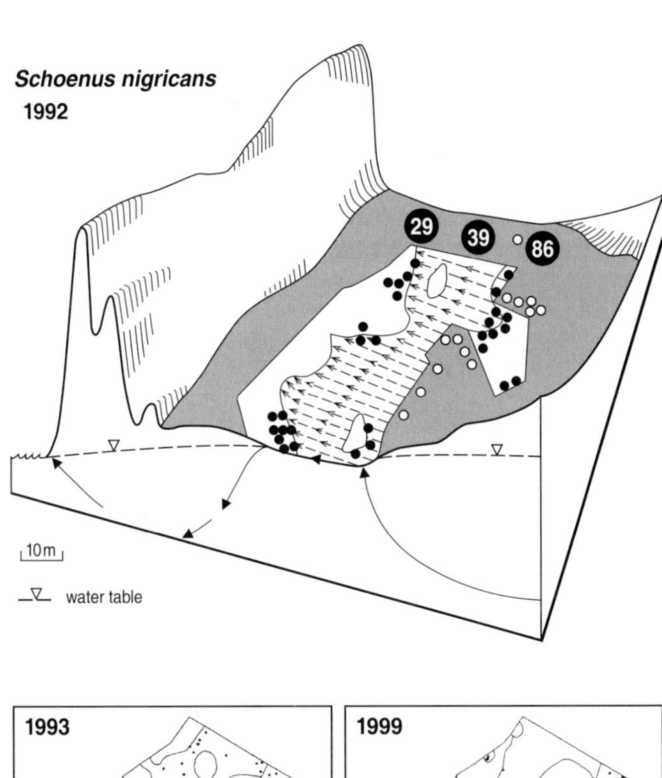

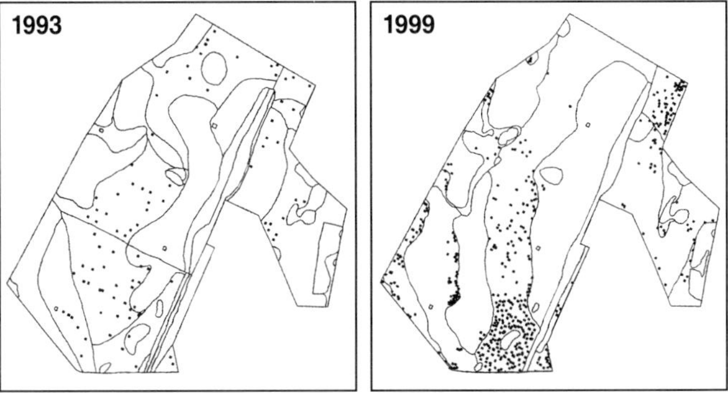

Fig. 6.3. Distribution of the pioneer species *Schoenus nigricans* in 1992, 1993 and 1999 in a sod-cut dune slack (Koegelwieck) on the Wadden Sea island of Terschelling. This part of the slack had been sod cut in 1991. Mature plants in 1992 are indicated by *open dots*, juveniles in 1992 and 1992 by *closed dots*. *Encircled numbers* indicate the calcium concentration (mg/l) measured in the surface water in May 1992. The *arrows* represent water flow (both groundwater and surface water flow)

from scarcely distributed individuals in the close vicinity of the sod-cut area (Fig. 6.3). One of the target species, for instance, *Schoenus nigricans*, established small populations, within 1 year after sod cutting, although no viable seeds were found in the soil, not even in the stage where *Schoenus nigricans* was abundant. So, apparently *Schoenus nigricans* had to establish a new population from the limited number of adult plants surviving in the uncut vegetation which was more than 75 years old (Lammerts et al. 1999). The distribution of young *Schoenus* plants in the slack suggests that the young plants establish themselves either close to the mature plants at the exfiltration side of the slack or were transported by surface water flow to the infiltration part of the slack. One year later, the distribution of juvenile *Schoenus* plants is more or less random in the sod-cut area, except for the very wet and the very dry parts. After 8 years the distribution of *Schoenus* plants has changed considerably. The species is now most abundant in the western part of the slack with slightly less wet conditions. The sites that were favourable for germination in 1991/1992 were evidently not the sites that are favourable for more adult plants.

6.3.3 Stability of Pioneer Stages

The influx of anoxic and iron-rich groundwater is important for vegetation succession in yet another way. Lammerts et al. (1995) showed that pioneer stages were much more stable in groundwater-fed dune slacks compared to slacks that were situated in infiltration areas. They hypothesised that the discharge of groundwater in spring and early summer keeps the soil moist so that laminated microbial- and algal mats do not dry out (Van Gemerden 1993). It was already known that algal mats can stabilise sandy substrates during the very early stages of dune slack formation (Pluis and De Winder 1990). When growth leads to the formation of visible layers these are called microbial mats. Prerequisites for the growth of microbial mats are the availability of water, much light, and the absence of excessive erosion and consumption by animals. Optimal growth conditions occur on bare soils that are regularly flooded or attain sufficient moisture by capillary water supply. Cyanobacteria in microbial mats can fix nitrogen (Stal et al. 1994) and the mats may develop in a relatively short period. They may, therefore, assist in the colonisation by phanerogams.

Photosynthesis is concentrated in the algal layer of the microbial mat. Respiration by heterotrophic bacteria rapidly depletes oxygen that is produced in this top layer. Consequently, oxygen penetration in microbial mats is shallow ranging from less than 2-mm depth in the dark to 5–6 mm during active photosynthesis (Van Gemerden 1993). In the absence of oxygen, alternative electron acceptors, such as sulphate, are used by heterotrophic bacteria, which use organic matter as carbon source. Well-developed mats in sulphate-rich environments are therefore characterised by an intense sulphur cycling. Sulphate-reducing bacteria produce sulphide which is partly oxidised again by either phototrophic sulphur bacteria in the presence of light or by colourless sulphur bacteria when sulphide diffuses to the oxic layer. Sulphide concentrations may reach toxic levels for higher plants depending on the amount of available sulphate and the amount of sulphide-fixing iron minerals.

Adema et al. (2002) measured sulphide and oxygen concentrations in a dune slack on the Wadden Sea island Texel in the Netherlands, using micro-electrodes (Fig. 6.4). They found that on the infiltration side of the slack where the topsoil had been decalcified, the sulphide concentrations reached toxic levels (30–90 μmol/l) for some higher plants, in particular sedges (Lamers et

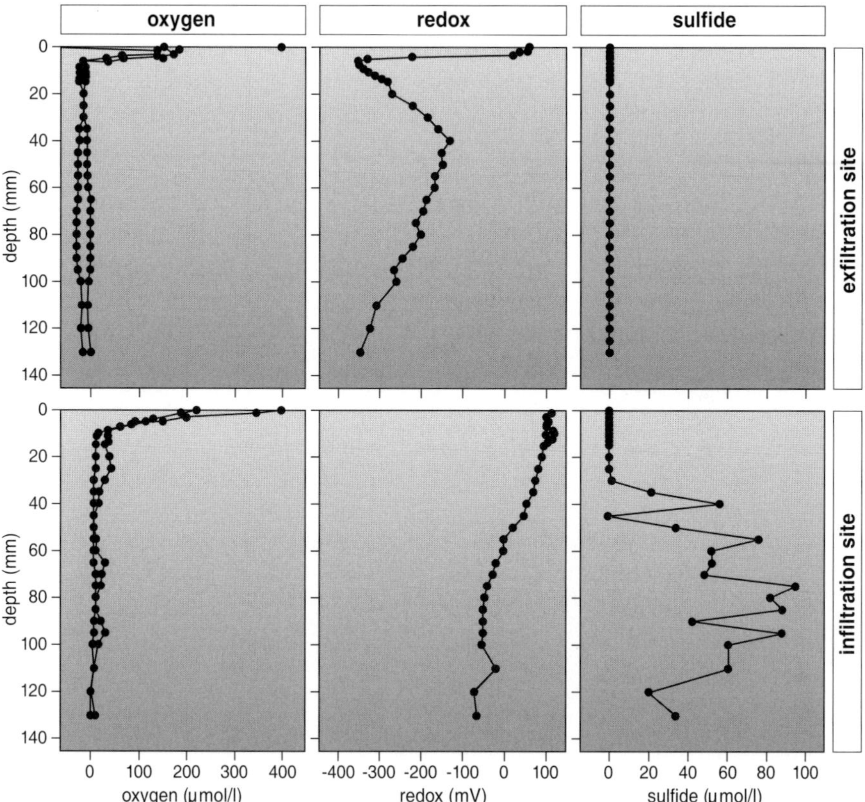

Fig. 6.4. Oxygen content, redox-potential, and sulfide concentrations measured in soil profiles in exfiltration and infiltration sites of 'De Buiten Muy' on the island of Texel in The Nertherlands (after Adema 2002). No free sulfide was measured in the exfiltration despite lower redox potentials. Apparently, the toxic sulfide is bound by regular supply of iron in the discharging groundwater

al. 1998). Plants that could grow at the infiltration side of the slack were common reed (*Phragmites australis)* and small pioneer species, such as *Samolus valerandi* and *Littorella uniflora*. At the exfiltration side, no sulphide was measured, although the redox potentials were much lower than in the infiltration site, due to continuous inflow of anaerobic and iron-rich groundwater (Fig. 6.5). The authors argued that the iron-rich groundwater fixed the free sulphide produced by the microbial mats by forming FeS. At the infiltration side, however, no iron was present any more and free sulphide could accumulate. These relatively high sulphide concentrations did not harm common reed, nor the pioneer species, since they are capable of oxidising sulphide to sulphate by releasing oxygen from their roots (radial oxygen release). The sulphide production in the infiltration areas can, however, release phosphates in the iron-depleted topsoil due to binding of sulphides with iron (Lamers et al. 1998). The infiltration side of such a slack, therefore, would not be able to maintain a pioneer vegetation for a long time and tall reed vegetation would soon take over.

A stable pioneer vegetation existed for over 60 years between the exfiltration and central parts of the slack because the pH is buffered here; sulphide production is neutralised by iron and acidification is prevented by discharge of calcareous groundwater. Sival et al. (1998) found that in exfiltration sites of dune slacks also secondary, in situ, carbonate deposition occurred in early

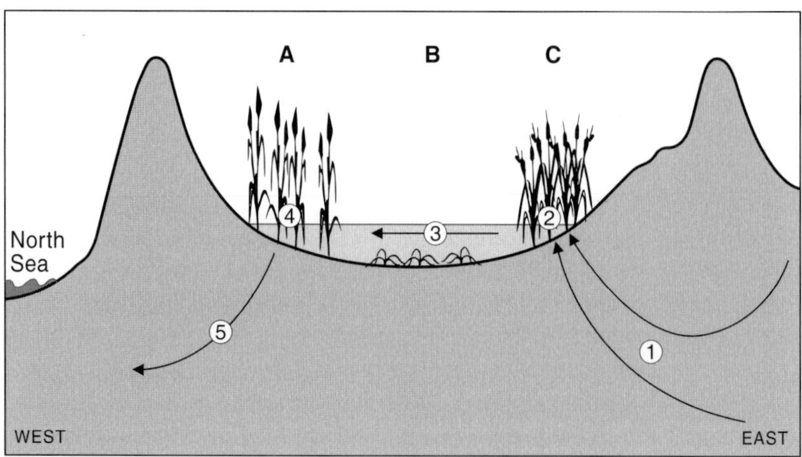

Fig. 6.5. Schematic presentation of a flow-through dune slack on a Dutch Wadden Sea island. In this particular case the vegetation zonation was derived from the dune slack 'De Buiten Muy' on the island of Texel. *A* Stand of common reed (*Phragmites australis*; *B* pioneer stage with *Littorella uniflora*; *C* tall sedges with *Carex riparia*. *1* Incoming calcium and iron-rich groundwater; *2* exfiltration of groundwater; *3* precipitation of iron and calcium; *4* infiltration of iron- and calcium-poor surface water; *5* sulfate reduction during infiltration. (After Adema 2002)

stages of dune slack succession. The carbonate was deposited in a very thin layer on the mineral soil. Loss of CO_2 from the calcareous groundwater resulted in carbonate precipitation at the soil-air interface (Chafetz 1994), thus counteracting soil acidification in a very significant way. Such carbonate precipitation occurs when water tables remain and temperatures are also high. Under such conditions CO_2 escapes from the discharging calcareous groundwater or is taken up by algae, mosses or small water plants. Calcium carbonate is then deposited as a thin silty layer on the soil or even on the leaves of plants. At the exfiltration side of the slack, therefore, the groundwater discharge contributes to maintaining a high pH, low nutrient availabilities, in particular phosphate, and preventing toxic sulphide conditions.

Summarising, the hydrological regime of a dune slack is essential for a good functioning of the dune slack ecosystem. Factors that stabilise the longevity of pioneer stages comprising many Red List species are always associated with a regular supply of groundwater.

6.4 Impact of Human Disturbances on Ecosystem Functioning

The high biodiversity noticed shortly before World War II often marked the beginning of a recovery process after a period of over-exploitation of the dune environment. Sod cutting of slacks occurred frequently but little documentation is available on the reasons why the slacks were sod cut. Fertilisation of gardens, were mentioned, roof material (Beinker 1996), but no real historical evidence has been presented. One reason for sod cutting during Word War II appears to be clear: the material was used to cover the fortifications of the German Atlantic Wall. These activities served one purpose very well. Dune slack succession was set back on a large scale and this is one of the reasons why many pioneer and typical dune slack species can still be found along the NW European coast.

Factors that contributed to the dramatic decline of biodiversity in wet dunes during the second half of the 20th century were lowering of the water levels in the adjacent polder areas, reclamation for agricultural use and afforestation with pine plantations. In the Netherlands large-scale disturbances of dune slack environment started around 1853 when the vast stock of fresh dune water became a major source of drinking water production for the large cities. Large dune areas actually became drinking water catchments. The exploitation of dune water resulted in a large-scale lowering of the water table by 2–3 m on average (Bakker and Stuyfzand 1993). At the same time, large parts of the dune area were saved as a landscape in a time of rapid industrialisation and rapid urbanisation in this densely populated area. Munoz-Reinoso (2001) also reported on the negative effects of groundwater abstraction in the

Doñana National Park in Spain. He found that the impact of a groundwater abstraction facility near a large tourist resort, was obscured by large fluctuations in precipitation from year to year. Using aerial photographs he assessed a clear increase of scrubs and trees (notably *Pinus pinea* from surrounding pine plantations) along the shores of dune ponds over a period of 23 years. He suggested that the following mechanism is responsible for this increase in trees: The invasion of trees and large shrubs, under natural conditions, is prevented by the occurrence of exceptionally high floods during the spring, killing most of the tree and shrub species from dryer habitats. This flooding frequency stabilises the open grassy and heath vegetation that is adapted to temporary very wet conditions. The abstraction of groundwater from the main aquifer prevents some discharge of groundwater in the dune ponds, thus preventing the exceptional high flood even in areas situated more than 5 km away (González Bernáldez et al. 1993).

The very negative effects of drinking water extraction in the Dutch dune areas led to the development of new production techniques. To increase the water tables in the dried out dune areas, surface water from the rivers Rhine and Meuse was transported into the dunes and through an extensive network of ponds and canals was infiltrated in the soil. For nature conservation this technique was disastrous. The input of polluted river water led to increased water tables in the dune slacks, but at the same time promoted eutrophication in practically all dune slacks (Van Dijk and Grootjans 1993).

In less-affected dune areas, as the Wadden Sea coast, 70 years of dune fixation, in which every spot of bare soil had to be covered with branches or hay due to legislation, this resulted in an almost complete stop in natural dune formation. This has led to rapid vegetation succession in dry dunes, but also in slacks. For economical reasons grazing by cattle has stopped in most of the European coastal areas. This has led to enhanced grass encroachment (Veer 1997) and the development of woodland. A positive feedback mechanism exists between increased biomass production and decreased groundwater levels. Tall vegetation types, such as shrubs and forests intercept more nutrients from atmospheric deposition than relatively open and short vegetation types, which leads to increased growth and a higher evapotranspiration. The result is an increased drop in water tables during the summer and consequently in a decreased discharge of groundwater in the dune slacks (Stuyfzand 1993). If the supply of groundwater decreases, shrubs and tall grass species invade the site and pioneer communities lose the competition due to increased availability of nutrients. The succession was stimulated even more by increased atmospheric N-deposition during the last 50 years. The total amount of nitrogen which was deposited on the vegetation via precipitation and dust particles increased from ca. 10 kg N ha^{-1} $year^{-1}$ in 1930 to ca. 25 kg N ha^{-1} $year^{-1}$ in 1980 (Stuyfzand, 1993) and has stabilised between 25 and 35 kg N ha^{-1} yr^{-1} in the late 1990s (Ten Harkel and Van der Meulen 1996; Van Wijnen 1999). All these human disturbances during the last 50 years have led to enormous loss of biodiversity along the European coasts.

At present, the pioneer stages in dune areas are rare in most parts of western Europe. Large blowing dune systems still occur in, for instance, Spain in the Coto Doñana, in Poland in the Wolinski National Park, in France near Bordeaux, on the German Wadden Sea island of Sylt, in Denmark and on the Dutch Wadden Sea island of Texel.

6.5 Restoration of Dune Slacks

During the last decade many restoration projects have been initiated to restore the biodiversity in European dune slacks. The first attempts, some 40–50 years ago, were not very successful. Ponds with rather steep slopes were dug in some convenient site to compensate for lost species-rich dune slacks. The number of target species establishing in such sites was rather low (Fig. 6.6). Better results were obtained by sod cutting in already existing dune slacks where relics of target species were still present in the grass- or shrub-dominated vegetation. Actually, the best results were obtained unintentionally. Very species-rich pioneer slacks developed behind the artificial sand dikes on large sand flats. These dykes were constructed for coastal defence purposes. The reason for these unexpected developments is that this situation resembles the natural dune slack formation by enclosure of sandy beaches by growing dune ridges. The vegetation development starts under near-natural conditions here. Usually, endangered pioneer species can easily colonise such areas since small populations are practically always present in older stages nearby, and dispersal mechanisms (wind, water, animals) appear to be very effective. If a well-developed soil seed bank is present, the pioneer stages are rich in Red List species immediately after the restoration measures.

In dune slacks, where the top soil has not yet been decalcified, mowing or grazing are also suitable restoration measures to restore or maintain species-rich dune slacks. However, when old dune slack soils have been decalcified mowing and grazing will not suffice to conserve the basiphilous dune slack species, because the acidification will proceed and organic material accumulates in the topsoil rapidly. These accumulated nutrient stocks in the soil compartment appear to act as a threshold. First, an endo-organic layer is formed in the top soil, but in decalcified soils also ecto-organic layers are formed on top of the surface. The surface is elevated in this way, preventing a good contact with the calcareous groundwater. Under such conditions basiphilous Red List species are no longer able to establish new populations. The soil seed bank underneath the organic layer may still be rich in Red List species, but when time passes the soil seed bank becomes depleted due to viability loss. When buffer mechanisms (blowing sand or groundwater fluxes) can be

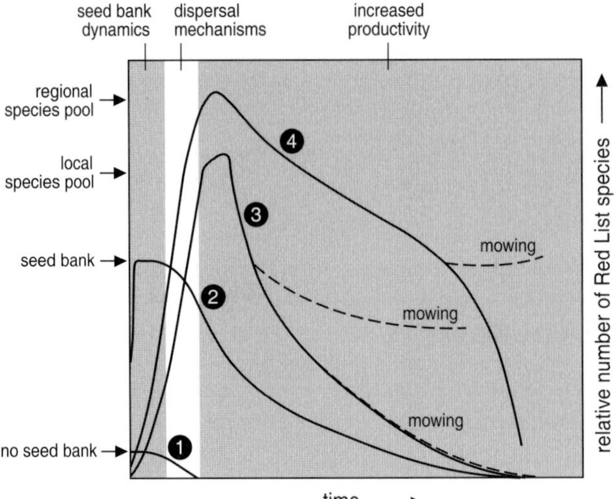

Fig. 6.6. Conceptual model of occurrence of endangered dune slack species (Red List species) after restoration measures have been carried out. *1* Unsuccessful projects where measures were carried out in unsuitable sites and where seed banks are depleted. *2* Temporary success, followed by rapid decrease in target species is encountered in slacks where environmental conditions are unfavourable, but where seed banks were still present. *3* Successful, but short-lived, reconstruction of pioneer vegetation with many Red List species. Dispersal mechanisms are effective, but environmental conditions are sub-optimal. Mowing may sometimes retard a rapid spread of later successional species and a rapid decline in Red List species. *4* Very successful projects where many typical dune slack species establish in large numbers and persist for many decades. Natural processes retard the succession towards late successional stages. A mowing regime may stabilise the pioneer stage even longer. (After Grootjans et al. 2002)

restored in acidified slacks, sod cutting is the best option here, since it removes the organic layer entirely.

Apart from man-made dune slacks there should be room for stimulating dynamic natural processes (action of wind and water) that form and sustain natural dune slacks. In this way endangered species can survive without human interference.

References

Adema EB (2002) Alternative stable states dune slacks succession. Thesis, University of Groningen

Adema EB, Grootjans AP, Petersen J, Grijpstra J (2002) Alternative stable states in a wet calcareous dune slack in The Netherlands. J Veg Sci 13:107–114

Armstrong W (1975) Waterlogged soils. In: Etherington JR (ed) Environment and plant growth. Wiley, London, pp 181–218

Bakker TWN, Stuyfzand PJ (1993) Nature conservation and extraction of drinking water in coastal dunes. In: Vos CC, Opdam P (eds) Landscape ecology of a stressed environment, Chapman and Hall, New York, pp 224–260

Beinker O (1996) Zur Vegetationskunde der Dünen im Listland der Insel Sylt. Kieler Notizen zur Pflanzenkunde in Schleswig-Holstein und Hamburg 25/26:128–166

Bekker RM, Bakker JP, Grandin U, Kalamees R, Milberg P, Poschlod P, Thompson K, Willems JH (1998) Seed size, shape and vertical distribution in the soil: indicators of seed longevity, Funct Ecol 12:834–842

Bekker RM, Lammerts EJ, Schutter A, Grootjans AP (1999) Vegetation development in dune slacks: the role of persistent seed banks, J Veg Sci 10:45–54

Berendse F, Lammerts EJ, Olff H (1998) Soil organic matter accumulation and its implication for nitrogen mineralisation and plant species composition during succession in coastal dune slacks, Plant Ecol 137:71–78

Boorman LAG, Londo G, Van der Maarel E (1997) Communities of dune slacks. In: Van der Maarel, E. (ed) Dry coastal ecosystems, part C. Ecosystems of the world, Elsevier, Amsterdam, pp 275–293

Chafetz HS (1994) Bacterially induced precipitates of calcium carbonate and lithification of microbial mats. In: Krumbein WE, Paterson DM, Stal LJ (eds) Biostabilisation of sediments. BIS Verlag, Oldenburg, pp 149–163

Crawford RMM, Wishart D (1966) A multivariate analysis of the development of dune slack vegetation in relation to coastal accretion at Tentsmuir Fife. J Ecol 54:729–744

Ernst WHO, Van der Ham NF (1988) Population structure and rejuvenation potential of *Schoenus nigricans* in coastal wet dune slacks. Acta Bot Neerl 37:451–465

Ernst, WHO, Slings QL, Nelisssen HJM (1996) Pedogenesis in coastal wet dune slacks after sod-cutting in relation to revegetation, Plant Soil 180:219-230

González Bernáldez F, Rey Benayas JM, Martínez A (1993) Ecological impact of groundwater extraction on wetlands (Douro Basin Spain). J Hydrol 141:219–238

Grootjans AP, Sival FP, Stuyfzand PJ (1996) Hydro-geochemical analysis of a degraded dune slack. Vegetatio 126:27–38

Grootjans AP, Ernst WHO, Stuyfzand PJ (1998) European dune slacks: strong interactions between vegetation, pedogenesis and hydrology, Trends Evol Ecol 13:96–100

Grootjans AP, Van den Ende FP, Walsweer AF (1997) The role of microbial mats during primary succession in calcareous dune slacks: an experimental approach. J Coastal Conserv 3:95–102

Grootjans AP, Geelen, Jansen AJM, Lammerts EJ (2002) Dune slack restoration in the Netherlands; successes and failures. Hydrobiologia 487:181–302

Kennoyer GJ, Anderson MP (1989) Groundwater's dynamic role in regulating acidity and chemistry in a precipitation lake. J Hydrol 109:287–306

Koerselman W (1992) The nature of nutrient limitation in Dutch dune slacks. In: Carter RWG, Curtis TGF, Sheehy-Skeffington MJ (eds) Coastal dunes. Ballgame, pp 189–199

Kooijman AM, Besse M (2002) On the higher availability of N and P in lime-poor than in lime-rich coastal dunes in the Netherlands. J Ecol 90:394–403

Lamers LPM, Tomassen HBM, Roelofs JGM (1998) Sulphate induced eutrophication and phytotoxicity in freshwater wetlands. Environ Sci Technol 32:199–205

Lammerts EJ, Grootjans AP, Stuyfzand PJ, Sival FP (1995) Endangered dune slack gastronomers in need of mineral water. In Salman AHPM, Berends H, Bonazountas M (eds) Coastal management and habitat conservation. EUCC, Leiden, pp 355–369

Lammerts EJ, Pegtel DM, Grootjans AP, Van der Veen A (1999) Nutrient limitation and vegetation change in a coastal dune slack. J Veg Sci 10:11–122

Munoz-Reinoso JC (2001) Vegetation changes and groundwater abstraction in SW Doñana, Spain. J Hydrol 242:197–209

Olff H, Huisman J, Van Tooren BF (1993) Species dynamics and nutrient accumulation during early primary succession in coastal sand dunes. J Ecol 81:693–706

Petersen J (2000) Die Dünentalvegetation der Wattenmeer-Inseln in der südlichen Nordsee. Eine pflanzensoziologische und ökologische Vergleichsuntersuchung unter Berücksichtigung von Nutzung und Naturschutz. Diss Univ Hannover, Husum Verlag (English summary)

Piotrowska H (1988) The dynamics of the dune vegetation on the Polish Baltic coast. Vegetatio 77:169–175

Pluis JLA, De Winder B (1990) Natural stabilisation. In: Bakker TWM, Jungerius PD, Klijn PA (eds) Dunes of the European coasts. Catena (Suppl) 18:195–208

Ranwell DS (1972) Ecology of salt marshes and sand dunes. Chapman and Hall, New York

Schat H, Bos AH, Scholten M (1984) The mineral nutrition of some therophytes from oligotrophic dune slack soils. Acta Oecol Oecol Plant 5:119–131

Sival FP, Grootjans AP (1996) Dynamics of seasonal bicarbonate supply in a dune slack: effects on organic matter, nitrogen pool and vegetation succession. Vegetatio 126:39–50

Sival FP, Mücher HJ, Van Delft SPJ (1998) Carbonate accumulation affected by hydrological conditions and their relevance for dune slack vegetation. J Coastal Conserv 4:91–100

Smith SE (1966) Physiology and ecology of orchid mycorrhizal fungi with reference to seedling nutrition. New Phytol 65:488–499

Stal LJ, Villbrandt M, De Winder B (1994) Nitrogen fixation in microbial mats. In: Krumbein WE, Paterson DM, Stal LJ (eds) Biostabilisation of sediments, BIS Verlag, Oldenburg, pp 384–399

Studer-Ehrensberger K, Studer C, Crawford RMM (1993) Competition at community boundaries: mechanisms of vegetation structure in a dune slack complex. Funct Ecol 7:156–168

Stuyfzand PJ (1993) Hydrochemistry and hydrology of the coastal dune area of the western Netherlands. PhD Thesis, Free Univ of Amsterdam

Ten Harkel MJ, Van der Meulen F (1996) Impact of grazing and atmospheric deposition on the vegetation of dry coastal dune grasslands. J Veg Sci 7:445–452

Thompson K, Bakker JP, Bekker RM (1997) Soil seed banks of north west Europe. University Press, Cambridge

Van Beckhoven K (1995) Rewetting of coastal dune slacks: effects on plant growth and soil processes, PhD Thesis, Free Univ of Amsterdam

Van Dijk HWJ, Grootjans AP (1993) Wet dune slacks: decline and new opportunities. Hydrobiologia 265:281–304

Van der Maarel E, Van der Maarel-Versluys M (1996). Distribution and conservation status of littoral vascular plant species along the European coasts. J Coastal Conserv 2:73–92

Van der Maarel E, Boot RGA, Van Dorp D, Rijntjes J (1985) Vegetation succession on the dunes near Oostvoorne, The Netherlands: a comparison of the vegetation in 1959 and 1980. Vegetatio 58:137–187

Van Gemerden H (1993) Microbial mats: a joint venture. Mar Geol 113:3–25

Van Wijnen H (1999) Nitrogen dynamics and vegetation succession in salt marshes. PhD Thesis, Univ of Groningen

Veer MAC (1997) Nitrogen availability in relation to vegetation changes resulting from grass encroachment in Dutch dry dunes. J Coastal Conserv 3:41–48

Willis AJ (1963) Braunton Burrows: the effects on the vegetation of the addition of mineral nutrients to the dune soils. J Ecol 51:353–374

Zoladeski CA (1991) Vegetation zonation in dune slacks on the Leba Bar, Polish Baltic Sea coast. J Veg Sci 2:255–258

Zunzunegui M, Diaz Barradas M, García Novo F (1998) Vegetation fluctuation in mediterranean dune ponds in relation to rainfall variation and water extraction. Appl Veg Sci 1:151–160

7 Coastal Dune Forest Rehabilitation: A Case Study on Rodent and Bird Assemblages in Northern Kwazulu-Natal, South Africa

R.J. van Aarde, T.D. Wassenaar, L. Niemand, T. Knowles and S. Ferreira

7.1 Introduction

Coastal dune forests in northern KwaZulu-Natal, South Africa, are continually exposed to natural and man-induced disturbances that usually initiate ecological succession (van Aarde et al. 1996a; Mentis and Ellery 1994). This succession is associated with temporal and spatial changes in vegetation structure that influence habitat suitability and ultimately the structure of vertebrate communities living there. For example, in the case of birds, we know from studies conducted elsewhere that species richness and diversity correlates with vegetation structural heterogeneity (see Kritzinger and van Aarde 1998 for references). Vegetation succession is also known to affect small mammals (Foster and Gaines 1991), though the patterns recorded in coastal dune forests are less obvious than those for birds (see Ferreira and van Aarde 1999 for references).

Ecological rehabilitation programmes often aim at minimising the compositional, structural and functional differences between undisturbed reference sites and rehabilitating sites (see van Aarde et al. 1996a, b). In the present chapter, we aim to characterise species traits that affect a species' occurrence, abundance and persistence at particular stages of the regenerating sere of coastal dune forest. For rodents and birds, as with other taxa, we expect that rehabilitation would result in community characteristics converging towards those of benchmark sites. Species present in newly regenerating sites should be pioneers and r-selected, while those inhabiting later stages of the regeneration sere should be K-selected (Smith and MacMahon 1981). Based on earlier studies on vertebrates, we also expect mean species-specific reproductive output to decrease as a result of increasing environmental stability with increasing successional age (May 1984; Mönkkönen and Helle 1987). Furthermore,

Ecological Studies, Vol. 171
M.L. Martínez, N.P. Psuty (Eds.)
Coastal Dunes, Ecology and Conservation

habitats developing towards benchmarks in response to rehabilitation are transient. Generalists therefore, should inhabit developmental gradients for longer periods than specialists.

In the present contribution, we describe the development of rodent and bird communities in our study area north of Richards Bay (South Africa), where dune forest rehabilitation commenced in 1977. About 24 years of continued rehabilitation gave rise to the development of a range of known-aged transient habitats converging onto an undisturbed coastal dune forest (van Aarde et al. 1996a,b; Kritzinger and van Aarde 1998; Ferreira and van Aarde 1997). The questions we will be addressing here are as follows: Do the life history traits of early vertebrate colonisers differ from those of later colonisers? Do these life history variables follow the traditional r–K dichotomy?

7.2 Study Area

7.2.1 Indian Ocean Coastal Dunes

The Indian Ocean coastal belt supports a distinctive vegetation system with 40 % endemicity among woody plants (Moll and White 1978). The coastal belt can be divided into four regions, namely the northern Swahili Centre of Endemism, the central Maputaland-Swahili Transitional Zone with little endemism, and the southern Maputaland and Pondoland Centres of Endemism (van Wyk 1996). Our study area is located within the southern Maputaland region. The Maputaland vegetation has been classified into 15–21 ecotypes, most of which include many endemic or localised plants usually associated with sandy soils (van Wyk 1996). One of these ecotypes includes dune forests, which occupy a narrow belt along the coastline (Moll and White 1978; Eeley et al. 1999) from Maputaland southwards where it becomes patchy and floristically impoverished (Moll and White 1978).

Coastal dunes support a relatively high diversity of vertebrates but limited endemism (McLachlan 1991). This may be ascribed to the relative narrowness of coastal dunes, allowing vertebrates from adjoining habitats free access in their search for additional food and shelter. Coastal dunes are also geologically relatively young and have thus had little time for the evolutionary development of unique species or subspecies (McLachlan 1991).

7.2.2 The Coastal Sand Dune Forests of KwaZulu-Natal

Tinley (1985) distinguished four vegetation zones in South African coastal dunes, one of these being forests. Coastal dune forests in a relatively undisturbed state form a narrow belt of potential habitat for vertebrates between the sea and the hinterland where it seldom extends further than 2 km from the coast in northern KwaZulu-Natal.

The dune forests of KwaZulu-Natal are located on Pleistocene and Recent sands and are exposed to relatively high rainfall (Tinley 1985). High leaching of soil minerals may limit soil fertility. These dunes have been covered by forest for approximately 8000 years (see references in Eeley et al. 1999). Human activities in the region have had a major influence on coastal dune plant communities in KwaZulu-Natal since the early Iron Age (Conlong and van Wyk 1991). By 1939 most of the dunes in the area were covered with small scrub (Stephens 1939 in Conlong and van Wyk 1991), suggesting that agricultural and pastoral activities of semi-permanent settlers dramatically degraded coastal dune forests. By 1974 the protective policies adopted by the then Department of Forestry against fire, woodcutting, shifting cultivation and grazing resulted in the recovery of some of these indigenous forests (Weisser 1978). More recently, these forests have been fragmented through the establishment of commercial exotic plantations and by opencast dune mining followed by ecological rehabilitation. Here, the withdrawal of man-induced disturbances usually initiates habitat age-related changes in vegetation composition and structure through successional processes (see van Aarde et al. 1996b). Such development provides transient habitats for vertebrates and invertebrates typical of coastal dunes of the region, all of which are colonising such areas on their own accord (see van Aarde et al. 1996a, b; Kritzinger and van Aarde 1998; Ferreira and van Aarde 1997, 2000).

7.2.3 The Post-Mining Rehabilitation of Coastal Dunes

Most of the present discussion is based on information collected over a ten-year period from 1991 to 2000 along a 40-km stretch of regenerating and mature coastal dune forest between Richards Bay (28°43′S; 32°12′E) and the Mapelane Nature Reserve (32°25′S; 28°27′E) (see Fig. 7.1). Richards Bay Minerals has been extracting heavy metals (zircon, ilmenite and rutile) from some of the dunes northeast of Richards Bay since July 1977. During these operations, a 400-m-wide shoreline strip of dune vegetation is preserved to reduce slumping as well as to preserve a species pool from which potential colonisers of the regenerating habitats may originate.

Before dune mining commences, the surface vegetation is cleared away and the topsoil is collected for later use in rehabilitation. A floating dredger and separation plant collects sand and separates the heavy metals by a gravitational process, after which the minerals are pumped to a stockpile on land. The remaining sand (>94 %) is pumped to an area behind the dredging pond, where new dunes are formed and shaped to resemble the topography of the dunes prior to mining. Topsoil collected prior to mining, is then spread over

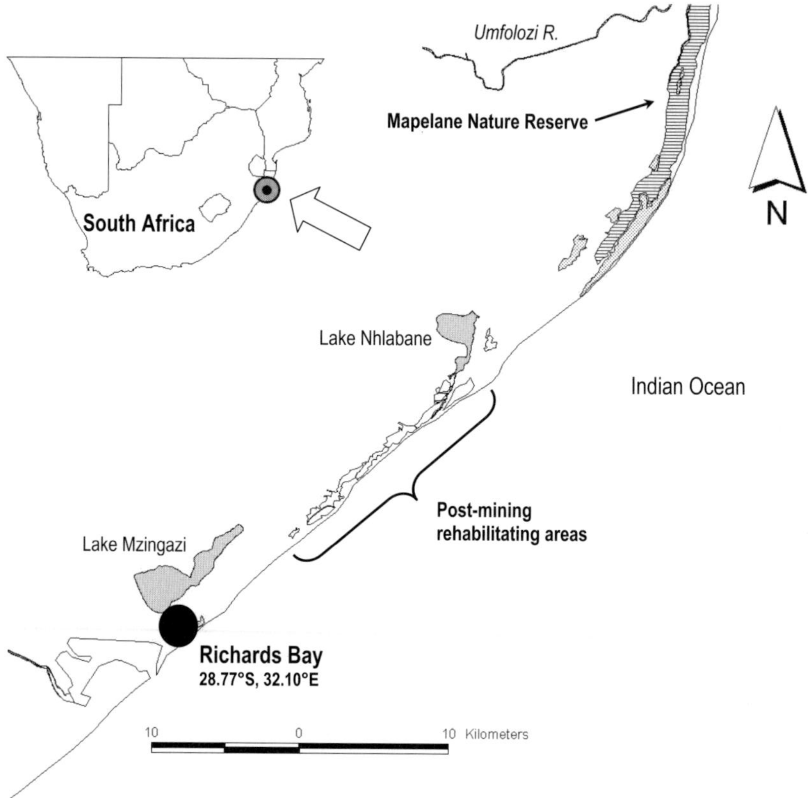

Fig. 7.1. A map of the study area showing the location of each particular regenerating site and the unmined mature coastal dune forest

the reshaped dunes. A seed mixture of annuals consisting of *Pennisetum americanum*, *Sorghum* sp. and *Crotalaria juncea* are incorporated in the topsoil. To reduce wind erosion and surface evaporation, 1.5-m-high hessian windbreaks are erected across the dunes. Within a month of the start of rehabilitation, this management programme gives rise to a dense plant cover that prevents erosion and apparently ameliorates the surface microclimate for the germination and subsequent establishment of indigenous species.

Within 3 to 6 months after the die-off of the annuals these areas are densely covered with grass and *Acacia kosiensis* seedlings, while at 2 years the canopy cover of low *A. kosiensis* trees is approaching 70–80 %. Further successional development is associated with the self-thinning of *A. kosiensis* (van Dyk 1996) and unassisted colonisation by plant and animal species, typical of mature dune forests in the region (see van Aarde et al. 1996a, b). None of the rodent or bird species recorded here are endemic to coastal dune forests.

7.3 Materials and Methods

The collection of information and the reduction of data have been described elsewhere (for rodent studies see Ferreira and van Aarde 1996, 1997, 1999, 2000; Koekemoer and van Aarde 2000; for bird studies, see Kritzinger and van Aarde 1998; Niemand 2001). Rodent and bird community sampling occurred on the same regenerating sites of known age (1, 5, 13, 17, 20 and 23 years old at the time of the present study) and in a neighbouring mature coastal dune forest site (see Fig. 7.1).

7.3.1 Rodents

Trapping took place during summer (December–January) over a 10-year period from 1991 to 2001 on three to six permanent replicate trapping grids located on each of the six regenerating sites and an unmined dune forest site. Trapping grids consisted of 49 trapping stations arranged in a 7x7 configuration with 15 m between trapping stations. A single Sherman live trap (75x90x230 mm), baited with peanut butter and raisins, was placed at each station. Grids were placed at least 200 m apart to ensure independent sampling. Trapping on a grid continued for three nights and traps were checked and rebaited each day at dawn. Trapped animals were identified to species level, marked (by toe clipping), sexed and weighed prior to release. Abundance was calculated as the minimum number alive (MNA) per grid (Krebs 1999).

7.3.2 Birds

Data was collected during December and January of 1994, 1996, 1998 and 2000 using line transect surveys following the methods of Kritzinger and van Aarde (1998). The number of transects per site was affected by the size of the regenerating site and varied from two to four transects per site. Transect starting points were randomised, but transects did not overlap and were at least 200 m apart. Surveys continued for approximately four hours, beginning 30–60 min after sunrise when birds are most active and conspicuous. Transect lines ranged from 250–500 m in length and line length was incorporated in the calculation of density to correct for variable transect length.

Data for each transect was analysed separately using the programme DISTANCE (Laake et al. 1993). Mean and error values for total density per hectare for each sampled site are based on these individual estimates. Site-specific relative density (rD) for each species was calculated using the equation rD=rN/rV, where rN is the number of a given species relative to the total number seen on the transect and rV is relative visibility (Buckland et al. 1992). rV values were calculated according to Buckland et al. (1992).

For each of the sites, typical species were identified as the most consistent species of a site, that is, those species contributing to the first 50 % of similarity within sites (Clarke and Warwick 1994). The present analysis is limited to the sets of typical species identified for the chronosequence of dune forest regeneration. Information on species body and clutch sizes was obtained from Maclean (1993).

The study was conducted over a ten-year period and most sites along the regenerating sere have been surveyed repetitively. A given site therefore contributed to more than one data point and data from sites of increasing age are not independent, thus not allowing for refined statistical analyses and curve fitting.

7.4 Results and Discussion

Coastal dunes in our study area are inhabited by vertebrates ranging in diet from granivores (rodents, birds) to frugivores (birds), insectivores (shrews, birds, reptiles, amphibians) and carnivores (reptiles, mammals). The presence of vertebrate species on specific sites or seral stages within such a successional sere conceivably depends on the presence of their resources, while the number of species and their absolute densities will depend on resource availability, area, and interspecific interactions.

7.4.1 Rodents

As mining of these dunes followed by dune rehabilitation may be considered a discrete disturbance event, the colonisation of such areas may be considered a recovery or regeneration of the relevant assemblages or communities. This implies that the undisturbed assemblage on stands of mature forest represents an entity towards which the disturbed assemblages can develop. Live-trapping (about 35,000 trap nights and 11,000 captures over a 10-year period of our study) on developing dune forests yielded eight rodent and three shrew species. All species recorded in unmined and relatively mature dune forests also occurred in regenerating dune forests with no clear change in the number of species inhabiting regenerating sites with increasing regeneration age (Fig. 7.2). Species inhabiting young regenerating stages were the same as those of mature forests and we may conclude that the post-disturbance recovery of rodent and shrew assemblages is not structured by habitat age-related factors.

Species richness at a given point along the chronosequence in our study, as in most habitats, is ultimately a consequence of the balance between local

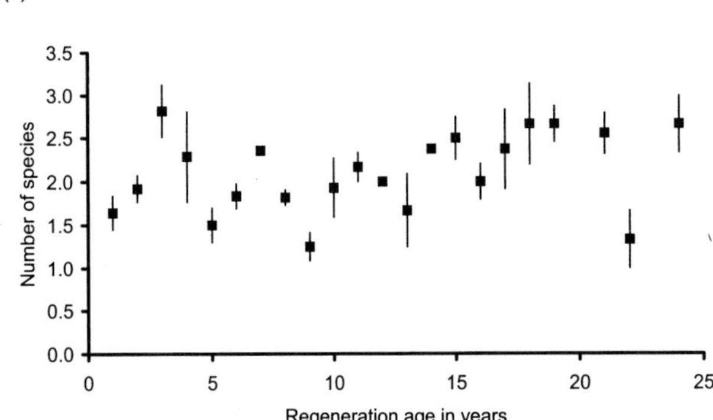

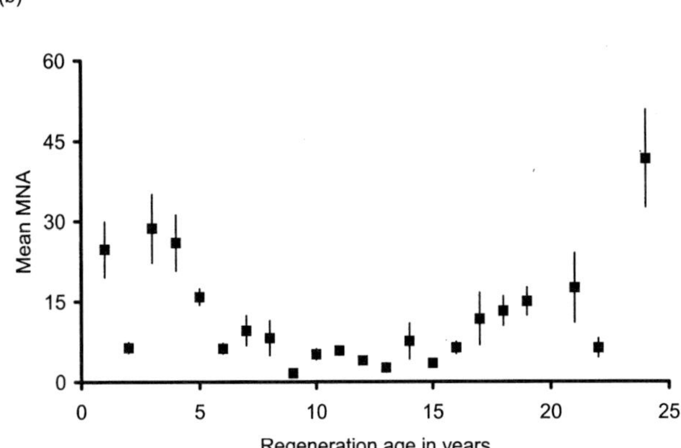

Fig. 7.2. **a** The mean (±SE) number of rodent species and **b** mean (±SE) minimum number of rodents alive (MNA) as a function of regeneration age along a chronosequence of dune forest development

colonisation and extinction. The lack of a clear age-related pattern in species richness might thus reflect a lack of a clear pattern in colonisation and extinction. This may also be true for mature dune forests in the region where natural disturbances evoked, e.g. by tree-falls, create gaps within the forest (Ferreira and van Aarde 2000). Tree-falls at different times create patches at different stages of recovery, leading to a variety of habitats and rodent assemblages. It may thus be argued that these coastal dune forests are predisposed to disturbance events – a given patch may be colonised by whichever species are available in the surrounding areas. The pioneer species are probably always present at all sites and ready to exploit any opportunity.

Our results suggest that rodent communities in these forests are extremely flexible with temporal changes in the composition of assemblages not always being unidirectional (see Ferreira and van Aarde 1996, 2000). Ferreira and van Aarde (1999) also showed that for some rodents, habitat features, rather than inter-specific interactions, might explain species-specific densities. Spatial and temporal variability in habitat features such as vegetation height, area covered by shrubs, volume of shrubs and litter depth (Ferreira and van Aarde

1999) appear to determine the occurrence of rodents in coastal dune forests. Rodent assemblages should thus be seen as loose collections of species, rather than tightly structured communities.

Rodent numbers (minimum number alive), expressed as a function of habitat regeneration age, followed a pattern of high numbers on regenerating sites less than five years of age, after which numbers remained relatively low until a site regeneration age of 15 years, thereafter steadily increasing (Fig. 7.2). All assemblages along the successional sere were dominated by the multi-mammate mouse (*Mastomys natalensis*) and the pouched mouse (*Saccostomus campestris*) (Ferreira and van Aarde 1996), although numbers for both species varied considerably (see Fig. 7.3). Studies by Foster and Gaines (1991) and Ferreira and van Aarde (1996) had shown that successional changes in rodent communities are characterised by species additions as well as changes in abundance. However, long-term data, such as ours, is conceivably affected by inter-annual differences in rainfall that may affect productivity that may overshadow the influence of regeneration age on assemblage parameters. Yet, the present findings are in agreement with the successional sere noted earlier by Ferreira and van Aarde (1996, based on data collected

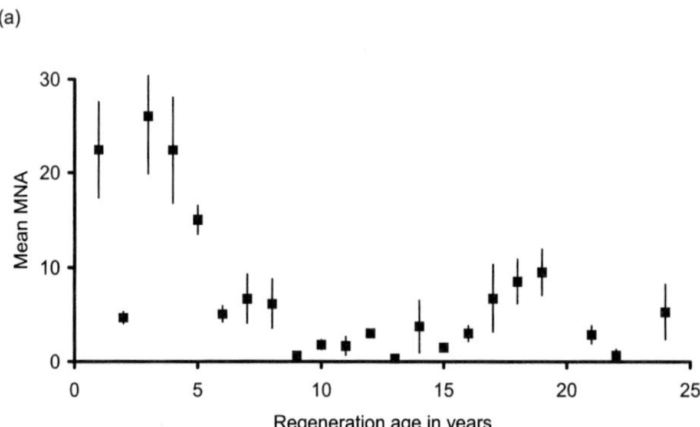

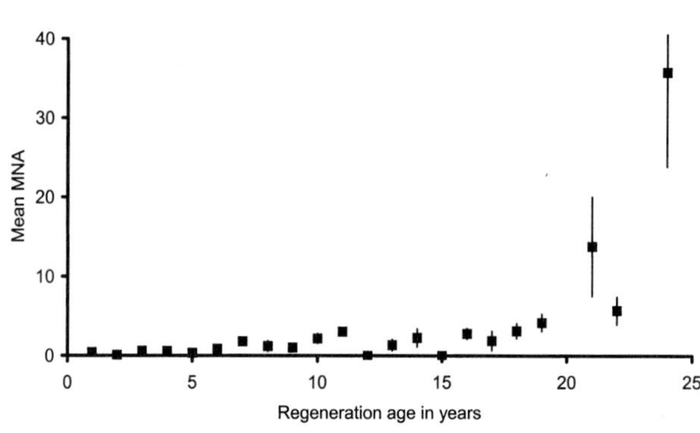

Fig. 7.3. The mean (±SE) minimum number alive (MNA) of **a** *Mastomys natalensis* and **b** *Saccostomys campestris* as a function of regeneration age along a chronosequence of dune forest development

between 1991 and 1993), where *M. natalensis* is replaced by *S. campestris* (Fig. 7.3). This pattern differed from that recorded later by Ferreira and van Aarde (2000), when between 1993 and 1995, *S. campestris* did not feature in the rodent successional sere. When considering all the data collected over a ten-year period from these dunes it is apparent that the pattern is determined by only two species. *Mastomys natalensis* is the most abundant species during the first few years of vegetation regeneration and *S. campestris* is dominant in sites older than 15 years of age (see Fig. 7.3).

During the first few years of forest regeneration, the vegetation is dominated by grasses and sedges (Conlong and van Wyk 1991). These conditions favour colonisation by a generalist pioneer such as *M. natalensis*, known to flourish in disturbed environments (Meester et al. 1979). *Saccostomus campestris*, a more specialised hoarder, especially of *Acacia* species seeds (Swanepoel 1972), could conceivably thrive on the 12-year and older regenerating sites where the dominant mature *A. kosiensis* trees are producing seed. These sites are, however, also characterised by disturbance caused by natural tree-falls that create patches ideal for pioneers, such as *M. natalensis*, thus explaining the continuing occurrence of this species along our successional sere.

Our studies suggest that the composition of rodent assemblages may be best explained by movement of animals between disturbed and undisturbed patches (see Ferreira and van Aarde 1996). High community dominance (only a few species dominating community structure) may also be due to limited interspecific competition and selective resource advantages at specific stages during forest regeneration. For instance, by experimentally manipulating food availability for rodents in early post-mining habitats, we have previously shown that community dominance increases (Shannon diversity decreases) with an increase in food availability (Koekemoer and van Aarde 2000). This was the result of an increase in the absolute numbers of the pioneer species rather than a change in the abundance of other species (Koekemoer and van Aarde 2000). It is thus fair to say that in our study area, the unstable environmental conditions that give rise to habitat changes, rather than interspecific interactions, result in temporal trends in rodent species richness and diversity (Koekemoer and van Aarde 2000).

How do these results reflect on our questions? Most species are present at the onset of rehabilitation, but typical pioneer species, such as *M. natalensis*, with high reproductive output and generalist feeding requirements (Meester et al. 1979) numerically dominates. In contrast, *S. campestris* (most prevalent on later stages) exhibits variable reproductive output (Westlin and Ferreira 2000) and more specialised feeding requirements (Swanepoel 1972). Early rodent dominants have different life history traits than those of later dominants, while these traits appear to follow the traditional r-K dichotomy.

7.4.2 Birds

We recorded 105 bird species during the four annual transect surveys completed between 1994 and 2000. Most (42 %) of the species on mature dunes also occurred on dunes regenerating in response to dune rehabilitation. The species (33 %) noted on regenerating sites but not in the mature forests were all grassland species typical of the region, e.g. the Rattling Cisticola (*Cisticola juncidis*), Grassveld Pipit (*Anthus cinnamomeus*) and Common Waxbill (*Estrilda astrid*). For birds, the rehabilitating habitats were similar to those typical of the region, with the result that regenerating habitats did not support any unique species (see Niemand 2001). This is to be expected, as local assemblages are probably dependent on regional species pools.

By reducing our data base and concentrating on species typical of known-aged and mature sites along the chronosequence of developing dune forest, it became clear that the number of typical species increases with regeneration age (Fig. 7.4). This increase coincided with a decrease in the contribution of each species to each of the site assemblages, as well as a decrease in their abundance and variability in abundance (Fig. 7.4). Successional development was further associated by an increase in body size, particularly after 10 years

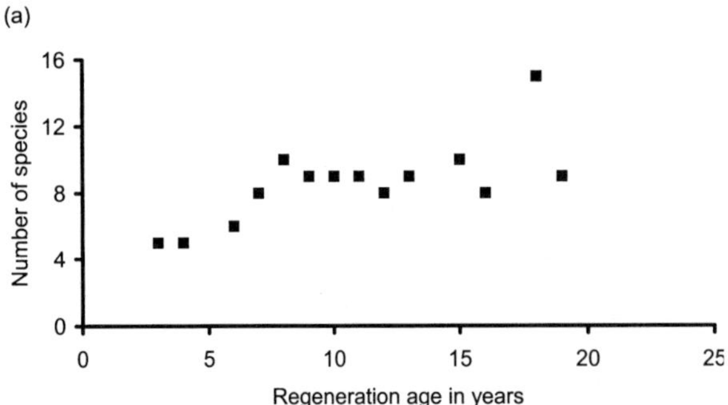

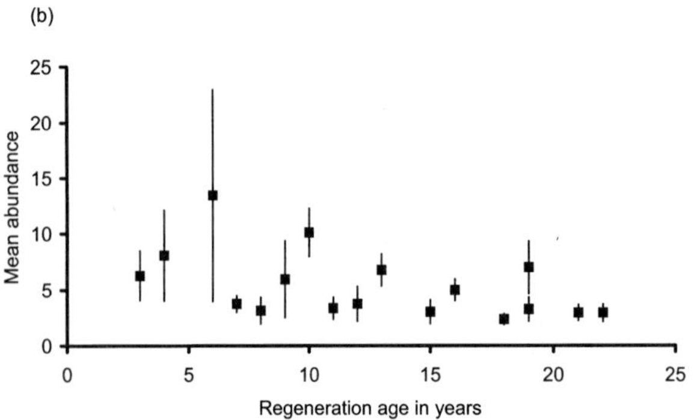

Fig. 7.4. a The total number of typical bird species along a known-aged chronosequence of dune forest development. **b** Mean (±SE) abundance for typical bird species based on summer censuses conducted along transect lines surveyed during 1994, 1996, 1998 and 2000

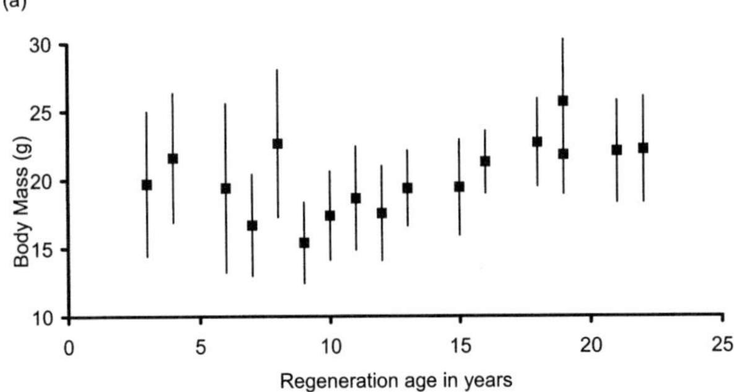

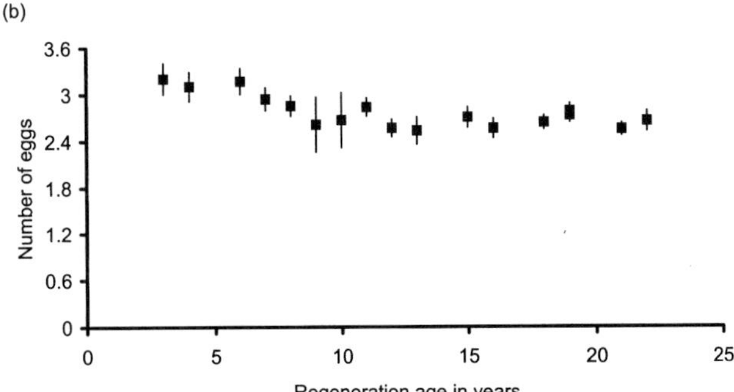

Fig. 7.5. a Mean (±SE) body mass in grams for birds identified as typical of a chronosequence of coastal dune forest development. **b** Mean (±SE) clutch size for birds identified as typical of a chronosequence of coastal dune forest development

of regeneration, and a decrease in clutch size, especially during the first 10 years of regeneration (Fig. 7.5). Therefore, the bird community of the regenerating sites followed developmental patterns typical of ecological succession (May 1984; Mönkkönen and Helle 1987).

In our studies, changes in the species composition of birds were closely associated with succession-induced changes in vegetation composition and structure (see also Kritzinger and van Aarde 1998). As expected (May 1984; Mönkkönen and Helle 1987), species typical of later successional stages were more K-selected (relatively large-bodied with relatively smaller clutches and occurring at relatively lower densities), than the r-strategists of earlier successional stages.

7.5 Conclusion

In response to the questions asked at the onset of the study we conclude that the life history traits of early colonisers differ from those of later colonisers for birds but not for rodents as all species colonise early. However, for rodents, life history traits of early dominants differed from those of later dominants.

Changes in these life history variables for birds appear to follow the classic r–K dichotomy associated with successional changes, but do not imply that rehabilitation has been successful or will succeed in the future. Trends in both bird and rodent communities do, however, indicate that rehabilitation is at least a management tool that could potentially reverse the ubiquitous trends of habitat loss and fragmentation that threatens the viability of species populations.

References

Buckland ST, Anderson DR, Burnham KT, Laake JL (1992) Distance sampling: Estimating abundance of biological populations. Chapman & Hall, London

Clarke KR, Warwick RM (1994) Change in marine communities: An approach to statistical analysis and interpretation. The Natural Environment Research Council, UK

Conlong DE, van Wyk RF (1991) Current understanding of grasslands of the dune systems of the Natal north coast. In: Everard DA, von Maltitz GP (eds) Dune forest dynamics in relation to land-use practices. FRD Report, National Research Foundation, Pretoria, pp 81–105

Eeley HAC, Lawes MJ, Piper SE (1999) The influence of climate-change on the distribution of indigenous forest in KwaZulu-Natal, South Africa. J Biogeogr 26:595–617

Ferreira SM, van Aarde RJ (1996) Changes in community characteristics of rodents in rehabilitating coastal dune forests in northern KwaZulu-Natal. Afr J Ecol 34:113–130

Ferreira SM, van Aarde RJ (1997) The chronosequence of rehabilitating stands of coastal dune forests: do rodents confirm it? S Afr J Sci 93:211–214

Ferreira SM, van Aarde RJ (1999) Habitat associations and competition in *Mastomys-Saccostomus-Aethomys* assemblages on coastal dune forests. Afr J Ecol 37:121–136

Ferreira SM, van Aarde RJ (2000) Maintaining diversity through intermediate disturbances: evidence from rodents colonising rehabilitating dunes. Afr J Ecol 38:286–294

Foster J, Gaines MS (1991) The effects of a successional habitat mosaic on a rodent community. Ecology 72:1358–1373

Koekemoer AC, van Aarde RJ (2000) The influence of food supplementation on a coastal dune rodent community. Afr J Ecol 38:343–351

Krebs CJ (1999) Ecological methodology 2nd edn. Addison-Wesley, Sydney

Kritzinger JJ, Van Aarde RJ (1998) The bird communities of rehabilitating coastal dunes at Richards Bay, KwaZulu-Natal. S Afr J Sci 94:71–78

Laake JL, Buckland ST, Anderson DR, Burnham KP (1993) DISTANCE User's Guide v 2.0. Colorado Cooperative Fish and Wildlife Research Unit, Colorado State Univ, Fort Collins, CO

Maclean GL (1993) Roberts' Birds of Southern Africa, 6th edn. John Voelcker Bird Book Fund, Cape Town

May PG (1984) Avian reproductive output in early and late successional habitats. Oikos 43:277–281

McLachlan A (1991) Ecology of coastal dunes. J Arid Environ 21:229–243

Meester J, Lloyd C.N.V., Rowe-Rowe D.T. (1979) A note on the ecological role of *Praomys natalensis*. S Afr J Sci 75:183–184

Mentis MT, Ellery WN (1994) Post-mining rehabilitation of dunes on the north-east coast of South Africa. S Afr J Sci 90:69–74

Mönkkönen M, Helle P (1987) Avian reproductive output in European forest succession. Oikos 50:239–246

Moll EJ, White F (1978) The Indian Ocean coastal belt. In: Werger MJA, van Bruggen AC (eds) Biogeography and ecology of southern Africa. Junk, The Hague, pp 561–598

Niemand LJ (2001) The contribution of the bird community of the regenerating coastal dunes at Richards Bay to regional diversity. MSc Thesis, Univ of Pretoria

Smith KG, MacMahon JA (1981) Bird communities along a montane sere: community structure and energetics. Auk 90: 62–77

Swanepoel P (1972) The population dynamics of rodents at Pongola, Northern Zululand, exposed to dieldrin cover spray. MSc Thesis, Univ of Pretoria

Tinley KL (1985) The coastal dunes of South Africa: a synthesis. South African National Scientific Programme Report. Council for Scientific and Industrial Research, Pretoria

Van Aarde RJ, Ferreira SM, Kritzinger JJ (1996a) Successional changes in rehabilitating coastal dune communities in northern KwaZulu/Natal, South Africa. Landscape Urban Plann 34:277–286

Van Aarde RJ, Ferreira SM, Kritzinger JJ, Van Dyk PJ, Vogt M, Wassenaar TD (1996b) An evaluation of habitat rehabilitation on coastal dune forests in northern KwaZulu-Natal, South Africa. Rest Ecol 4:334–345

Van Dyk PJ (1996) The population biology of the sweet thorn *Acacia karroo* in rehabilitating coastal dune forests in northern KwaZulu-Natal, South Africa. MSc Thesis, Univ of Pretoria

Van Wyk AE (1996) Biodiversity of the Maputaland centre. In: van der Maesen LJG, van der Burgt XM, van Medenbach de Rooy JM (eds) The biodiversity of African plants. Proc XIVth AETFAT Congress. Kluwer, Dordrecht, pp 198–207

Weisser PJ (1978) Changes in the area of grasslands on the dunes between Richards Bay and the Mfolozi River, 1937 to 1974. Proc Grassland Society of South Africa, vol 13, pp 95–97

Westlin LM, Ferreira SM (2000) Do pouched mice alter litter size through resorption in response to resource availability? S Afr J Wildlife Res 30:1–4

III Living in a Stressful Environment

8 Burial of Plants as a Selective Force in Sand Dunes

M.A. Maun

8.1 Introduction

Burial of plants is a recurrent event in coastal dunes because of the activity of waves and wind. Waves dump large quantities of sand on the beach that is later moved inland by the action of wind velocities exceeding about 16 km/h. Plants growing on the foredunes not only have to contend with burial by sand but also with a wide variety of other environmental stresses such as desiccation, nutrient shortage, and salt spray along sea coasts. Perhaps the most important stress is burial in sand because burial alters all aspects of the plant and the soil micro-environment, such as soil temperature, soil moisture, bulk density, nutrient status, soil pH and oxygen levels. This physical alteration of the micro-environment may increase soil microorganisms, change the ratio between aerobic and anaerobic microbes, decrease mycorrhizal fungi, increase the rate of respiration and curtail photosynthesis. Burial stress occurs with such regular frequency that it has strong selective consequences to fitness and organisms must make physiological adjustments in succeeding generations in order to survive. I define stress according to Grime (1979) "the external constraints that limit the rate of dry matter production of all or part of vegetation". Could burial in sand be defined as a stress? It depends on the amount of burial. Small amounts of burial specific to a species do not cause any stress. Actually, it is beneficial and plants exhibit a stimulation response. However, above a certain threshold level of burial, specific to each species, it becomes a stress.

At the community level, if burial occurs on a regular basis in a habitat, there is selection against species with a conservative growth habit. Burial acts as a filter that eliminates species when burial exceeds their threshold of survival. Eventually the community consists of plant species that have become functionally adapted to grow and prosper under conditions that deny survival opportunities to other species. These adaptations or physiological adjustments are the most successful means of coping with the encountered environmental constraints and may range from changes to allocation patterns of

Ecological Studies, Vol. 171
M.L. Martínez, N.P. Psuty (Eds.)
Coastal Dunes, Ecology and Conservation

metabolic resources, and/or modification of structural components. As shown in other chapters, plants adapted to live in sand dunes play a major role in dune formation and dune morphology because of their different growth forms and their significant abilities to grow through the sand deposits and utilise the meagre resources of sandy habitats.

In this chapter, I would like to emphasise three major objectives. The first major objective is to demonstrate relationships between vegetation and burial as an environmental force. It is important to know how individual species of foredunes are distributed in relation to the gradient in sand deposition. The second major objective is to show the responses of seeds, seedlings, and adult plants to burial episodes because each stage in the life cycle of a plant has a slightly different mechanism to cope with this stress. The third major objective is to examine the process of stimulation of growth under continued burial conditions and degeneration when burial ceases. It would be useful to critically evaluate the hypotheses generated over the past century and present experimental evidence in favour or against these hypotheses.

8.2 Storm Damage of Foredunes – A Case History

The major problem faced by foredune plants along coasts is disruption of habitat by wind and wave action. I present a case history of a storm that occurred in1986–1987 along the Lake Huron shoreline. The lake levels started to rise in 1984 and by 1986 had risen by about 1.25 m above the long-term average. Wave storms in fall of 1986 and early spring of 1987 eroded the middle beach, upper beach and approximately half of the first dune ridge (Fig. 8.1). All populations of *Ammophila breviligulata, Calamovilfa longifolia* and other annual and biennial species of the beach and foredune were completely destroyed and a bare area was created. In addition, major changes occurred in the physiography and re-arrangement of the foredune terrain. However, within about 1 month, some of the sand began to return and was deposited on the beach along with the flotsom and jetsom from the lake. The deposited material consisted of seeds of annuals, biennials, perennials and rhizome fragments of grasses especially *A. breviligulata*, cuttings of herbaceous plants such as *Potentilla anserina* and *Tussilago farfara*, and twigs and branches of different trees and shrubs. Seeds of almost all species germinated and fragments and cuttings of grasses, shrubs and trees started to grow, however, a large majority of the species were short lived. The main reasons for their mortality were desiccation, erosion of sand, burial in sand, sand blasting and insect attack. Since there was little or no vegetation on the beach to arrest the movement of sand, a large proportion of sand was deposited on the crest of the first dune ridge through openings in the dune ridge caused by pedestrian traffic. Perumal (1994) installed 96 steel stakes at different places on the first dune ridge and measured the amount of sand depo-

Fig. 8.1. The erosion of first dune ridge during the wave storms of fall 1986 and early spring 1987

sition at each stake at regular intervals for two years. He showed that sand accretion ranged from 0 to 74 cm. By the end of 2 years the original complement of sand dune species consisting of *Cakile edentula, Corispermum hyssopifolium, Euphorbia polygonifolia, Artemisia caudata, A. breviligulata* and *Calamovilfa longifolia* had re-established on the upper beach and started to arrest sand movement and re-build the foredune. These species reclaimed the habitat because of two traits, (1) ability to disperse back to the habitat (return) and (2) then show high rates of re-establishment.

8.2.1 Return

A species must have a mechanism of dispersing back to the habitat. I will elaborate two main mechanisms. First, the most important trait used by propagules of plant species is *dispersal in water*. According to Ridley (1930), to be successful in water dispersal, the propagules (1) should be able to float in water without being waterlogged, (2) should not imbibe water while afloat and (3) should not lose viability while being transported in water. The worldwide distribution of some species of several genera such as *Cakile, Ammophila, Ipomoea, Calystegia, Sesuvium, Honckenya, Crambe* and several others can be attributed to the ability of seeds, and fragments of plants to meet all three criteria. *Wind dispersal* also contributed to the establishment of some plant species particularly annuals. Second, a large number of most successful species on coastal foredunes are grasses and vines that expand into the open areas of the beach by producing *creeping rhizomes or stolons* (Table 8.1). The storm waves that destroyed their habitat also fragmented these rhizomes or stolons and transported them back to the same shoreline or to another shoreline where they quickly regenerated and established new populations (Maun 1984, 1985). Many species of different genera and families listed in Table 8.1 along coasts of the world exhibit this mode of vegetative regeneration. This trait provides an efficient solution to the demand exerted by frequent destructive storms along shorelines and may be an example of parallelism or convergence. *Parallelism* may be defined as independent acquisition of similar phenotypic traits in species with a common heritage in response to similar selective pressures imposed by the environment. When species do not have a common heritage, evolutionary parallelism is called *convergence.* According to Mayr (1977), "if there is only one efficient solution for a certain functional demand, very different gene complexes will come up with the same solution, no matter how different the pathway by which it is achieved".

Table 8.1. Partial list of species of taxa that produce rhizomes, stolons or suckers along different shorelines of the world. These rhizomes or stolons are fractured by storm waves in autumn and early spring months and transported back to the same shoreline or to new shorelines where they establish new populations. This is a convergent trait exhibited by many families of plants

Name of species and family	Occurrence
Ammophila breviligulata, A. arenaria (Gramineae)	North America and Europe
Calamophila baltica (Gramineae)	Europe
Leymus arenarius, L. mollis (Gramineae)	North America and Europe
Elymus farctus, (Gramineae)	Europe
Panicum racemosum (Gramineae)	South America (Brazil)
Phragmitis communis (Gramineae)	North America and Europe
Ischaemum anthrefroides (Gramineae)	Japan
Spinifex hirsutus, S. sericeus (Gramineae)	Australia
Spinifex littoreus (Gramineae)	India, Malay Peninsula
Distichlis stricta (Gramineae)	North America and Europe
Ehrharta villosa (Gramineae)	South Africa
Thinopyrum distichum (Gramineae)	South Africa
Carex arenaria, C. kobomugi, C. eriocepahala (Cyperaceae)	Europe, Japan, America
Ipomoea pes-caprae, Ipomoea stolonifera (Convolvulaceae)	Tropics
Calystegia soldanella (Convolvulaceae)	Europe
Sesuvium portulacastrum (Portulacaceae)	Tropics
Honkenya peploides (Caryophyllaceae)	Europe and North America

Vegetative regeneration along coasts is adaptive for three main reasons:

1. It takes less time for a species to establish, become adult and reach reproductive stage because of large carbohydrate reserves in these fragments. For example, normally *Crambe maritima* plants establishing from seeds take about 5 to 8 years before they come to flower but plants establishing from fragments flower within one year (Scott and Randall 1976).
2. The rate of establishment of plants is much higher from fragments compared with seeds. In a comparison between survivorship of *A. breviligulata* from rhizome fragments and seedlings, more than 85 % of plants established from rhizome fragments compared to only 4 % from seedlings (Maun 1984).
3. It provides the fastest way of re-occupying the habitat. Populations of *A. breviligulata, Ipomoea pes-caprae, Spinifex hirsutus* expand towards the shoreline by forming an advancing front consisting entirely of rhizomes or stolons. According to Woodhouse (1982), planted stands of *A. breviligulata* were capable of 50-fold increase in area per year. Similarly, many other beach species expand very fast into the open bare areas along the coast.

However, all foredune perennials also allocate resources into sexual reproduction. There are three main advantages of seed production. First, even though the establishment from seeds is a stochastic event in dunes (Maun 1985; Lichter 1998), it incorporates genetic variability into the population and eliminates the major disadvantage of vegetative regeneration in that the offspring is genetically identical to that of its parents. Second, the seeds of most species possess enforced or innate dormancy that allows them to prolong their life. Third, seeds are able to disperse to more distant shorelines than rhizome or stolon fragments.

8.2.2 Re-Establishment

The problems for re-establishment of plant species are similar to those of other habitat types but foredune species have to contend with additional stresses imposed by (1) burial by sand, (2) sand blasting, (3) salt spray and (4) very low nutrient levels. Periodic observations over the years showed that within about ten years all traces of storm damage were completely obliterated and the plant community recovered approximately to its former levels (Fig. 8.2). The most important observation was that in spite of the invasion by a large number of species after the storm only the original complement of species prior to the storm reestablished and reclaimed the habitat. The species that contributed to reclamation of the habitat were the perennial rhizomatous grasses, *A. breviligulata* and *Calamovilfa longifolia*. Their success was primarily due to their ability to re-establish and grow vertically in response to burial by sand. The species on other shorelines of the world may be different

Fig. 8.2. The recovery of plant community to approximately its former levels. Note the gradual foredune slope of the first dune ridge formed by *Ammophila breviligulata* and *Calamovilfa longifolia*. Photograph taken after 8 years of recovery in June 1995

but the process of re-colonization and dune formation by lateral and upward extension of plants in response to burial is the same.

8.3 A Conceptual Model of Plant Response to Burial

As mentioned above, all foredune species returned after the storm but many aliens also dispersed to the habitat. Maun (1998) showed that plants exhibited three types of responses to burial (Fig. 8.3).

1. A *"negative inhibitory response"* in which the plant is unable to withstand burial and dies soon after the episode. For example, large trees of *Quercus velutina* and *Pinus resinosa* are readily killed by burial in sand. As shown earlier, propagules of many plant species are cast on the shorelines by waves. They may produce seedlings or ramets but the conditions on the beach and foredunes are not suitable for their survival and they succumb to burial and other unfavourable conditions.
2. A *"neutral and then negative response"* in which the plant shows little or no visible response initially because burial depth is within its limits of tolerance. However, as the level of sand accretion increases the response

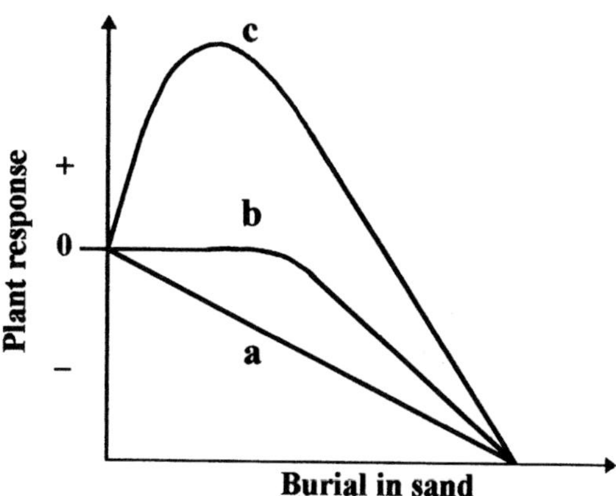

Fig. 8.3. Three possible response curves of plant species: *a* a negative inhibitory response, *b* a neutral and then negative response and *c* a positive stimulatory response, to being buried in sand on coastal dunes. (Adapted from Maun 1998)

becomes negative and the plant eventually dies (Maun 1998). Some tree species such as *Juniperus virginiana* may survive burial for a few years but are eventually killed depending on the amount and rate of sand accretion.

3. A "*positive stimulatory response*" in which the species exhibit enhancement of growth following a certain threshold level of burial. This is by far the most common response shown by all sand dune species (Maun and Baye 1989). Different life forms vary, by many orders of magnitude, in the relative amounts of burial at which they exhibit a stimulatory response but their response curves are similar. For example, the amount of burial may only be a few mm in lichens and mosses but a meter or more in some grass and tree species (Maun 1998). However, plant species vary in their maximum tolerance limits above which they start to show a negative response and are eventually killed.

8.4 Post-Burial Responses of Plants

Each stage in the life history of a plant has its own mechanism to tolerate the stress caused by burial. We will examine the response of seeds, seedlings, adult plants and communities to burial episodes.

8.4.1 Seeds and Seed Bank

Following dispersal, the seeds of plants accumulated in depressions in the sand surface where they were covered with leaves of deciduous trees and herbaceous plants (Maun 1981). During the autumn and winter months, these micro-sites are buried to various depths by sand depending on their location and wind velocities. The seeds respond to burial in three ways. First, the seeds will germinate and the seedlings will emerge if they are situated at an optimal depth specific for the species. Second, the seeds will germinate but the seedlings are unable to emerge because the seed is buried too deep and it does not have enough stored energy to take the growing point above the sand surface. These seedlings eventually decay in the soil and are lost. Third, the seeds may not germinate because they undergo enforced or innate dormancy and become part of the seed bank. This was by far the most common response shown by deeply buried seeds of seven dune species (Zhang and Maun 1994). The seeds were forced to become dormant as a result of burial and as the depth of burial increased, the degree of enforced dormancy increased.

The emergence of a seedling is dependent on the energy contained in the seeds and there is clear evidence of a positive relationship between seed mass and depth of emergence (Maun 1998). The larger the seed, the greater was the maximum depth from which its seedlings emerged (Maun and Lapierre 1986). This relationship held true for seed mass both within and between species. However, since the variability in burial depths in foredune habitats is very high and even though a certain proportions of seeds may be buried too deep, there is always a certain proportion of seeds buried at optimum depths (Maun 1981).

Most species on sand dunes possess a transient *seed bank* and do not have a significant carryover of seeds from one year to the next (Rowland and Maun 2001; Planisek and Pippen 1984; Barbour 1972; Mack 1976; Watkinson 1978). Overall, the number of seeds in the seed banks of sand dunes is very low as shown by Baptista and Shumway (1998). They determined seed bank composition of four coastal dunes along Cape Cod National Seashore by collecting sand samples and germinating seeds in a greenhouse. Seedlings emerged from only 20 % of the sand samples indicating that the seeds were highly clumped. A total of 254 seedlings emerged from all sand samples of which 85 % belonged to *Artemisia caudata*, 5 % to *A. breviligulata* and 3 % to *Solidago sempervirens*. Several other species, *Chenopodium rubrum, Hudsonia* spp., *Artemisia stelleriana, Cakile edentula* and *Polygonella articulata* contributed less than 2 % to the seedling population. Moreover, as shown earlier, many species of foredunes have the potential to form a seed bank (Zhang and Maun 1994). Another source of seeds is a temporary seed bank on the above ground plant parts. Several species of foredunes retain seeds in cones or inflorescences as an above ground seed bank and release them gradually at appropriate environmental cues thus ensuring the dispersal of some seeds into safe sites (Zhang and Maun 1994).

8.4.2 Seedlings

The survival of seedlings is usually extremely low in sand dunes because of various environmental stresses such as desiccation, erosion of sand, insect attack and excessive burial in sand. However, there is evidence that partial burial stimulates the growth of seedlings. For example, partially buried seedlings of *A. breviligulata* and *Calamovilfa longifolia* showed higher net CO_2 uptake (Yuan et al. 1993), *Uniola paniculata* seedlings responded to burial by an increase in tillering (Wagner 1964), and *Cakile edentula* exhibited greater production of flowers and seeds per plant compared to control (Maun 1994). Similarly, seedlings of all six tropical species, *Chamaecrista chamaecristoides, Palafoxia lindenii, Schizachyrium scoparium, Trachypogon plumosus* (formerly *gouini*), *Canavalia rosea* and *Ipomoea pes-caprae* responded to burial by an increase in biomass and leaf area (Martínez and Moreno-Casasola 1996). All species except *T. plumosus* allocated greater biomass to aboveground plant parts.

Burial beyond a certain threshold level proved fatal to the seedlings. Young plants of *Cakile edentula, C. maritima, Corispermum hyssopifolium, Salsola*

Fig. 8.4. A plant of *Cakile edentula* var. *edentula* being buried on a beach along the sea coast of Prince Edward Island, Canada, on the Gulf of St. Lawrence. Note the formation of a shadow dune on the lee of the plant. Complete burial of annual or biennial plants usually kills them

kali, Honckenya peploides and many others often grow in clumps at the location of last year's plants or as single plants on the midbeach where they may be partially buried and form shadow dunes (Fig. 8.4). With few exceptions complete burial almost always killed the seedlings unless they were re-exposed within a few days. According to Harris and Davy (1987), the seedlings of *Elymus farctus* survived if they were re-exposed after about 1 week, but died if left buried for 2 weeks. The energy for this short term survival came from stored reserves in the roots and stems (Harris and Davy 1988). Brown (1997) showed that upon burial the normal source-sink relationship was reversed and the stored material was mobilized and transferred to existing photosynthetic tissues. A morphological examination of buried seedlings showed etiolation of leaves and stems within about 10 days of burial. Under field conditions seedling survival may also be affected by the depth from which it emerged. Seedlings of *Panicum virgatum* emerging from deeply buried seeds survived significantly lower post-emergence burial depths than those emerging from shallow depths (Zhang and Maun 1991).

8.4.3 Adult Plants

Below a certain threshold level of burial specific for each dune species, plants show an increase in vigour by exhibiting higher net CO_2 uptake (Yuan et al. 1993), higher density, percent cover, and biomass per plant and per unit area (Maun 1998). For example, dominant foredune grasses of the Great Lakes, *A. breviligulata, Calamovilfa longifolia, Agropyron psammophilum*, and *Panicum virgatum*, showed an increase in density at burial depths ranging between 5 and 20 cm but started to decline at higher levels of burial (Maun and Lapierre 1984; Perumal 1994; Maun 1996). Seliskar (1994) also showed a similar relationship between the number of panicles and burial depth. However, even though the density decreased there was an increase in biomass per shoot after their emergence above the sand surface. Similar conclusions were reported by Eldred and Maun (1982) and Disraeli (1984) in natural stands of *A. breviligulata* and by Sykes and Wilson (1990) who artificially subjected 30 New Zealand sand dune species to different burial treatments.

8.4.4 Plant Communities

In natural foredune communities the distribution of plants is related to variability in burial depths in the habitat. Moreno-Casasola (1986) showed that there was a close relationship between natural sand movement, topography, and spatial distribution of plant communities. In habitats with high levels of sand mobility *Croton punctatus, Palafoxia lindenii* and *Chamaecrista chamaecristoides* survived and reproduced successfully. Similarly, in a study on primary succession on mobile tropical dunes, Martínez et al. (2001) showed that the spatial distribution, coverage, diversity and relative frequency of early colonizers, *Chamaecrista chamaecristoides* and *Palafoxia lindenii*, were positively correlated with the amount of burial in sand. In contrast, the later colonizing species, *Schizachrium scoparium* and *Trachypogon plumosus*, were less tolerant of sand deposition and were abundant only in areas where sand movement had decreased substantially. Although sand mobility was probably the most important factor, other factors such as soil moisture, soil temperature, biotic interactions and plant life histories also played a role in spatial and temporal variability (Martínez et al. 2001). In an artificial burial experiment Maun and Perumal (1999) showed that the number of plant species in the community decreased with an increase in burial. As the burial depth increased beyond the level of tolerance of a species, the plants started to deteriorate and eventually died. Indeed, sand dune species may be classified as non-tolerant, tolerant and sand dependent. Annual species were eliminated first followed by biennials and then perennials (Maun and Perumal 1999). Eventually, however, a stage was reached when the amount of sand accretion exceeded the tolerance limits of even the sand-dependent species and a bare area was created. The survival of plants is also dependent on the rate at which a plant is buried in sand. In an experiment on *Cirsium pitcheri* plants buried gradually recovered within a few days probably because their leaves were still above the sand surface and had continued to function normally (Maun et al. 1996). In contrast, one time burial of plants significantly delayed the emergence and recovery of plants.

8.5 Burial – The Primary Cause of Zonation

The differential tolerance of sand dune species to burial may be one of the principal causes of zonation of plant species on coastal foredunes (Maun and Perumal 1999). Martin (1959) showed that as one moved inland from the shoreline along the Atlantic coast of North Carolina, the total deposition of sand decreased and the species occurrence was related to the amount of sand burial. For example, *A. breviligulata* and *Carex kobomugi*, were very vigorous in areas with average sand deposition of about 17 to 28 cm/year in the first 40 m from the beginning of the primary foredune. When the sand deposition in the next 20 m (41–60 m) decreased to about 3–5 cm, the two species became sparse. Farther inland deflation exceeded sand deposition and both species degenerated and exhibited a significant decline in vigour. Burial also retards sand dune succession (Poulson 1999). He showed that abiotic forces such as burial by sand, high winds and substrate instability along Lake Michigan continuously modified the local environment and did not allow species of the next stage in succession to gain a foothold. Similarly Olff et al. (1993) showed that earliest dune stage deviated from the general successional pattern because of sand deposition. This constant disturbance of the habitat did not allow the species to converge to permanent plant communities. Similar observations were made by Houle (1997) in a sub-arctic foredune along Hudson Bay, where productivity was low and plant-plant interactions were non-existent because of high disturbance caused by wind and wave action.

8.6 Degeneration Response

Burial of plants has both a positive and a negative aspect. It has a stimulating positive effect on plant growth up to a certain level of burial in sand. However, in plant communities with little or no sand deposition there is a decline in density, plant height, net CO_2 uptake, flowering, tillering and biomass per unit area. Several possible causes of degeneration such as deficiency of nutrients (Willis 1965), increase in competition (Marshall 1965; Huiskes and Harper 1979), desiccation of growing point (Olson 1958), accumulation of organic matter (Waterman 1919), decortication of roots (Marshall 1965), and harmful soil organisms (Van der Putten et al. 1988), have been proposed over the years. However, strong counter arguments, as shown below, have been advanced against each hypothesis. For example, *nutrient deficiency* can not be a factor because even burial by leached sand or acid washed sand (no nutrients) increased growth (Hope-Simpson and Jefferies 1966; Maze and Whalley 1992). Decline due to *increase in competition* is not relevant because debilitated stands of *Ammophila* do not contain any other species (Hope-Simpson and Jefferies 1966; Baye 1990; Poulson 1999). Actually, *A. breviligulata* retards natural sand dune succession (Poulson 1999) and heterospecific removal of two species showed no effect on either species (Houle 1998). *Desiccation of growing point* is a good possibility because even in the absence of sand burial, *A. breviligulata* continues to elongate its internodes into the dry surface sand where it desiccates. However, the hypothesis has not been tested. *Accumulation of organic matter* cannot be responsible for the decline because experimental addition of organic matter did not inhibit growth of plants (Zaremba 1982). Actually, it stimulated growth. Hope-Simpson and Jefferies (1966) found no evidence for the *decortication of roots* because wiry decorticated roots frequently terminated in a fully functional fleshy distal end. *Nematodes and harmful soil organisms* do destroy functional roots (De Rooij-van der Goes 1996), however, the injurious effects of pathogenic fungi (Newsham et al. 1995) and nematodes (Little and Maun 1996) were mitigated by mycorrhizal fungi.

What then are the possible reasons for the degeneration of plants? In spite of the many studies conducted to answer this question, the reasons are obscure and any suggestions must remain tentative. However, there are strong indications that this decline is caused by an interaction of several factors. For example, five factors, (1) complete exploitation of the soil volume by roots, (2) decline in the formation of new roots thereby causing a decrease in colonization by mycorrhizal fungi, (3) desiccation of the growing point as it continues to grow upward even in the absence of sand deposition (4) physiological deterioration in plant functions, and (5) soil microorganism activity, may interact to cause a decline in plant growth.

8.7 Stimulation Response

What are the possible causes of stimulation? There is strong evidence that single factors are inadequate to explain the enhancement of plant vigour. I would therefore propose a *"Multifactor hypothesis"* composed of four major biotic and physical variables, (1) increased soil volume, (2) increased soil nutrients, (3) increased activity of mycorrhizal fungi, and (4) reactive growth by the plant to burial (Fig. 8.5). Burial increases soil volume and creates more new space for the growth and expansion of the plant and its roots. The apical meristems of the plant grow through the burial deposits probably because of the etiolation response and emerge above the new sand surface. The new sand deposit increases the amount of soil nutrients that cannot be used by the plant until new roots develop in the sand deposit that may take 2–4 weeks. However, mycorrhizal fungi, ubiquitous in sand dune systems (Perumal 1994), expand into the new deposit almost immediately and exploit the soil resources to the benefit of the plant. For more details on the occurrence of mycorrhizal fungi in sand dunes refer to Koske et al. (Chap. 11, this Vol.). The mycorrhizal fungi

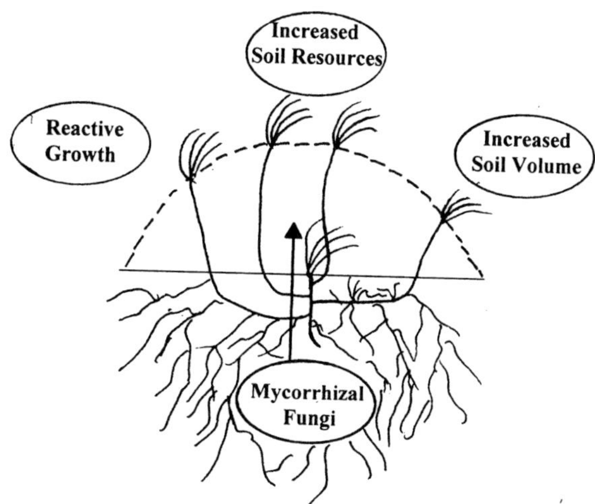

Fig. 8.5. A "Multifactor hypothesis" to explain the stimulation response of sand dune species following burial in sand

not only enlarge the nutrient absorbing surface of the roots, but also compete with harmful soil organisms for space on the roots. In addition, the plant reacts to the burial stress by mobilizing all its resources that allows it to overcome the burial episode. The response may be mediated by hormone production whereby the plant directs all its energy towards making physiological adjustments, changing its morphology, and finally emerging from the burial deposit.

8.8 Summary

Coastal sand dune systems are highly dynamic because of the activity of wind, waves and tides. The most important stress encountered by plant species growing here is probably the burial by sand. Burial acts as a very strong selective force that alters the composition of plant communities by selective elimination of species with a conservative growth habit. Burial curtails the photosynthetic capacity of the plant, increases the respiration rate and alters the microclimate around the plant. However, all foredune plant species have developed adaptations that allow them to withstand certain threshold levels of sand burial. Other traits such as dispersal in water, vegetative growth by rhizomes and stolons and lateral and vertical growth also allow them to occupy this habitat permanently. Deep burial of seeds induces enforced dormancy of seeds. The response of seedlings and adult plants to burial is similar and emergence of plants is related to the amount of stored energy reserves in their roots and rhizomes. They not only survive certain amounts of burial but their growth is also stimulated by it. There is mounting evidence that the principal causes of this stimulation are an increase in soil volume around the base of the plant that contains nutrients, probably in small amounts, which are exploited by the mycorrhizal fungi already associated with roots of dune plants. A plant also exhibits a reactive growth response to burial. Conversely, as soon as the sand dune stabilizes and sand accretion ceases, there is a marked decline in vigour and density of foredune populations. Several possible hypotheses and counter arguments have been advanced in the past century but no consensus has been reached. In all likelihood, the phenomenon of decline in sites with no sand deposition is caused by an interaction of several environmental factors.

Acknowledgments. I would like to thank the Natural Sciences and Engineering Council of Canada for supporting my research program on "Adaptations of Plants to the Sand Dune Environment" over the last 25 years.

References

Baptista TL, Shumway SW (1998) A comparison of the seed banks of sand dunes with different disturbance histories on Cape Cod National Seashore. Rhodora 100:298–313

Barbour MG (1972) Seedling establishment of *Cakile maritima* at Bodega Head, California. Bull Torr Bot Club 99:11–16

Baye PR (1990) Comparative growth responses and population ecology of European and American beachgrasses (*Ammophila* spp.) in relation to sand accretion and salinity. PhD Thesis, Univ Western Ontario, London, Ontario

Brown JF (1997) Effects of experimental burial on survival, growth, and resource allocation of three species of dune plants. J Ecol 85:151–158

De Rooij-van der Goes PCEM (1996) Soil borne plant pathogens of *Ammophila arenaria* in coastal foredunes. PhD Thesis, Landbouw Univ Wageningen, The Netherlands

Disraeli DJ (1984) The effect of sand deposits on the growth and morphology of *Ammophila breviligulata*. J Ecol 72:145–154

Eldred RA, Maun MA (1982) A multivariate approach to the problem of decline in vigour of *Ammophila*. Can J Bot 60:1371–1380

Grime JP (1979) Plant strategies and vegetation processes. Wiley, New York

Harris D, Davy AJ (1987) Seedling growth in *Elymus farctus* after episodes of burial with sand. Ann Bot 60:587–593

Harris D, Davy AJ (1988) Carbon and nutrient allocation in *Elymus farctus* seedlings after burial with sand. Ann Bot 61:147–157

Hope-Simpson JF, Jefferies RL (1966) Observations relating to vigour and debility in marramgrass, *Ammophila arenaria* (L) Link. J Ecol 54:271–274

Houle G (1997) Interaction between resources and abiotic conditions control plant performance on subarctic coastal dunes. Am J Bot 84:1729–1737

Houle G (1998) Plant response to heterospecific neighbour removal and nutrient addition in a subarctic coastal dune system (northern Quebec, Canada). Ecoscience 5:526–533

Huiskes AHL, Harper JL (1979) The demography of leaves and tillers of *Ammophila arenaria* in a dune sere. Oecol Plant 14:435–446

Lichter J (1998) Primary succession and forest development on coastal Lake Michigan sand dunes. Ecol Monogr 68:487–510

Little LR, Maun MA (1996) The '*Ammophila* problem' revisited: a role for mycorrhizal fungi. J Ecol 84:1–7

Mack RN (1976) Survivorship of *Cerastium atrovirens* at Abberffraw Anglesey. J Ecol 64: 109–312

Marshall JK (1965) *Corynephorus canescens* (L) P. Beauv. as a model for the *Ammophila* problem. J Ecol 53:447–463

Martin WE (1959) Vegetation of Island Beach State Park. Ecol Monogr 29:1–46

Martínez ML, Moreno-Casasola P (1996) Effects of burial by sand on seedling growth and survival in six tropical sand dune species from the Gulf of Mexico. J Coastal Res 12:406–419

Martínez ML, Vázquez G, Salvador SC (2001) Spatial and temporal variability during primary succession on tropical coastal sand dunes. J Veg Sci 12:361–372

Maun MA (1981) Seed germination and seedling establishment of *Calamovilfa longifolia* on Lake Huron sand dunes. Can J Bot 59:460–469

Maun MA (1984) Colonizing ability of *Ammophila breviligulata* through vegetative regeneration. J Ecol 72:565–574

Maun MA (1985) Population biology of *Ammophila breviligulata* and *Calamovilfa longifolia* on Lake Huron sand dunes. I. Habitat, growth form, reproduction and establishment. Can J Bot 63:113–124

Maun MA (1994) Adaptations enhancing survival and establishment of seedlings on coastal dune systems. Vegetatio 111:59–70

Maun MA (1996) The effects of burial by sand on survival and growth of *Calamovilfa longifolia*. Ecoscience 3:93–100

Maun MA (1998) Adaptations of plants to burial in coastal sand dune systems. Can J Bot 76:713–738

Maun MA, Baye PR (1989) The ecology of *Ammophila breviligulata* Fern. on coastal dune systems. CRC Crit Rev Aquat Sci 1:661–681

Maun MA, Lapierre J (1984) The effects of burial by sand on *Ammophila breviligulata*. J Ecol 72:827–839

Maun MA, Lapierre J (1986) Effects of burial by sand on seed germination and seedling establishment of four dune species. Am J Bot 73:450–455

Maun MA, Perumal J (1999) Zonation of vegetation on lacustrine coastal dunes: effects of burial by sand. Ecol Lett 2:14–18

Maun MA, Elberling H, D'Ulisse A (1996) The effects of burial by sand on survival and growth of Pitcher's thistle (*Cirsium pitcheri*) along Lake Huron. J Coastal Conserv 2:3–12

Mayr E (1977) Populations, species and evolution. Harvard University Press, Cambridge, MA

Maze KM, Whalley RDB (1992) Effects of salt spray and sand burial on *Spinifex sericeus* R. Br. Aust J Ecol 17:9–19

Moreno-Casasola P (1986) Sand movement as a factor in the distribution of plant communities in a coastal dune system. Vegetatio 65:67–76

Newsham KK, Fitter AH, Watkinson AR (1995) Arbuscular mycorrhiza protect an annual grass from root pathogenic fungi in the field. J Ecol 83:991–1000

Olff H, Huisman J, van Tooren BF (1993) Species dynamics and nutrient accumulation during early succession in coastal sand dunes. J Ecol 81:693–706

Olson JS (1958) Rates of succession and soil changes on southern Lake Michigan sand dunes. Bot Gaz (Chicago) 119:125–170

Perumal J (1994) Effects of burial in sand on dune plant communities and ecophysiology of component species. PhD Thesis, Univ Western Ontario, London, Ontario

Planisek SL, Pippen RW (1984) Do sand dunes have seed banks? Mich Bot 23:169–177

Poulson T (1999) Autogenic, allogenic and individualistic mechanisms of dune succession at Miller, Indiana. Nat Areas J 19:172–176

Ridley HN (1930) The dispersal of plants throughout the world. L Reeve, Kent, UK

Rowland J, Maun MA (2001) Restoration ecology of an endangered plant species: establishment of new populations of *Cirsium pitcheri*. Restoration Ecol 9:60–70

Scott GAM, Randall RE (1976) Biological flora of British Isles: *Crambe maritima* L. J Ecol 64:1077–1091

Seliskar DM (1994) The effect of accelerated sand accretion on growth, carbohydrate reserves and ethylene production in *Ammophila breviligulata* (Poaceae). Am J Bot 81:536–541

Sykes MT Wilson JB (1990) An experimental investigation into the response of New Zealand sand dune species to different depths of burial by sand. Acta Bot Neerl 39:171–181

Van der Putten WH, Van Dijk C, Troelstra SR (1988) Biotic soil factors affecting the growth and development of *Ammophila arenaria*. Oecologia 76:313–320

Wagner RH (1964) The ecology of *Uniola paniculata* in the dune-strand habitat of North Carolina. Ecol Monogr 34:79–96

Waterman WG (1919) Development of root systems under dune conditions. Bot Gaz 68:22–53

Watkinson AR (1978) The demography of a sand dune annual: *Vulpia fasciculata* II. The dynamics of seed populations. J Ecol 66:35–44

Willis AJ (1965) The influence of mineral nutrients on the growth of *Ammophila arenaria*. J Ecol 53:735–745

Woodhouse WW Jr (1982) Coastal sand dunes of the US. *In* Creation and restoration of coastal plant communities. Lewis RR III (ed) CRC Press, Boca Raton, pp 1–44

Yuan T, Maun MA, Hopkins WG (1993) Effects of sand accretion on photosynthesis, leaf-water potential and morphology of two dune grasses. Funct Ecol 7:676–682

Zaremba RE (1982) The role of vegetation and overwash in the landward migration of a northern barrier beach: Nauset Spit-Eastham, Massachusetts. PhD Thesis, Univ of Massachusetts, Amherst, MA

Zhang J, Maun MA (1991) Establishment and growth of *Panicum virgatum* L. seedlings on a Lake Erie sand dune. Bull Torrey Bot Club 118:141–153

Zhang J, Maun MA (1994) Potential for seed bank formation in seven Great Lakes sand dune species. Am J Bot 81:387–394

9 Physiological Characteristics of Coastal Dune Pioneer Species from the Eastern Cape, South Africa, in Relation to Stress and Disturbance

B.S. Ripley and N.W. Pammenter

9.1 Introduction

A characteristic of coastal dune systems is that species diversity and total plant biomass is less than that of the adjoining inland areas, and furthermore, there is generally pronounced zonation of the species present on the dunes (Clements 1916; Lubke 1983). Why are the foredunes inhabited by only few species, and dominated by even fewer?

There are obviously a number of factors controlling plant distribution and productivity, but for the foredunes in particular, adaptations to resource stress and/or disturbance are likely to be very important (Barbour 1992). The resources required by plants for growth include light, water and nutrients, and stress could be a consequence of either deficiencies or excesses of these resources. Adaptations to acquire resources present in low amounts, or to resist those present in excess, may also play a role in competition between species. Disturbance is the process by which part or all of a plant is damaged or destroyed, generally by physical processes. Disturbance is particularly high in coastal foredunes and includes wind, salt-spray, occasional inundation by seawater and sand movement. The latter can lead to burial, exposure, or physical damage from sand-blasting. Salt spray could be argued to be an environmental stress, but because it can lead to direct physical damage, we prefer to consider it a disturbance. However the distinction is unimportant in the discussion that follows. Further complications arise because of adaptations to a combination of stresses and/or disturbance, and because plant response may vary with stage in the life cycle (particularly seedlings in comparison with established plants).

Ecological Studies, Vol. 171
M.L. Martínez, N.P. Psuty (Eds.)
Coastal Dunes, Ecology and Conservation

9.2 A Conceptual Model of Resource Limitation and Plant Performance

To consider the adaptations to resource stresses and disturbances, a conceptual model of foredune plant performance is presented (Fig. 9.1). The model describes the relationship between the supply of resources, plant function in relation to the acquisition and use of these resources, and the link to the resultant productivity. This gives insight into the potential limitations created by the particular resources. The effect that disturbance may have is also indicated. Differences among species may explain why species survive on the foredunes and why particular species are more abundant within the particular microhabitats of the foredunes.

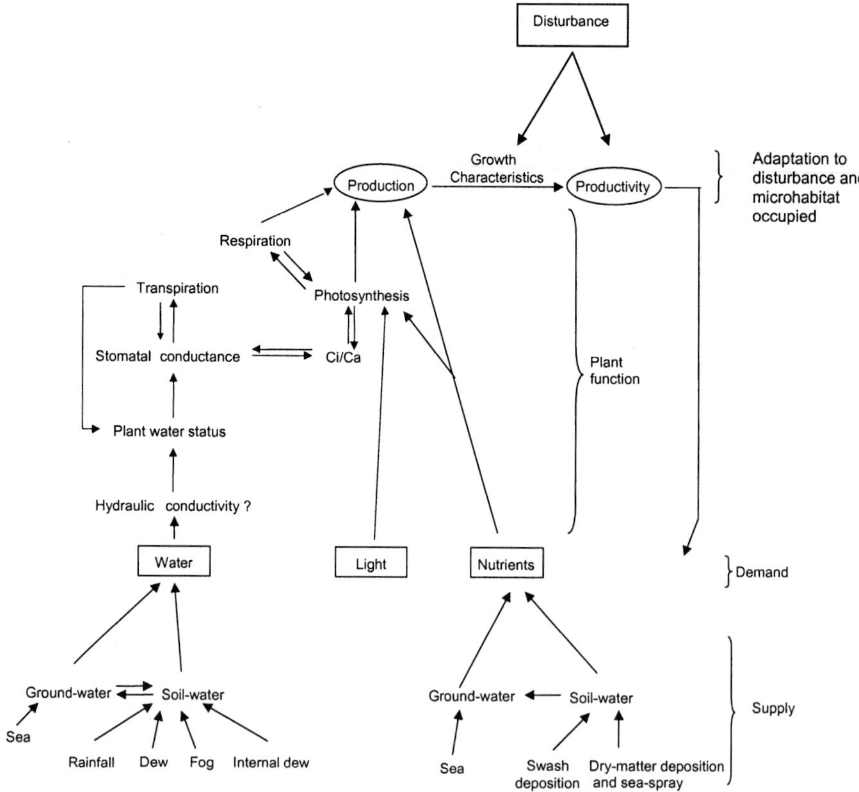

Fig. 9.1. A conceptual model of how plant production and ultimately productivity (model outputs) are related to the acquisition and use of water, nutrients and light (model inputs) and how, in turn, productivity determines the demand for resources. The environment supplies resources and the possible origins of these resources are indicated. Disturbance can influence growth characteristics and productivity, and hence demand. See text for explanation. (Atmospheric fallout refers to the input of nutrients by atmospheric fall-out, such as dust etc.)

Water availability and transpiration rate, in conjunction with plant hydraulic conductance, determine leaf water status (Passioura 1982), which in turn affects stomatal conductance (Mansfield and Davies 1981). Stomatal conductance affects the diffusion of CO_2 into the leaf and influences the ratio of intercellular (C_i) to atmospheric (C_a) CO_2 concentration (Farquhar and Sharkey 1982). C_i, in conjunction with light absorption and the intrinsic capacity of the photosynthetic apparatus (which is dependent *inter alia* upon mineral nutrients) determines the photosynthetic rate, and photosynthetic rate in turn influences C_i. Photosynthate production and respiration determine net carbon gain. Carbon gain, biomass allocation patterns and the acquisition of mineral nutrients determine production. Production per unit land area, as determined by the species growth characteristics, determines productivity. Biomass allocation patterns are likely to differ among species and are also modified by the prevailing stress and disturbance (microhabitat) conditions. Disturbance can reduce standing biomass and hence resource demand.

The relationship between plant growth and resource availability can be complex, although it is often thought that plant growth is limited by the supply of resources. The amount of plant material present as biomass per unit land area or leaf area index (LAI, leaf area per unit ground surface area) is obviously important when considering resource supply and demand. A low standing biomass could be considered as a direct consequence of low resource supply. Alternatively, it could be viewed as a phenomenon that ensures that each unit of biomass is supplied with adequate resources and, consequently, that physiological stress does not occur. This is not a facile argument: restriction of standing biomass by resource supply implies that the plants present exploit the resources available down to some minimum level; restriction of biomass that avoids stress implies availability of resources in excess of those utilised. Avoidance of stress would permit high growth rates per unit plant mass, and concomitant death and loss of plant parts would prevent biomass accumulation, i.e. a low standing crop but high turnover rates. In the study described in this chapter the highest growth rate per unit plant mass actually occurred in the species with the lowest standing biomass. There is considerable biomass turnover, such that these high rates of dry matter production do not lead to high biomass accumulation. To distinguish between the two possibilities is not simple, but adaptations that enable more efficient resource acquisition or reduce resource demand may be taken as indicators that resources are limiting. Similarly, symptoms of stress, such as reduced stomatal conductance, net assimilation rates, photosynthetic electron transport, all leading to reduced growth rates, may also indicate the inadequacy of a resource.

To assess the importance of resource limitation and physiological stress in terms of Fig. 9.1 requires a considerable data set. Certainly with respect to coastal dunes, comprehensive data sets of that nature are not available. This chapter largely describes a study on the foredunes of the coast of the Eastern

Cape, South Africa, in which physiological and growth measurements were made over a number of years (Ripley 2001). The data set is such that it makes it possible to assess the extent and possible consequences of physiological stress in relation to resource availability.

9.3 Study Site, Species and Parameters Measured

The study was conducted at the mouth of Old Woman's River, 30 km northeast of Port Alfred (27°08′49″E; 33°28′59″S) over a period of 3 years, with measurements being taken in every season. Mean annual rainfall is 618 mm, with a peak in summer but winter rains are not unusual. Mean maximum and minimum temperatures for summer are 24.6 and 17.8 °C, and for winter 20.2 and 10.2 °C, respectively.

The species selected for study were *Arctotheca populifolia* (Berg.) Norl. (Asteraceae), *Ipomoea pes-caprae* (L.) R. Br. (Convolvulaceae) and *Scaevola plumieri* (L.) Vahl (Goodeniaceae) (Fig. 9.2). *S. plumieri* is a classic 'dune builder' with semi-succulent leaves, branching and growing up through sand trapped by the expanding above-ground parts of the plant. *A. populifolia*, also with semi-succulent leaves, has a more compact growth form, and occurs as individual plants towards base of the seaward slope of the foredune. *I. pes-caprae* occurs on the more stable sand at the top of the foredune and further inland. The growth form is prostrate, with long stems running across the top of the sand; the leaves, although occasionally semi-succulent, are far less so than the leaves of the other species. These species were chosen because they are the major foredune species along the entire east coast of southern Africa. Water and mineral nutrients are often considered to be potentially limiting factors in sand dune ecosystems (because of the nature of the substrate) (Willis 1965; Morris et al. 1974; Rozema et al. 1985; Hesp 1991). Consequently, the water relations and nutrient status of the plants were studied on a seasonal basis, and these were related to photosynthetic characteristics and growth rates (measured as leaf and stem production).

To assess the availability of resources and to scale up resource utilisation from leaf to ground surface area requires knowledge of the LAI. This was fairly simple to measure for *I. pes-caprae* and *S. plumieri* as these species form fairly uniform 'stands' on well-vegetated foredunes. Individual plants of *A. populifolia* are small and more widely scattered than the other species, and measurement of LAI was more difficult and less reliable. What was important was to assess the area of soil that the roots could exploit, and so the approach used was to randomly throw small quadrats where the plants of this species were growing, ensuring that plants occurred in every quadrat.

Various measurements of physiological and growth characteristics were made (see below for details) on a total of six visits over a three year period,

Fig. 9.2. **A** *Arctotheca populifolia*; **B** *Ipomoea pes-caprae*; **C** *Scaevola plumieri*, pioneer plants found growing on the foredunes on the east coast of southern Africa

two in each of the seasons: summer (1997 and 1998), winter (1996 and 1998) and spring (1997 and 1998). Replicate data (see Tables for details) were statistically compared both between experimental days and between species, although only the significance (at the 95 % confidence level) of the species comparison is given in the Tables. For a more complete comparison see Ripley (2001).

9.4 Water Relations

Stomatal conductance and transpiration rates were measured by porometry, stem or leaf water potential with a pressure chamber, and pressure-volume curves were established to assess points of turgor loss and the osmotic component of water potential.

Stomatal conductances for all species showed normal diurnal responses, with only *I. pes-caprae* sometimes showing midday reductions (Table 9.1). Mean maximum values for *A. populifolia* were particularly high, being over

Table 9.1. Interspecific differences in water-relations and transpiration rates measured for *A. populifolia, I. pes-caprae* and *S. plumieri*

No.	Parameters measured		Species		
			A. populifolia	*I. pes-caprae*	*S. plumieri*
1	Diurnal stomatal conductance, N=15		One-peaked for 15 replicates	Two-peaked for 9 replicates	One-peaked for 15 replicates
2	Maximum leaf conductance (mol m^{-2} s^{-1}), N=15		1.3±1.1[a]	0.17±0.05[b]	0.33±0.19[b]
3	Midday leaf or shoot water potential (MPa), N=15		−1.01±0.04[a]	−0.93±0.05[a]	−1.56±0.08[b]
4	Leaf water potential where turgor potential=zero (MPa), N=18		−1.31±0.54[a]	−1.65±0.55[a]	−1.61±0.41[a]
5	Osmotic potential when tissue is fully hydrated (MPa), N=18		−1.24±0.53[a]	−1.48±0.47[a]	−1.46±0.35[a]
6	Maximum transpiration rates (mmol H_2O m^{-2} s^{-1}), N=15		4.2±1.5[a]	3.7±1.4[a]	4.1±1.5[a]
7	Annual water budgets in an average rainfall year (1978–1998)	Rainfall (l m^{-2} $year^{-1}$)	618±168	618±168	618±168
		Transpiration (l m^{-2} $year^{-1}$)	208±12	593±34	430±30
		Difference (l m^{-2} $year^{-1}$)	410	25	188
8	Number of months in 1997 (rainfall =546 mm) when predicted volumes of water transpired exceeded volumes of rainfall per unit dune surface-area.		2	6	5

Values with the same superscript letter are not significantly different at the 95 % probability level (*post hoc* ANOVA and Tukey multiple range test)

1 mmol m^{-2} s^{-1}, while those of *I. pes-caprae* were much lower (Table 9.1). *A. populifolia* possesses an indumentum of silvery hairs that reduces leaf conductance (Ripley et al. 1999). Mean midday leaf water potentials were characteristic of mesic plants, ranging from –0.93 MPa in *I. pes-caprae* to –1.56 MPa in *S. plumieri*, and values seldom dropped below the water potential corresponding to loss of turgor (Table 9.1). Osmotic potential at full turgor did not differ among the species, with an average of about –1.4 MPa (Table 9.1), and there were no indications of seasonal osmotic adjustment. Mean maximum transpiration rates were high (Table 9.1, with values up to 6.5 mmol m^{-2} s^{-1} being recorded). These values are similar to those of temperate dune species (Pavlik 1983), and higher than those reported from some plants from semi-arid areas (Caldwell 1985; von Willert et al. 1989). In *S. plumieri* there was a strong linear relationship between transpiration rate and evaporative demand (measured as vapour pressure difference, VPD), suggesting little stomatal control of transpiration (see also Peter and Ripley 2000). A similar response was shown by *A. populifolia*, although the relationship was not as good as for *S. plumieri*, possibly as a consequence of fewer data points. In *I. pes-caprae* the relationship was non-linear, suggesting stomatal limitation of transpiration at high VPD. No distinct seasonal patterns were observed in plant water relations; rather, the plants responded to the immediate environmental conditions.

It is possible to make some rough calculations of the water budget of the foredune system. Using the measured relationships between E and VPD, and by calculating VPD from air temperature and relative humidity supplied from weather stations, transpiration rates can be estimated for any point in time (Peter and Ripley 2000). These estimated transpiration rates can be integrated over the year to give annual water use, and from LAI, this can be converted to units of mm water per unit ground area and compared with average annual rainfall. For all three species estimated annual average water utilisation by transpiration was less than annual average rainfall (1.7; although, because of the difficulty of assessing LAI for *A. populifolia*, calculated water use for this species may be an underestimate). Rainfall for 1997 was below average, and these rough calculations indicated a water use exceeding rainfall for a number of months during the year (Table 9.1). Either the plants were not transpiring as rapidly as the model based on measured E and VPD would suggest or they were accessing some other source of water, such as ground water, water condensed within the cold upper depths of the sand dunes (termed "internal dew" by Olsson-Seffer 1909; see Salisbury 1952 for details of this process), or the water stored in the soil column. Isotopic measurements were unable to distinguish between rain- and groundwater as a water supply, because of the rapidity with which the isotopic composition of the ground water reflected that of a particular rain event. The budgeting procedure was altered to incorporate the soil water. From knowledge of rooting depth the volume of soil available to the plants could be calculated, and from this and the water content of the soil at field capacity, the amount of water available to plants could be calculated. Soil water could then be modelled by estimates of subsequent transpiration and input from rain. When this was done at no point did soil water drop below 0.5 %, the value suggested by Salisbury (1952) to be the lower limit at which plants could extract water from the sandy dune soils.

9.5 Mineral Nutrients

Soil nutrient contents and nutrient contents of successive leaves on a stem (thus generating a 'time series'), were measured using standard analytical procedures.

Sand dune soils are generally low in mineral nutrients, and these dunes were no exception, with concentrations of N, P and K being 1.6±1.4, 0.32±0.02 and 2.57±0.41 mg 100 g^{-1} dry sand, respectively (N=6). Despite this, leaf nutrient concentrations were not particularly low (Table 9.2). These concentrations are well within the range measured on plants growing on less oligotrophic soils (Allen 1989). Measurements on consecutive leaves on a stem generally show that total leaf content of P and K initially rose as the leaf expanded and then declined after full expansion. This indicated that young leaves were net importers of P and K, and older leaves net exporters (Table 9.2); considerable internal recycling was occurring. Similar recycling of N and K has been shown in *S. plumieri* (Harte and Pammenter 1983). Ca accumulated in the leaves of *A. populifolia* and *I. pes-caprae* but remained relatively unchanged in the leaves of *S. plumieri*. In all three species Mg and Na were accumulated throughout the life of the leaves (Table 9.2). Na was accumulated to particularly high levels in *S. plumieri* (on a dry mass, but not wet mass, basis the Na levels in old leaves of this species were higher than those in the leaves of the mangrove *Avicennia marina*; Harte and Pammenter 1983). This accumulation was probably a consequence of a combination of high transpiration rates and long leaf life span. Cl concentrations increased slightly or remained relatively unchanged with increasing leaf age (Table 9.2).

9.6 Photosynthetic Characteristics

Photosynthetic rates were measured using a portable infrared gas analyser (LCA-II and PLC leaf chamber, ADC, Hoddeson, UK, or LI-6400, LI-COR, Lincoln, USA); potential photoinhibition was assessed by chlorophyll fluorescence (PEA, Hansatech, King's Lynn, UK); night-time respiration was mea-

Table 9.2. Interspecific differences in nutrient-relations measured for *A. populifolia*, *I. pes-caprae* and *S. plumieri*. Data from Ripley 2001

No.	Parameters measured		Species		
			A. populifolia	*I. pes-caprae*	*S. plumieri*
1	Leaf nitrogen content (mg g^{-1} dry wt.), N=6		14.1 ± 2.3^{ab}	15.3 ± 1.3^{a}	10.2 ± 2.4^{b}
2	Leaf phosphorus content (mg g^{-1} dry wt.), N=6		1.2 ± 0.3^{a}	1.9 ± 0.4^{b}	1.6 ± 0.2^{b}
3	Leaf potassium content (mg g^{-1} dry wt.), N=6		7.1 ± 2.5^{a}	19.4 ± 4.0^{b}	16.8 ± 6.8^{b}
4	Remobilisation of indicated nutrients from older to younger leaves, based on a comparison of the slopes of changes in leaf weight and nutrient concentration in response to increasing leaf age. N=6	Phosphorus	Yes	Yes	No
		Potassium	Yes	Yes	Yes
5	Accumulation of indicated nutrients. Considered accumulating if average concentration of leaf 20 exceeded that to leaf 5. N=6	Calcium	Yes	Yes	No
		Magnesium	Yes	Yes	Yes
		Sodium	Yes	Yes	Yes
		Chloride	Yes	No	Yes

Values with the same superscript letter are not significantly different at the 95 % probability level (*post hoc* ANOVA and Tukey multiple range test)

sured by enclosing shoots in a chamber and conducting standard gas exchange measurements.

The three species showed a range of photosynthetic characteristics (Table 9.3). *A. populifolia* had a high photosynthetic capacity, measured in terms of CO_2 saturated rates of ribulosebisphosphate (RuBP) regeneration and carboxylation efficiency (Table 9.3; related to rubisco activity; Farquhar and Sharkey 1982). It also had low stomatal limitations and high efficiency of radiation utilisation, the latter probably being related to the high chlorophyll content per unit area (Table 9.3). The hair layer reduced transpiration more than photosynthesis, giving rise to a higher leaf-level water use efficiency than was measured for the other species (Table 9.3). There was only a small reduction in the maximum quantum yield of photosynthesis (Table 9.3), which recovered quickly (ca. 30 min) on darkening, indicating little photoinhibition of photosynthesis. The hair layer contributed to this low photoinhibition (Ripley et al. 1999). Consequent upon the high capacity as well as high stomatal conductance, maximum light saturated rates of photosynthesis were high (up to 34 $\mu mol\ m^{-2}\ s^{-1}$; Table 9.3). *S. plumieri* showed characteristics similar to those of *A. populifolia*, except that capacities and rates were not as high (Table 9.3). Photoinhibition at midday was slightly more marked, but recovery was rapid (ca. 60 min).

I. pes-caprae showed lower capacities and slower CO_2 assimilation rates than the other species, and this was associated with the lowest leaf chlorophyll concentration (on an area basis, Table 9.3). Photoinhibition was also more marked in this species and recovery of maximum quantum yield of photosynthesis took approximately 120 min. *I. pes-caprae* was the only species that showed stomatal control of water loss, and these reductions in stomatal conductance may have contributed to the generally low photosynthetic performance of the species (Table 9.3).

9.7 Growth Rates

Aboveground growth was estimated by tagging individual leaves on several shoots and counting and measuring new leaves and stems produced subsequent to tagging. Belowground growth was not measured. It is extremely difficult to do this in the field, and our unpublished observations suggest that data derived from pot experiments would be unreliable; growth rates and growth patterns of these species in pots are considerably different from that of material in the field. However, root:shoot ratios are carefully regulated (Farrar 1999) and under constant conditions it is likely that aboveground growth reflects belowground growth. Biomass was measured by harvesting quadrats and structural carbohydrates were measured according to Buysse and Merckx (1993).

Table 9.3. Interspecific differences in photosynthetic characteristics measured for *A. populifolia*, *I. pes-caprae* and *S. plumieri*. Data from Ripley 2001

No.	Parameters measured	Species		
		A. populifolia	*I. pes-caprae*	*S. plumieri*
1	Peak assimilation rate from diurnal curves (μmol CO_2 m^{-2} s^{-1}), *N*=15	25.9±6.9[a]	9.7±5.2[b]	20.1±4.5[c]
2	RuBP regeneration rate (μmol m^{-2} s^{-1}), *N*=10	37.0±3.4[a]	23.4±5.4[b]	24.9±6.0[b]
3	Carboxylation efficiency (mmol CO_2 m^{-2} s^{-1}), *N*=10	118.3±20.6[a]	71.1±14.4[b]	95.8±17.6[c]
4	Efficiency of utilisation of incident PPFD (μmol mol^{-1}), *N*=15	55.7±5.0 [a]	39.1±3.0 [b]	42.7±12.0 [ab]
5	Chlorophyll content a+b (μg cm^{-2} leaf surface area), *N*=36	43.0±11.8 [a]	19.0±6.1 [b]	37.0±10.5 [a]
6	Stomatal limitation (%), *N*=10	16.5±3.4 [a]	45.7±11.0 [b]	26.2±11.4 [a]
7	Instantaneous water use efficiency (mmol CO_2 mol^{-1} H_2O), *N*=15	5.9±2.4 [a]	2.3±1.1 [b]	4.7±1.8 [a]
8	Midday reduction in maximum quantum yield of primary photochemistry (%), *N*=15	18.7±9.6 [a]	28.0±21.5 [ab]	32.3±16.5 [b]
9	Light saturated assimilation rate (μmol CO_2 m^{-2} s^{-1}), *N*=15	33.6±2.0 [a]	16.4±3.4 [b]	19.0±3.6 [b]

Values with the same superscript letter are not significantly different at the 95 % probability level (*post hoc* ANOVA and Tukey multiple range test)

Amongst the species there was a general relationship between CO_2 assimilation rates and growth rates (compare Tables 9.3 and 9.4, parameter 1), and also between whole shoot dark respiration per unit leaf area (Table 9.4) and growth. Interestingly, dark respiration rates as a percentage of maximum photosynthetic rates (1–5 %, depending on species and season) were lower than recorded in either temperate dune species (4.5–10.3 %; Pavlik 1983) or savanna grasses or trees (9–18 %; Scholes and Walker 1993). As growth rates are high, this low respiration rate is probably a consequence of low maintenance respiration. This is a characteristic that may be expected from short-lived leaves (Table 9.4)

Growth rates may be expressed as new leaf and stem production per shoot, biomass produced per unit land area (primary productivity) or per unit initial biomass (relative growth rate on a biomass basis; Table 9.4). The standing above-ground biomass that gave rise to the measured production varied among the three species (Table 9.4). These values are similar to, although a little higher, particularly for *S. plumieri*, than those reported for temperate dune systems (Barbour and Robichaux 1976). *A. populifolia* showed the highest growth rate in terms of production per shoot, and per g dry matter initially present (Table 9.4). Because of rapid leaf and stem death, there was no marked accumulation of dry matter and mean leaf life span was short (Table 9.4). Total non-structural carbohydrates (TNC) were low, presumably newly assimilated material was allocated directly to growth, with little reserve material being deposited (Table 9.4). Growth in this species was surprisingly non-seasonal. *I. pes-caprae* had the lowest growth rate per shoot, with mean life span being about 176 d (Table 9.4). Leaf and stem TNC was similar to that of *A. populifolia*, but more biomass was allocated to leaves; presumably the creeping habit of *I. pes-caprae* makes considerable allocation to stems unnecessary (Table 9.4). *I. pes-caprae* growth rate expressed per unit biomass was lower than *A. populifolia* (Table 9.4) and most of that initial biomass was leaf, rather than stem and so growth efficiency per unit leaf area would have been low. The production per shoot of *S. plumieri* was intermediate between the other species (Table 9.4), and leaves were longer lived (ca. 217 d). This species had the highest leaf and stem TNC (Table 9.4), suggesting that reserves may be available for non-seasonal growth, or to support the substantial reproductive investment (numerous large fruit) characteristic of this species. Production per unit biomass was lower than in *I. pes-caprae* (Table 9.4). However, *S. plumieri* has substantial stems, and growth efficiency per unit leaf area would be higher than that of *I. pes-caprae*.

Because the three species had different LAI, differences in production at the stem level tended to even out when expressed as a primary productivity (Table 9.4; the caveat about estimating LAI for *A. populifolia* remains). Because the productivity data are strictly gross primary productivity (based on new biomass, rather than increment in biomass, i.e. death and loss of plant parts was not accounted for), comparisons with published values of net pri-

Table 9.4. Interspecific differences in production and productivity measured for *A. populifolia*, *I. pes-caprae* and *S. plumieri*. Data from Ripley 2001

No.	Parameters measured	Species		
		A. populifolia	*I. pes-caprae*	*S. plumieri*
1	Shoot primary production (g dry mass shoot^{-1} year^{-1}), *N*=20	42.4 ± 18.7^{a}	17.8 ± 15.0^{b}	25.3 ± 10.3^{c}
2	Whole shoot respiration rate (µmol CO_2 m^{-2} s^{-1}), *N*=6	0.47 ± 0.15^{a}	0.15 ± 0.03^{b}	0.22 ± 0.01^{b}
3	Leaf production (g dry wt. shoot^{-1} month^{-1}), *N*=20	1.08 ± 0.48^{a}	1.09 ± 0.76^{a}	1.34 ± 0.61^{b}
4	Stem production (g dry wt. shoot^{-1} month^{-1}), *N*=20	2.43 ± 1.47^{a}	0.33 ± 0.10^{b}	0.76 ± 0.29^{c}
5	Aboveground primary productivity (g dry wt. m^{-2} year^{-1}), *N*=10	479.5 ± 211.6^{a}	386.5 ± 326.2^{b}	541.4 ± 220.6^{a}
6	Aboveground biomass (g dry wt. m^{-2}), *N*=5	63.1 ± 17.7^{a}	134.6 ± 40.9^{ab}	271.4 ± 133.3^{b}
7	Relative growth rate (g g^{-1} average initial biomass year^{-1}), *N*=5	7.6 ± 3.4^{a}	2.9 ± 2.4^{b}	2.0 ± 0.8^{b}
8	Leaf longevity (days), *N*=5	ca. 69	ca. 176	ca. 217
9	Total non-structural leaf carbohydrates (mg g^{-1} dry wt.)	1.63 ± 0.70^{a}	1.81 ± 0.70^{a}	3.06 ± 1.76^{b}
10	Leaf area index (LAI), *N*=5	0.23 ± 0.06^{a}	0.78 ± 0.26^{ab}	1.05 ± 0.51^{b}
11	Specific leaf area (SLA, cm^2 g^{-1}), *N*=6	60.6 ± 6.9^{a}	77.3 ± 5.2^{b}	35.8 ± 6.2^{c}

Values with the same superscript letter are not significantly different at the 95 % probability level (*post hoc* ANOVA and Tukey multiple range test)

mary production are difficult. Nevertheless, productivity for the species on this dune system (ranging from 380 g m^{-2} $year^{-1}$ for *I. pes-caprae* to 540 for *S. plumieri*) compare favourably with the 250 m^{-2} $year^{-1}$ recorded for *Ammophila arenaria* on a Scottish mobile dune system (Deshmukh 1977), and the range of 280 to 500 m^{-2} $year^{-1}$ for *Spartina alterniflora* in a North American salt marsh (Stroud and Cooper 1968), and are similar to the value of 440 g m^{-2} $year^{-1}$ reported for a South African savanna (Scholes and Walker 1993).

9.8 Stress and Disturbance

To what extent do the data gathered in the investigation described here 'explain' the occurrence of the foredune species? We pointed out the general perception that plants occupying the foredune may be subjected to stresses associated with limitations on resources, as well as to disturbance. Resource utilisation generated by plant growth obviously cannot exceed supply, but the requirement for resources in a resource-limited environment can be generated in different ways. A particular species may have a high standing biomass and low growth per unit biomass, or a species may have a low standing biomass and exhibit high growth and turnover rates: the resource requirement in each case may be similar. Physiological symptoms of stress or adaptations that increase access to resources may indicate a potential resource limitation.

Two of the species investigated (*A. populifolia* and *S. plumieri*) showed no evidence of water stress or limitation at the leaf level: stomatal conductances were high and transpiration responded primarily to atmospheric demand, and leaf water potentials did not drop below the turgor loss point. On a per unit land basis water use by the dune plants would have been lower than that of a well vegetated area, simply because of the low LAI on the foredune. *S. plumieri* had a higher LAI than *A. populifolia*, but rooted deeper, and may on occasions have tapped ground water. *I. pes-caprae* had lower stomatal conductances and did occasionally show a midday depression in conductance, indicating an inadequate water supply. However, leaf water potentials were always higher than those of *S. plumieri* (perhaps indicating a lower hydraulic resistance in *I. pes-caprae*) and did not drop below the turgor loss point.

It is more difficult to assess limitations imposed by nutrient shortages. Although dune soils are low in nutrients, there is a continual input from salt spray, as well as dry-matter deposition, and although the pool size of nutrients in the soil may be low, there may be a fairly rapid through-put. It is possible that high transpiration rates are necessary for the acquisition of sufficient nutrients from the dilute soil solution. Additionally, there is evidence of internal recycling of nutrients, which would obviously be important in the case of high leaf turnover rates. Intuitively, the high rates of leaf production and growth suggest that, at the shoot meristem level, cell division and growth are not limited by nutrients. More detailed studies of nutrient cycling, with particular emphasis on rates of nutrient input into the system, loss by leaching and utilisation by the plants (on a per plant and per unit ground area basis) are required. Nutrient supplementation experiments were considered, but Barbour et al. (1985) have criticised such experiments, questioning their use in nutrient ecology.

Rates of net CO_2 assimilation were high (although that of *I. pes-caprae* was lower than the other species), suggesting that there were no immediate shortages of resources at the leaf level. Allocation of photosynthate differed in that the proportion of stem in newly produced above-ground dry mass was lower in *I. pes-caprae* than the other species. This reflects the growth habits of the species. The high rates of photosynthesis at the leaf level translated into high rates of growth at the level of individual stems. *A. populifolia* showed particularly high rates of growth, coupled with high rates of death and loss of plant parts, giving rise to very high turnover rates.

The three species have different, although related, growth strategies, and these strategies allow the species to tolerate disturbance. Standing biomass and LAI on the foredune is low, and at the level of the individual leaf or stem, the plants do not appear to be suffering physiological stress or resource limitation. *A. populifolia* occurs predominantly at the seaward base of the foredune, where rates of sand movement are highest. It has the highest growth rate on a stem or initial dry mass basis, but the lowest leaf life span, and shows rapid biomass turnover. A single plant can soon become fragmented into several individual 'daughter' plants. This rapid turnover permits rapid adjustment to changes in the local topography of the dune that occurs with unpredictable sand movement. *S. plumieri* colonises the main dune itself. It is a classic 'dune builder' trapping and growing through accreting sand, and casual observations indicate that the species becomes moribund in stable sand. (Sand burial has been shown to enhance photosynthetic rates and affect leaf morphology in *Ammophila breviligulata* and *Calamovilfa longifolia* as well; Yuan et al. 1993). The high growth rate of *S. plumieri* permits the maintenance of a (relatively) high LAI, and also the production of stem to grow through the accreting sand. In the system studied *I. pes-caprae* is almost an anomaly: it occurs predominantly on the more stable crest and landward slope of the dune, and does not suffer frequent large-scale disturbance. It is the species that shows some signs of water limitation in that stomatal conductances are low, and this appears to negatively impact on its photosynthetic and growth rates. However, seedlings of this species are capable of responding to disturbance, re-allocating resources from roots to shoots (Martinez and Moreno-Casaola 1996) on sand burial. It is not known why it is absent from the front of the foredune in the system we studied.

For all three species, because of limited nutrient and water availability, the high rates of growth can be sustained only because plant cover is low. What is the mechanism that maintains this low cover? It does not seem to be an immediate resource limitation at the leaf or stem level. There are two other possibilities. Firstly, disturbance could limit the establishment of new individuals (seedlings are the most vulnerable part of the life cycle) and perhaps could limit the extension of existing individuals (sand removal, particularly, would do this). Secondly, it is possible that each species has an inherent leaf turnover rate such that the production of new leaves is, in the long term, in balance with the death of old ones. Detailed long-term growth and demographic studies are required.

It is probably the necessity to tolerate disturbance (including salt spray) that prevents the invasion of other species onto the foredune. It is well established that salt spray has an influence on species zonation on coastal dune systems (Boyce 1954; Donnelly and Pammenter 1983; Davy and Figueroa 1993). The growth strategy adopted by foredune species, largely to tolerate sand movement (Maun 1998), is likely to be a disadvantage in stable dunes, where competition and thus resource sequestration by the plants is important. This probably explains why the foredune species do not occur further inland. It is interesting to note that *I. pes-caprae* does occur further from the high tide line than the other foredune species.

9.9 Conclusions

The success of the species inhabiting the foredune of the system we studied (and probably many other mesic tropical and subtropical dune systems) appears to lie in a combination of three factors. Firstly, the plants of the foredune must, of course, have physical characteristics that can resist salt spray and the physical assault of sand blasting. Secondly, plant cover is low, possibly as a consequence of disturbance, particularly sand movement, such that individual stems do not suffer stress imposed by severe resource limitations. Thirdly, growth patterns are such that there is rapid turnover of biomass, which permits accommodation to the disturbance associated with sand movement, as well as reducing biomass accumulation; growth patterns vary amongst species, and this may contribute to the suitability of a species for a particular microhabitat.

References

Allen SE (1989) Chemical analysis of ecological materials, 2nd edn. Blackwell, Oxford

Barbour MG (1992) Life at the leading edge: The beach plant syndrome. In: Seeliger V (ed) Coastal plant communities of Latin America. Academic Press, San Diego, pp 291–306

Barbour MG, Robichaux RH (1976) Beach phytomass along the Californian coast. Bull Torrey Bot Club 103:16–20

Barbour MG, DeJong TM, Pavlik BM (1985) Marine beach and dune plant communities. In: Chabot BF, Mooney HA (eds) Physiological ecology of North American plant communities. Chapman and Hall, New York, pp 296–322

Boyce SG (1954) The salt spray community. Ecol Monogr 24:29–69

Buysse J, Merckx R (1993) An improved colourimetric method to quantify sugar content in plant tissue. J Exp Bot 44:1627–1629

Caldwell D (1985) Cold desert. In: Chabot BF, Mooney HA (eds) Physiological and ecology of North American plant communities. Chapman and Hall, London, pp 198–212

Clements FE (1916) Plant succession. Carnegie Inst Washington. Publ 242

Davy AJ, Figueroa ME (1993) The colonization of strandlines. In: Miles J, Walton DWH (eds) Primary succession on land. Special publication series of the British Ecol Soc, No 12. Blackwell, Oxford , pp 113–131

Deshmukh IK (1977) Fixation, accumulation and release of energy by *Ammophila arenaria* in a sand dune succession. In: Jefferies RL, Davy AJ (eds) Ecological processes in coastal environments in a sand dune succession. Blackwell, London, pp 353–362

Donnelly FA, Pammenter NW (1983) Vegetation zonation on a Natal coastal sand-dune system in relation to salt spray and soil analysis. S Afr J Bot 2:46–51

Farquhar GD, Sharkey TD (1982) Stomatal conductance and photosynthesis. Annu Rev Plant Physiol 33:317–45

Farrar JJ (1999) Acquisition, partitioning and loss of carbon. In: Scholes JD, Barker MG, Press MC (eds) Physiological plant ecology. Blackwell, Oxford, pp 25–43

Harte JM, Pammenter NW (1983) Leaf nutrient content in relation to senescence in the coastal sand dune pioneer *Scaevola thunbergii* Eckl. & Zeyh. S Afr J Sci 79:420–422

Hesp PA (1991) Ecological processes and plant adaptations on coastal dunes. J Arid Environ 21:165–191

Lubke RA (1983) A survey of the coastal vegetation near Port Alfred, Eastern Cape. Bothalia 14:725–738

Mansfield TA, Davies WJ (1981) Stomata and stomatal mechanisms. In: Paleg LG, Aspinall D (eds) The physiology and biochemistry of drought resistance. Academic Press, London, pp 315–347

Martínez M L, Moreno-Casasola P (1996) Effects of burial by sand on seedling growth and survival in six tropical sand dune species from the Gulf of Mexico. J Coastal Res 12: 406–419

Maun MA (1998) Adaptations of plants to burial in coastal sand dunes. Can J Bot 76(5):707–712

Morris M, Eveleigh DE, Riggs SC, Tiffney WN (1974) Nitrogen fixation in the Bayberry (*Myrica pennsylvanica*) and its role in coastal succession. Am J Bot 61:867–870

Olsson-Seffer P (1909) Hydrodynamic factors influencing plant life on sandy sea shores. New Phytol 8:37–49

Passioura JB (1982) Water in the soil-plant-atmosphere-continuum. In: Lange OL, Nobel PS, Osmond CB, Zeigler H (eds) Encyclopedia of plant physiology, new series12B. Springer, Berlin Heidelberg New York, pp 5–33

Pavlik BM (1983) Nutrient and productivity relations of the dune grass *Ammophila arenaria* and *Elymus mollis*. I. Blade photosynthesis and nitrogen use efficiency in the laboratory and field. Oecologia 57:233–238

Peter CI, Ripley BS (2000) An empirical formula for estimating the water use of *Scaevola plumieri*. S Afr J Sci 96:593–596

Ripley BS (2001) The ecophysiology of selected coastal dune pioneer plants of the Eastern Cape. PhD Thesis, Rhodes University, South Africa

Ripley BS, Pammenter NW, Smith VR (1999) Function of leaf hairs revisited: The hair layer on leaves *Arctotheca populifolia* reduces photoinhibition, but leads to higher leaf temperatures caused by lower transpiration rates. J Plant Physiol 155:78–85

Rozema JP, Bijwaard G, Prast G, Broekman R (1985) Ecophysiological adaptations of coastal halophytes from foredunes and salt marshes. Vegetatio 62:499–521

Salisbury E (1952) Downs and dunes. Their plant life and its environment. G Bell & Sons, London

Scholes RJ, Walker BH (1993) An African savanna: Synthesis of the Nylsvley study. Cambridge University Press, Cambridge

Stroud LM, Cooper AW (1968) Colour-infrared aerial photographic interpretation and net primary productivity of a regularly flooded North Carolina salt marsh. Water Resource Res Inst, Univ of North Carolina, Rep No 14

von Willert DJ, Herppich M, Miller JM (1989) Photosynthetic characteristics and leaf water relations of mountain fynbos vegetation in the Cedarberg area (South Africa). S Afr J Bot 55(3):288–298

Willis AJ (1965) The influence of mineral nutrients on the growth of *Ammophila arenaria*. J Ecol 53:735–745

Yuan T, Maun MA, Hopkins WG (1993) Effects of sand accretion on photosynthesis, leaf-water potential and morphology of two dune grasses. Funct Ecol 7:676–682

10 Plant Functional Types in Coastal Dune Habitats

F. García Novo, M.C. Díaz Barradas, M. Zunzunegui,
R. García Mora and J.B. Gallego Fernández

10.1 Plant Functional Types

Most living organisms have evolved through a hazardous sequence of expansion to new areas, new environments and communities, coupled with extinction in the original habitats and in some of the newly settled areas. All current biological traits, either morphological or physiological, play or have played their role in the evolutionary line. Focusing on functional traits helps us to understand biological systems in ecological terms.

The use of functional types (FT), or functional groups (FG), as opposed to taxonomic entities, emphasises physiological vs. morphological characters as a basis for species association in "natural" groups. However, a proper conceptual construction of either FT or FG is not yet available, and definitions only address certain topics (Semenova and van der Maarel 2000).

Over the last two decades extensive research has been conducted to identify plant groups on the basis of association of structural and functional characters. Several problems arise: which set of characters or traits will be used and whether the emerging functional types represent sound ecological groups or are a mere numerical construct, depending solely on the original data set.

The association between the presence of certain characters and the ecological performance of the individual may be of a multiple nature: the same degree of fitness (recorded as survival, productivity, or reproductive success), may be attained by various strategies based upon a different association of characters. Response groups of species may thus be delineated on ecological grounds. But it is almost impossible to compare the strategies of annuals with those of perennials or trees with small plants, where contrasts go beyond assigning them to different functional groups.

Taxonomic affinities and functional types were all generated by the same evolutionary mechanism. However, it is only taxonomic affinities which have been preserved through evolution as opposed to ecological types which have

Ecological Studies, Vol. 171
M.L. Martínez, N.P. Psuty (Eds.)
Coastal Dunes, Ecology and Conservation

been reshaped time and again, through community assembling, ecosystem waning and waxing, and biogeographical shifting.

Functional types may embrace those species with a common life strategy, sharing some sets of characters enabling them to succeed within an ecosystem. Broader definitions of the strategy may reduce the common set of characters and widen the species number, producing naive results. More refined analysis will lead to the identification of some new relevant characters eventually leading to the subdivision of former FTs into smaller ones, or recognising narrower functional groups of a greater ecological interest, but difficult to compare with other areas.

Emphasising one ecological role against another may bias the selection of characters. The perception of plant cycles or strategies by the ecologist may favour particular sets of characters. Permanent morphological or functional characters may find a disproportionate presentation in the analysis against temporary characters or discontinuous interactions. Aerial organs yield most traits to FT analysis.

Gitay and Noble (1997) recognise three approaches to FT detection: (1) an inductive analysis of communities or ecosystems sorting out species into supposed functional types; (2) a deductive analysis based on some basic ecological processes and the set of functional categories linking to them; (3) the numerical analysis of sets of associated characters to seek clusters of species and characters.

Gitay and Noble also discuss some concepts related to FT which have been used in ecological literature such as strategy, syndrome, guild, functional type, league, ecological sector and others. The authors stress the differences between establishing common categories based on resource use or on response to perturbation. The authors identify a structural guild for species exploiting the same resource in different ways against a functional guild for those exploiting the same resource in the same way. A response group shares the same response to a perturbation by different mechanisms, as opposed to a functional group which shares both response and mechanism.

Semenova and van der Maarel (2000) clarify the redundant terminology in use in the FT literature. They suggest functional traits or states of plants which may be referred to as "trait states" abandoning the use of attributes. They propose syndromes be restricted to a set of plant traits connected to an environmental constraint. Strategy types usually arise from the three adaptive strategies described by Grime (1974), which relate species traits to resource availability.

Weiher et al. (1999) made a valuable attempt to focus on the "core list of plant traits" in an attempt to base a common ground to "tailor our own list of traits by adding details specific to our region and research agenda" (p. 617). A core list (see Table 10.1) containing 12 traits and related plant ecological functions is given as an example of the analysis. The contribution brings the FT discussion back to the ecosystem level and all driving interactions therein.

Table 10.1. Core list of plant traits of FT definitions. (Weiher et al. 1999; Table 10.2)

Seed mass	Leaf water content or surface leaf area
Seed shape	Height
Dispersal type	Aboveground biomass
Onset of flowering	Stem density
Life history	Resprouting ability
	Clonality

The core list may be simplified to a minimum set of traits best describing the plants' ability to cope with major ecological challenges. Weiher et al. (1999) propose: (1) a leaf measurement (specific leaf area SLA, leaf water content); (2) seed mass; (3) adult size (height or above-ground biomass). Westoby (1998) described a geometrical representation, the LHS (leaf, height, seed) scheme of plant strategies, using a plot on three axes: SLA, canopy height, and seed mass. The close agreement between Weiher et al. and Westoby suggests that the two similar concepts merely are running under different names.

The above suggestions are traced back to the triangular scheme of strategies proposed by Grime (1974). This line of research underlines the association of core plant traits, which pertain to long-standing evolutionary forces in plant organisation. Evolutionary strategies go beyond ecological FT analysis. But comparing both sets of results may help understanding plant groups and interpreting trait association.

Coastal dune plant communities survive in harsh environments with marked habitat constraints, causing the selection of functional groups and functional types that may be identified and linked to the ecological processes. At the local scale, FT analysis aids in interpreting community function.

At the regional scale, the use of FTs favours the comparison of communities in similar environments but having few species in common.

Hesp (1991) adding to previous work by Doing (1985) on beaches and coastal dune plants, described latitudinal patterns of biogeographic affinities, species richness, degree of endemism, desert species, and plant functional traits such as succulence, C4 and adaptation to stress. The association of succulents and desert plants to arid coastal climates, and C4 species to tropical latitudes, exemplifies the correspondence between plant traits and climatic type in dune vegetation at the regional scale.

10.2 Dune Habitats as Environmental Islands

In the coastal environment, the foredunes are that portion of the beach/dune profile that extends from the mean tide line to the top of the frontal dune

(Barbour 1992). Together with the coastal dunes they offer an array of geomorphologic features, soil types and environmental processes widely occurring in this sandy environment (Seelinger 1992). Plant populations survive in a harsh environment where adverse features include: salt spray, highly permeable substrate with a low field capacity, dry sand with occasional high temperatures near the surface, intense radiation both directed and reflected, high winds, and substrate mobility (Barbour et al. 1985; Hesp 1991). Dune fields further inland are not prone to wave action, and salt spray declines with distance to the shoreline (Brown and McLachlan 1990).

Favourable conditions include higher air humidity under sea wind conditions, and a limited plant competition for space, nutrients or other resources. The piezometric dome of the aquifer under the dune may provide a higher water supply to depressed surfaces, with temporary flooding and pond formation on slacks. Flooding becomes an unfavourable condition in wet dune slacks, controlling vegetation development.

Dune field environments appear to be homogeneous because of the dominant sand cover, when in fact they are not. Wind speed, sand advance and particle transport differ widely over the dune surface. This variation combined with soil moisture availability and the temperature regime at the soil surface produces a range of terrestrial environments. Temporal variation adds to spatial heterogeneity in providing what has been named a "stochastic environment". Air turbulence and ensuing sand transport and soil water scarcity represent the main driving interactions at the dune environment, and the main constraints to plant development.

Under irregular climate regimes the sand mantles may present pulses of high sand transport and front advancement, inserted among others of near stability. Slack and peridunar ponds fluctuate from dry beds under periods of scarce rainfall to large water bodies under heavy rainfall intervals. Dune vegetation closely follows precipitation patterns (van der Maarel 1981). Around dune depressions water level fluctuations start succession processes in vegetation from the long-term (centuries to millennia), long-scale (kilometres) examples of Lake Michigan (Cowles 1899), to the minute dimensions (10–50 m) and short intervals (1–5 years) of Doñana ponds (Zunzunegui et al. 1998).

The above examples, underline the fragmented character of dune field environments, where terrestrial habitats fluctuate in soil water availability, temperature, soil stability and sand grain transport. What is more, connectivity among the different pieces of the same habitat type may be low due to contrasted environmental characteristics. It is in this context that García-Mora et al. (1999) have addressed the patchy dune-field habitats as "environmental islands". Martínez et al. (2001) underlined the repeated process of plant succession occurring at the fragmented habitats due to repeated interruption caused by disruption pulses.

Resistance to adverse interactions, resource exploitation, and dispersal to favourable sites, are the main ecological issues for plant dune communities.

10.2.1 Adverse Interactions

Climate changes may drive the system beyond the fluctuation boundaries, either setting stable dune crests in motion or bringing sand displacement to a halt through changes in wind regimes (Carter 1995). Sand sheets may be set in motion after human disturbance: logging, fire, overgrazing, and marram grass collection favour the building up of dunes. Large maritime dunes may grow to high slip faces (precipitation dunes; Cooper 1958) which bury any forest as they move inland. The depositional processes or the alternation of sand erosion/accretion is a major hazard for most plant species (Seelinger 1992).

An active growth response to burial has been interpreted, long ago, as a key character for dune plants. *Ammophila arenaria* presents two growth patterns with emerging or spreading habits (Huiskes 1979). Davy and Figueroa (1993) report *Cakile maritima* seedlings emerging through 16 cm sand burial, *Atriplex laciniata* and *Salsola kali* emerging from a depth of 8 cm. *Elymus farctus* studied by Gimingham (1964) in Scotland, succeeds in emerging through 23 cm sand burial. *A. arenaria* and *E. farctus*, from European coasts, present a small increase in productivity under moderate sand engulfment (Ranwell 1972). *Chamaecrista chamaecristoides, Palafoxia lindenii, Trachypogon gouini* from the Gulf of Mexico, *Schyzachyrium scoparium* from North America and the tropical creeping runners *Canavalia rosea* and *Ipomea pes caprae* also present a positive growth response to moderate burial (Martínez and Moreno Casasola 1996). *Ammophila breviligulata* markedly increases vigour when buried by 2–35-cm sand layers (Disraeli 1984).

Intervals of high precipitation usually promote plant growth. Drought periods often coupled to high soil surface temperatures and increased sand transport are serious impairments to plant growth in dune habitats, causing plant mortality and plant cover disruption. Plant resistance to drought appears to be associated to several traits of dune plants. Danin (1996) describes a collection of dune plant types from the eastern Mediterranean Basin and California, with a careful morphological analysis of plant growth patterns and some traits involved in their resistance to substrate instability and drought.

10.3 Some Examples of Applications of Plant FTs to Dune Vegetation Analysis

It is difficult to compare plant FT covering widely differing life cycles or morphologies because they have few organs or traits in common to characterise a functional type. The application of plant FTs to dune vegetation needs a preliminary discussion to focus on certain comparable plant types, i.e. vascular plants, perennials, and some environmental relationships in order to overcome naive results. Diaz et al. (1998) list criteria for FT applications including easily measurable plant traits. Two recent examples will be shown, the former from stabilised dune fields and the latter from active foredunes.

Díaz-Barradas et al. (1999) studied the FT of perennial vegetation (largely scrub) from the stabilised coastal dunes of Doñana National Park (Atlantic coast of SW Spain), along a gradient of water availability from dune crests to depressions fed by aquifer discharges (see Fig. 10.1). For plant trait selection, 24 vegetative and functional characters were chosen, to include morphology, responses to resource availability and environmental stress, reproduction and dispersion (Table 10.2). The 24 traits x 20 species matrix was subjected to DCA and TWINSPAN to identify clusters of species with similar traits, which were assumed to represent FTs at the species level (Keddy 1992). These are shown in Fig. 10.2 with some representative species from TWINSPAN classification of traits x species matrix; six plant FTs are identified (see Fig. 10.3).

Two FTs will be singled out: FT E (pioneer shrubs, with small canopies, hairy leaves, able to withstand very negative leaf water potentials, dry dehiscent fruits and obligate seeders); and FT D (mature shrubs with larger canopies, longer life span, more sclerophyllous leaves, fleshy fruits and sprouting after fire).

Previous studies on Mediterranean woody plants have identified two main ecological groups, which match two of the plant FTs described above. Margaris (1981) suggested two "adaptation syndromes": phrygana, for garrigue species marked by seasonal leaf heteromorphism, and maquis, for sclerophyllous woody plants, keeping the same leaves throughout the whole year.

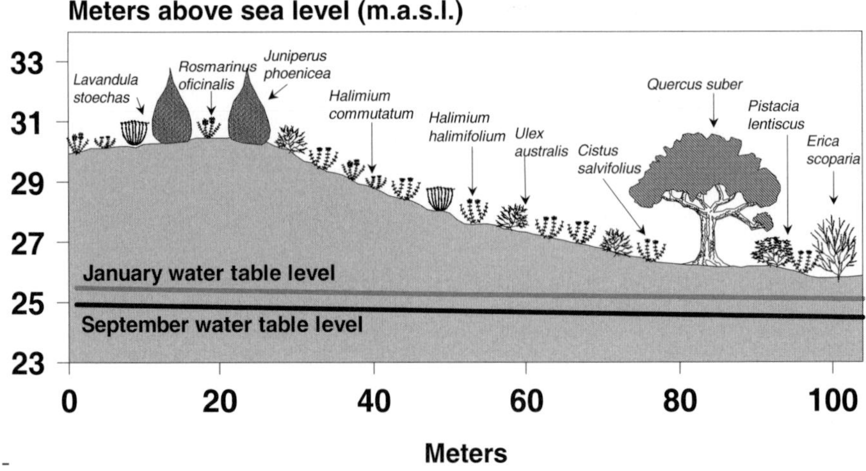

Fig. 10.1. A representation of stabilised dune scrub vegetation along a gradient of soil water availability at the stabilised dunes of Doñana N. Park (SW Spain). September level represents the maximum water table depth, at the end of Summer period. (Díaz Barradas et al. 1999)

Table 10.2. Traits recorded on the 20 scrub species of the stabilised sands of Doñana National Park. (Díaz Barradas et al. 1999)

Traits	Description	Classes in matrix[a]
Years	Average plant age	1: 5–25, 2: 25–50, 3: 50–100, 4: >100 (5)
Height	Average plant height	1:<60 cm, 2: 60–100 cm, 3: 100 cm–3 m, 4: 3–5 m, 5: >5 m (7)
Diameter	Average canopy diameter	1: <50 cm, 2: 50–100 cm, 3: 1–2 m, 4: 2–3 m, 5: >3.5 m (1, 5)
Leaf size	Average leaf area	1: <0.10 cm^2, 2: 0.10–0.25 cm^2, 3: 0.25–2.25 cm^2, 4: 2.25–12.25 cm^2 (7)
Leaf colour		1:all green, 2:green and white, 3:all white. (1, 5)
Leaf margin		1: entire, 2: revolute, 3:lobed. (5, 7)
Leaf texture		1: malacophyll, 2: semi-sclerophyll, 3: sclerophyll. (5, 7)
Leaf hairiness		1: non hairy, 2: hairy lower side, 3: hairy upper side (5, 7)
Leaf duration		1: <6 month, 2: 6–14 month, 3: 14–26 month, 4: 26–28 month (5, 7)
L/S	No. of leaves per 10 cm of stem	1: <10, 2: 10–20, 3: 20–50, 4: 50–100, 5: >100 (1, 5)
Stem diameter	Measured at ground level	1:<2 cm, 2: 2–5 cm, 3: 5–10 cm, 4: <20 cm (1, 5)
Bark consistency		1: smooth, 2: fibrous, 3: corky. (1, 5)
Spininess		1: absent, 2: with spine (8)
Underground stem		1: lignotubers, 2: others (1, 5)
Root morphology		1: tap root, 2: horizontal roots, 3: vertical-horizontal roots (1, 5)
Root depth		1: <25 cm, 2: 25–50 cm, 3: 50–100 cm, 4: >100 cm (3, 5)
Regeneration	Vegetative regeneration after fire	1: plant killed 2: from epicormic buds, 3: from non epicornic buds, 4: from epicornic and non epicornic buds (2, 5)
Fruit dehiscence		1: dry indehiscent, 2: dry dehiscent, 3: fleshy indehiscent, 4: fleshy dehiscent (8)
Fruit type		1: capsule, 2: cone, 3: nutlet, 4: legume, 5: achene, 6: drupe, 7:berry (8)
Pollination	Type of pollination	1: zoophyllous, 2: anemophyllous (8)
Slwp	Summer leaf water potential	1: <–2 MPa, 2: –2–3 MPa, 3: –3–4.5 MPa, 4: >–4.5 MPa (1, 4, 6)
Wlwp	Winter leaf water potential	1: <–0.5 MPa, 2: –0.5–0.6 MPa, 3: –0.6–0.7 Mpa, 4: >-0.7 MPa (1, 4, 6)
Cost of growth	Cost of leaf growth	1: <1.5, 2: 1.5–1.6, 3: >1.8 g glucose g^{-1} dry wt. day^{-1} (9)
Cost maintenance	Cost of leaf maintenance	1: <0.0130, 2: 0.0130–0.0140, 3: >0.0140 g glucose g^{-1} dry wt. day^{-1} (9)

[a] Numbers in parentheses: 1, Field observation; 2, García-Novo (1977); 3, Martínez et al. (1998); 4, J. Merino et al. (1976); 5, J. Merino in Specht (1988); 6, O. Merino et al. (1995); 7, Orshan (1989); 8, Valdés et al. (1987); 9, J. Merino (1987)

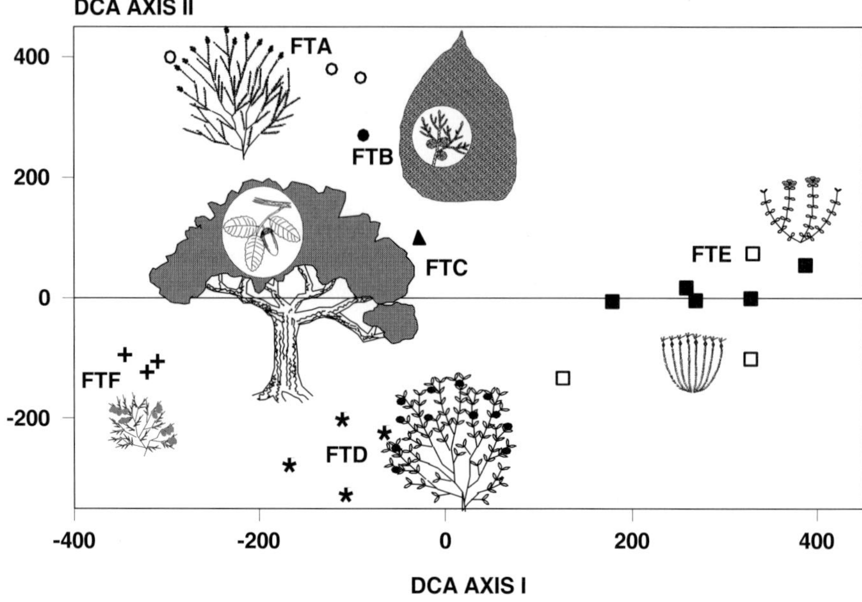

Fig. 10.2. Analysis of dune scrub vegetation along a gradient of water soil availability, at Doñana N. Park. (SW Spain). Approximate location of identified plant functional types A to F (see text) using a draft of a representative species on the plot of DCA of the traits x species matrix. Species symbols refer to Fig 10.1. FT symbols refer to Fig 10.3. (Díaz Barradas et al. 1999)

Orshan (1989) and Terradas (1991) back a similar division. In the Doñana dune scrub analysis, FT E plants may be ascribed to the phrygana group and FT D plants to the maquis group.

In a more refined analysis of plant FTs in 13 species of dune scrub vegetation (Díaz-Barradas et al. 2001), 14 traits were used. Seven traits (one half) were morphological characters, three were reproductive characters and another four were definite physiological characters[1]. Hierarchical classification of species, using the Pearson product moment as the dissimilarity coefficient, first separates spiny legumes (*Stauracanthus genistoides, Ulex australis*) and *Corema album*. The ten extant species are then subdivided into two groups matching the maquis vs. phrygana opposition. Further subdivision separates species groups in accordance to FT analysis previously described (Díaz Barradas et al. 1999), although using a different set of traits. The result

[1] The list of traits included: morphology (7): *canopy height, presence of spines or thorns, leaf area, leaf margin shape, leaf pilosity, leaf strength;* reproduction (3): *sprouting, dispersion, pollination;* physiology (4): *annual range of proline content in leaves (winter–summer values), range of chlorophyll content, range of photosynthetic efficiency in leaves (Fv/Fm), range of plant suction water potential.*

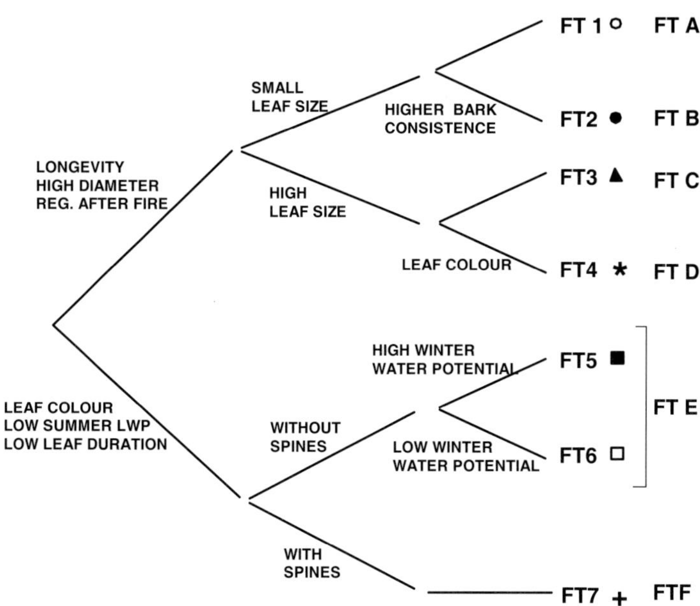

Fig. 10.3. Dendrogram of plant FT by TWINSPAN analysis of the traits x species matrix showing the main trait indicators at each division. (Díaz Barradas et al. 1999)

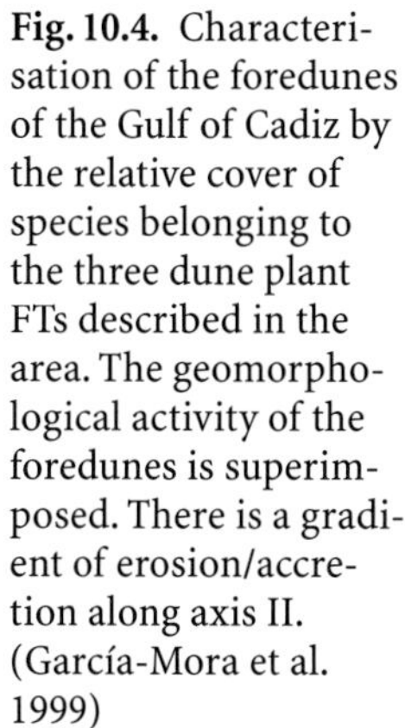
Fig. 10.4. Characterisation of the foredunes of the Gulf of Cadiz by the relative cover of species belonging to the three dune plant FTs described in the area. The geomorphological activity of the foredunes is superimposed. There is a gradient of erosion/accretion along axis II. (García-Mora et al. 1999)

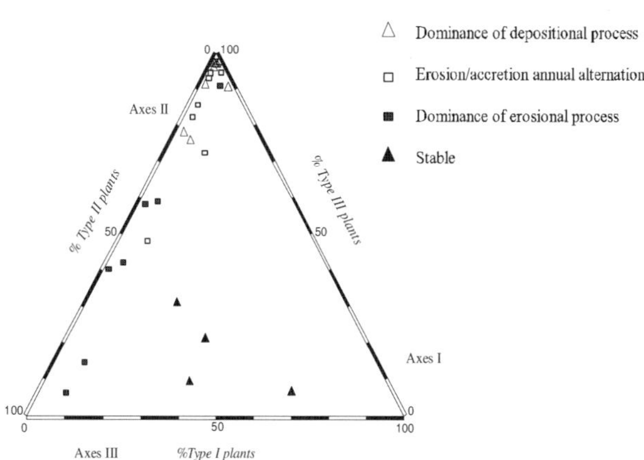

confirms the consistency of the groups of species that were identified through the FT analysis.

The results also showed that physiological traits can be useful for plant FT definition. The main advantage behind physiology is that it reveals the actual functioning of plants, helping to link in a causal manner the statistical associations of traits and the response to environmental variables.

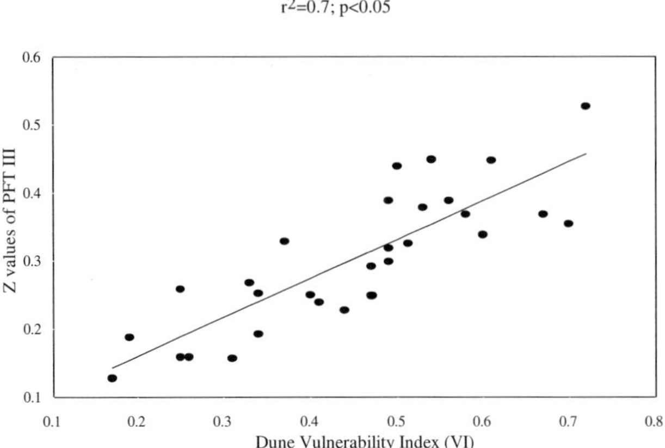

Fig. 10.5. Linear regression of FT III plant diversity (measured as *Z*, see text) against foredune condition estimated using the Vulnerability Index of Bodere et al. (1991). As the Vulnerability Index grows, due to increased disturbance, Z values also rise. (García-Mora et al. 2001)

García-Mora et al. (1999) in the foredunes of the Gulf of Cadiz (SW Spain and S Portugal) provides a second example of plant FT studies in coastal dunes in a fairly diverse vegetation (55 species of 49 genera) using 7 plant traits (4 morphological characters and 3 functional attributes). Three plant FTs were identified in the analysis. Then, the FTs were used to characterise the foredune vegetation, substituting in each inventory the abundance of each species for the abundance of its FT, and summing up the results. Making the sum of covers of the three FTs equal to 100 % in each site, it was possible to plot them in a triangular diagramme, each side representing the abundance of one of the dune plant FTs described above (Fig. 10.4). FTs abundance in a vegetation sample was shown to be associated with environmental conditions of the dune site. As the dune vulnerability index (Bordere et al. 1991) increases, the diversity of one FT of vegetation measured as Z (slope of species area plot) also rises (see Fig. 10.5).

The examples show some applications of FTs to the identification of ecological groups of species and their associated environments in dune vegetation studies.

10.4 Dune Habitat Confinement

Ecological literature highlights the spatial variability of coastal dune vegetation, relating it to geomorphologic processes (Ranwell 1972; Carter 1995). Detailed long-term studies of coastal dunes by van der Maarel et al. (1985), Studer-Ehrensberger et al. (1993), Crawford et al. (1997), Geelen (1997) have shown the strong fluctuation of both environment and vegetation types in these unstable systems. Each dune patch behaves like a short-term environmental island separated from others by high disturbance corridors and subject to extinction-invasion processes. As Harper (1977) pointed out, many dune annuals have no buried seed bank and need to be subject to a steady seed rain.

A repeated extinction model of community, focusing on temporal phenomena, and on the importance of history and chance, is common to environments with a strong recurrent disturbance, wiping out local populations from patches (Townsend 1989). Dune plants often exhibit an aggregate pattern with empty sites and vegetated ones with uneven plant composition through the operation of a series of environmental filters which produce a patchy environment. The filter set is not orderly sequenced as proposed for Woodward and Kelly (1997), but a random series of events for each patch, adverse interactions interwoven with favourable inputs of nutrients and rain spells.

Davy and Figueroa (1993) suggest, although do not formulate, a "rich patch" hypothesis, through an initial pulse of nutrients heterogeneously distributed over dune surfaces, thus permitting different vegetation assemblages to develop over time. The dune thus presents patches, each bearing a temporary vegetation facies, according to the previous history of environmental events, which acted as filters or provided resources and helped seeds and propagules to the site.

Looking at high-beach and coastal dune vegetation, there is a remarkable confinement of plant distribution. A small number of species share dune habitats rarely growing elsewhere. In turn, dune vegetation tends to exclude most species growing outside. Danin (1996) mentions an extreme confinement example: *Swallenia alexandrae,* a North American grass requiring sand accretion, which is dispersed by "gliding" fruits on sandy surfaces, is confined to a single dune.

The confinement also operates at local scales selecting the species composition of dune communities in each habitat, and inducing patchiness in vegetation with large areas void of any plant cover. Dune plants need powerful dispersion mechanisms to new suitable patches in order to survive in the area. The causes of littoral confinement may be different from one species to another. They may be due to specific requirements of seawater salt, sand accretion, or the existence of active dispersion processes. Another source for confinement is the limited ability for competition (nutrients, water, ground), or to withstand predation. Association to vertebrate dispersers may help the species to spread. The following examples illustrate this point.

The umbrella pine (*Pinus pinea*) was introduced to Doñana N. Park early in the 19th century. At present, pines survive in the mobile dune slacks, thanks to seed dispersal by birds (magpies, azured wing magpies). Cone dispersion by wind will not be able to cope with dune crest advance, averaging 5–6 m/year, and a time lag from tree seedling establishment to cone production of 15 years as a minimum (García-Novo and Merino 1997). Umbrella pine survival in Mediterranean coastal dune areas largely depends on the available mechanisms for seed dispersal.

The large reed (*Arundo donax*), a freshwater phreatophite, maintains large populations in the foredunes of the Gulf of Cadiz where they are supplied by freshwater seeping at the dune basis from local aquifer discharge. Every few years (last time March 2003) heavy storms erode reed stands, taking the rhizomes along the littoral and piling up litter deposits on the embryo dunes, where reed rhizomes germinate and rebuild the stands. *A. donax* populations, although confined to foredune upwelling sites, persist in the Gulf of Cadiz on behalf of a local dispersion process exchanging individuals among the metapopulations of the littoral dunes.

10.5 Conclusions

Coastal dunes offer a collection of habitats that may vary through time and space, where terrestrial habitats may undergo critical periods of soil instability, drought, high surface temperature and salt spray, limiting plant life therein.

Coastal dune species are often confined to the littoral fringe or to neighbouring dune fields, facing unstable and harsh environments but not spreading outside the dune area.

Dune plant confinement also produces a biogeographical substitution with a floristic change of vegetation along the coasts in a sequenced substitution of taxa: varieties, subspecies, species, give way to one another along the coastal stretch.

All dune species show a degree of fitness to local environments but do not share a common set of traits so as to define a single plant FT. Instead, several plant FTs and functional groups may be recognised within the community, and similar FTs may occur in dune communities with little species in common. PFTs offer a valuable tool for the study of coastal dune vegetation where species from several biogeographical and ecological origins coexist thanks to different life strategies and dispersion processes.

The study of plant dispersion mechanisms and dispersers, and the inclusion of physiological traits may be valuable for future FT studies in dune vegetation, recalling what Westoby and Leishman (1997) pointed out, FT classifications will necessarily be purpose-dependent at two levels: (1) the relevance of the set of chosen characters for the purpose of classification. (2) the assumption that final classification will depend on the previous character selection.

One final remark, taken from Dereck S. Ranwell, a forerunner in sand dune studies: "Perhaps we might end this chapter with a plea that ... we should turn our attention to the more detailed study of the unit "fragments" of the ecosystem, the better to eventually understand the whole" (Ranwell 1972, p. 181).

References

Barbour MG (1992) Life at the leading edge: the beach plant syndrome. In: Seelinger U (ed) Coastal plant communities of Latin America. Academic Press,, New York, pp 291–307

Barbour MG, De Jong TM, Pavlik BM (1985) Marine beach and dune plant communities. In: Chabot BF, Mooney HA (eds) Physiological ecology of North American plant communities, Chapman and Hall, New York, pp 296–322

Brown AC, McLachlan A (1990) Ecology of sandy shores, 2nd edn. Elsevier, Amsterdam

Carter RWG (1995) Coastal environments: an introduction to the physical, ecological and cultural systems of coastlines, 5th edn. Academic Press, London

Cowles HC (1899) The ecological relations of the vegetation on the sand dunes of Lake Michigan. Bot Gaz 27:95–117;167–202; 281–308; 361–391

Cooper WI (1958) Coastal sand dunes of Oregon and Washington. Geol Soc America Memoir 72. Baltimore

Crawford RMM, Studer-Ehresberger K, Studer C (1997) Flood induced change on a dune slack observed over 24 years. In: García-Novo F, Crawford RMM, Díaz-Barradas MC (eds) The ecology and conservation of European dunes. EUDC-Universidad de Sevilla, pp 27–40

Danin A (1996) Plants of desert dunes. In: Cloudsley-Thompson JL (ed) Springer, Berlin Heidelberg New York

Davy AJ, Figueroa ME (1993) The colonization of strandlines. In: Miles J, Walton DWH (eds) Primary succession on land. Blackwell, Oxford, pp 113–131

Díaz S, Cabido M, Casanoves F (1998) Plant functional traits and environmental filters at regional scale. J Veg Sci 9:113–122

Díaz-Barradas MC, Zunzunegui M, Tirado R, Ain-Lhout F, García-Novo F (1999) Plant functional types and ecosystem function in Mediterranean shrubland. J Veg Sci 10:709–716

Díaz-Barradas MC, Zunzunegui M, Ain-Lhout F, Clavijo A, García-Novo F (2001) To live or to survive in Doñana dunes. IAVS Symp. Munich, July 2001

Doing H (1985) Coastal fore-dune zonation and succession in various parts of the world. Vegetatio 61:65–75

Disraeli DJ (1984) The effect of sand deposits on the growth and morphology of *Ammophila_breviligulata*. J Ecol 72:145–154

García-Mora MR, Gallego-Fernández JB, García-Novo F (1999) Plant functional types in coastal foredunes in relation to environmental stress and disturbance. J Veg Sci 10:27–34

García-Mora MR, Gallego-Fernández JB, Williams AT, García-Novo F (2001) A coastal dune vulnerability classification. A case study of the SW Iberian Peninsula. J Coastal Res 17(4):802–811

García Novo F (1977) The effects of fire on the vegetation of Doñana National Park. In: Mooney HA, Conrad CE (eds) Proc Symp Environmental Consequences of Fire and

Fuel Management in Mediterranean Ecosystems, USDA For Serv Gen Techn WO-3. US Dept Agric, California, pp 318–325
García-Novo F, Merino J (1997) Pattern and process in the dune system of the Doñana National Park, SW Spain. In: Van der Maarel E (ed) Ecosystems of the world. 2C: Dry coastal ecosystems. Elsevier, Amsterdam, pp 453–468
Geelen LHWT (1997) Landscape succession in the Haasvelderduninen (in the past 50 years). In: García-Novo F, Crawford RMM, Díaz-Barradas MC (eds) The ecology and conservation of European dunes. EUDC-Univ de Sevilla. Sevilla, pp 267–276
Gimingham CH (1964) Maritime and sub-maritime communities. In: Burnett JH (ed) The vegetation of Scotland. Oliver and Boyd. Edinburgh, pp 67–142
Gitay H, Noble IR (1997) What are functional types and how should we seek them?. In: Smith TM, Shugart HH, Woodward FI (eds) Plant functional types: their relevance to ecosystem properties and global change. Cambridge University Press, Cambridge, pp 3–19
Grime JP (1974) Vegetation classification by reference to strategies. Nature 250:26–31
Harper JL (1977) Population biology of plants. Academic Press, London
Hesp PA (1991) Ecological processes and plant adaptations on coastal dunes. J Arid Environ 2:165–191
Huiskes AHL (1979) Biological flora of the British Isles. *Ammophila arenaria* (L.) Link. J Ecol 67:363–382
Keddy PA (1992) Assembly and response rules: two goals for predictive community. J Veg Sci 3:57–164
Margaris NS (1981) Adaptative plant strategies in plants dominating Mediterranean type ecosystems. In: Di Castri F, Goosdall DW, Specht RL (eds) Mediterranean type shrublands. Ecosystems of the world 11. Elsevier. Amsterdam, pp 309–315
Martínez F, Merino O, Martín A, García Martín D, Merino J (1998) Belowground structure and production in a Mediterranean sand dune shrub community. Plant Soil 201:209–216
Martínez ML, Moreno-Casasola P (1996) Effects of burial by sand and seedling growth and survival in six tropical sand dune species from the Gulf of Mexico. J Coastal Res 12(2):406–419
Martínez ML, Vázquez G, Sánchez-Colón S (2001) Spatial and temporal variability during primary succession on tropical coastal sand dunes. J Veg Sci 12:361–372
Merino J (1987) The cost of growing and maintaining leaves of Mediterranean plants. In: Tenhunen JD, Catarino FM, Lange OL, Oechel WC (eds) Plant response to stress: functional analysis in Mediterranean ecosystems. Springer, Berlin, Heidelberg New York, pp 553–564
Merino J, García-Novo F, Sánchez Díaz M (1976) Annual fluctuation of water potential in the xerophitic shrub of Doñana Biological Reserve (Spain) Oecol Plant 1:1–11
Merino O, Villar R, Martín A, García D, Merino J (1995) Vegetation response to climatic change in a dune ecosystem in Southern Spain. In: Moreno JM Oechel WC (eds) Global change and Mediterranean type ecosystems. Ecological Studies 117. Springer, Berlin Heidelberg New York, pp 225–238
Orshan G (1989) Plant phenomorphological studies in Mediterranean type ecosystems. Geobotany 12. Kluwer, Dordrecht
Ranwell DS (1972) Ecology of salt marshes and sand dunes. Chapman and Hall, London 7(4):433–455
Seelinger U (1992) Coastal foredunes of southern Brazil: physiography, habitats and vegetation. In: Seelinger U (ed) Coastal plant communities of Latin America, Academic Press, New York, pp 367–381
Semenova GV, van der Maarel E (2000) Plant functional types-a strategic perspective. J Veg Sci 11:917–922

Specht RL (1988) Mediterranean-type ecosystems a data source book. Kluwer, Dordrecht

Studer-Ehrensberger K, Studer C, Crawford RMM (1993) Flood induced change in a dune slack observed over 24 years. Funct Ecol 7:156–68

Terradas J (1991) Mediterranean woody plant growth forms, biomass and production in the Eastern part of the Iberian peninsula. In: Ros JD, Prat N (eds) Homage to Ramón Margalef. Oecol Aquat 10:337–349

Townsend CR (1989) The patch dynamics concept of stream community ecology JN Am Benthol Soc 8(1):36–50

Valdés B, Talavera S, Fernández Galiano E (eds) (1987) Flora Vascular de Andalucía Occidental. Ketres Editora, Barcelona

van der Maarel E (1981) Fluctuations in a coastal dune grassland due to fluctuations in rainfall: experimental evidence. Vegetatio 47:259–265

van der Maarel E, Boot R, van Dorp D, Rijntjes J (1985) Vegetation succession on the dunes of Oostvoorne, The Netherlands; a comparison of the vegetation in 1959 and 1980. Vegetatio 58:137–185

Weiher E, van der Werf A, Thompson K, Roderick M, Garnier E, Eriksson O (1999) Challenging Theophrastus. A common core list of plant traits for functional ecology. J Veg Sci 10:609–620

Westoby M (1998) A leaf-height-seed (LHS) plant ecology strategy scheme. Plant Soil 199:213–227

Westoby M, Leishman M (1997) Categorizing plant species into functional types. In: SmithTM, Shugart HH, Woodward FI (eds) Plant functional types: their relevance to ecosystem properties and global change. Cambridge University Press, Cambridge, pp 104–121

Woodward FI, Kelly CK (1997) Plant functional types: towards a definition by environmental constraints. In Smith TM, Shugart HH, Woodward FI (eds) Plant functional types:their relevance to ecosystem properties and global change. Cambridge Univ Press, Cambridge, pp 47–65

Zunzunegui M, Díaz Barradas MC, García-Novo F (1998) Vegetation fluctuation in Mediterranean dune ponds in relation to rainfall variation and water extraction. Appl Veg Sci 1:151–160

IV Biotic Interactions

11 Arbuscular Mycorrhizas in Coastal Dunes

R.E. Koske, J.N. Gemma, L. Corkidi, C. Sigüenza, and E. Rincón

11.1 Introduction

Sand dune systems are among the best studied of primary successional sites and have attracted the attention of plant ecologists for over a century (Cowles 1899). Surprisingly, the traditional explanation of dune succession overlooks the critical contribution of mutualistic fungi that facilitate the invasion of barren areas. In fact, many of the dominant, dune-building plants appear to be incapable of growing in the dune environment if their roots are not associated with arbuscular mycorrhizal (AM) fungi, the topic of this chapter.

The roots of dune species, like the vast majority of vascular plants, form symbiotic associations with fungi in the order Glomales (arbuscular mycorrhizas). The fungal mycelia provide mineral nutrients in exchange for carbon compounds of the host plant (Smith and Read 1997).

Arbuscular mycorrhizas were first reported in sand dunes in *Pancraticum maritima, Convolvolus soldanella* and *Cineraria maritima* in Italy (Stahl 1900). Since then, all the surveys of dunes throughout the world have shown that most of the plants distributed in embryo dunes and foredunes, mobile, semi-fixed and fixed dunes are heavily colonized by AM fungi ,e.g., in US mainland (Koske and Halvorson 1981, 1989), Canada (Dalpé 1989), Hawaii (Koske and Gemma 1996), Australia (Peterson et al. 1985; Logan et al. 1989), Brazil (Trufem 1995), Chile (Godoy and González 1994), India (Kulkarni et al. 1997), Italy (Giovannetti and Nicolson 1983), Mexico (Corkidi and Rincón 1997a), and Poland (Blaszkowski 1993, 1994; Tadych and Blaszkowski 2000a).

AM fungi are obligate symbionts, unable to complete their life cycle without a host plant (Smith and Read 1997). Less recognized is the fact that most plants grown under natural conditions (as opposed to greenhouse conditions with added fertilizer, water, and pesticides) cannot thrive without AM fungi (Trappe 1987). The AM association may be of considerable ecological significance for the establishment and growth of sand dune pioneer plants, because the fungi enhance plant nutrient uptake, particularly phosphorus (P), increase plant tolerance to drought and salt stress, and protect against soil

Ecological Studies, Vol. 171
M.L. Martínez, N.P. Psuty (Eds.)
Coastal Dunes, Ecology and Conservation

pathogens (Newsham et al. 1995; Little and Maun 1996, 1997; Perumal and Maun 1999; Tsang and Maun 1999; Augé 2001; Gemma et al. 2002).

While most studies of AM fungi in dunes have focused on maritime sites, arbuscular mycorrhizas are also well represented in lacustrine (Koske et al. 1975) and inland dunes (Al-Agely and Reeves 1995). Other kinds of mycorrhizas are formed by some dune species. Ericoid mycorrhizas are formed by Ericaceae, and Ectomycorrhizas by Pinaceae, Fagaceae, and some Salicaceae (see Read 1989).

The objectives of this chapter are to review a variety of studies on AM fungi and plants associated with AM fungi in coastal dunes and to point out some topics that are especially poorly known. Constraints on space prevented us from writing a more complete review.

11.2 Life History of AM Fungi in Coastal Dunes

As the roots of plants grow through the soil they are contacted by hyphae of AM fungi. The hyphae can originate from large spores (some up to 1 mm diameter), roots and rhizomes of plants already mycorrhizal, or from dead fragments of AM roots and rhizomes. The fungal hyphae invade the root cortex, producing intercellular hyphae and intracellular hyphal coils, vesicles, and arbuscules (the site of exchange between plant and fungus) (Smith and Read 1997; Figs. 11.1, 11.2). Hyphae grow out into the soil for several centimeters, greatly enlarging the absorbing capacity of the root system, and acquire phosphate and other nutrients that often are too distant to be available to the relatively coarse root system. A prodigious amount of hyphae grows from the roots into the soil. In Florida dunes, Sylvia (1986) found ca. 9 m of hyphae/g of sand and ca. 462 m of hyphae/cm of root length associated with *Uniola paniculata*. New spores are formed on the hyphae, in the soil or in roots, completing the life cycle of the fungus (Smith and Read 1997).

Unlike spores of AM fungi, the network of AM fungal hyphae in soil is easily and rapidly disrupted by a variety of abiotic and biotic factors. Destruction of this network typically leads to a significant reduction in the rate at which roots are colonized by the fungi (e.g., Read and Birch 1988; Koske and Gemma 1997), depriving transplants or seedlings the benefits of the association (Fig. 11.3). Hyphal networks in soil can be destroyed by shrinkage and expansion of soils (resulting from cycles of drought and hydration or freezing and thawing), by mechanical damage such as that caused by cultivation or compaction, by burrowing animals, by destruction of the hyphae by grazing microfauna or parasitic microorganisms, and by loss of vegetative cover (e.g., Read and Birch 1988; Jasper et al. 1989; Koske and Gemma 1997).

With the exception of *Gigaspora gigantea*, very little is known of the biology of individual species of AM fungi that can be isolated from dunes. *G.*

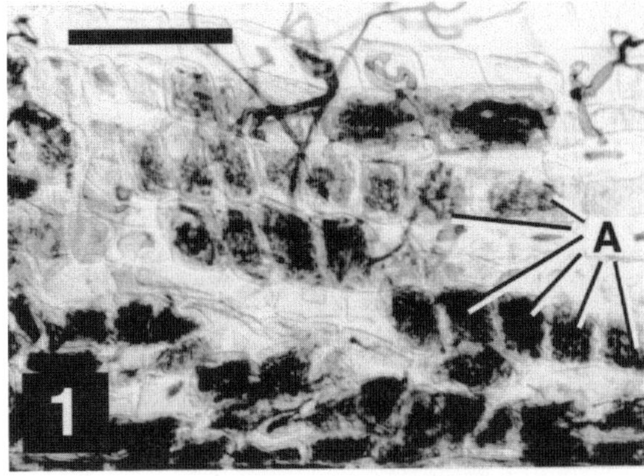

Fig. 11.1. Dark-staining hyphae and arbuscules of AM fungi in roots of American beachgrass (*Ammophila breviligulata*). Note high percentage of cortical cells with arbuscules (*A*). *Bar* 270 µm

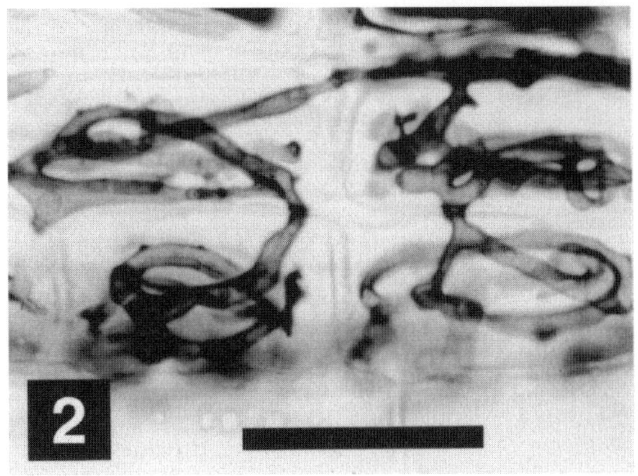

Fig. 11.2. Hyphal coils of AM fungi in roots of *A. breviligulata*

Fig. 11.3. Effect of disruption of the hyphal network of AM fungi in dune soil. Two cores were taken from the dune. Core on *left* was removed, shaken, and replaced into corning tube. Core on *right* was undisturbed. Corn plants (*Zea mays*) were grown for 8 weeks. Note reduced growth in plant growing in disturbed soil

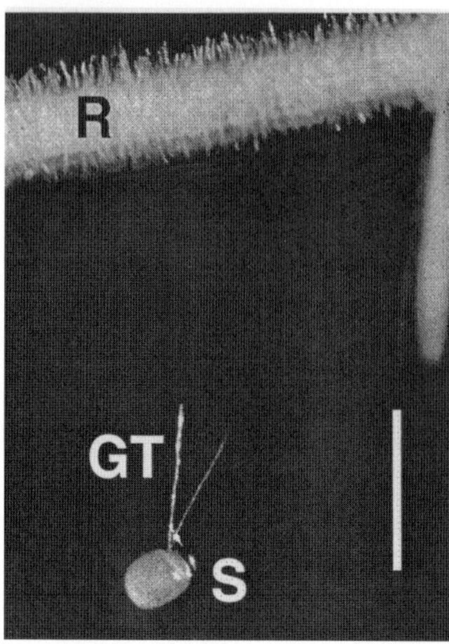

Fig. 11.4. Germinating spore (*S*) of *Gigaspora gigantea*. Germ tube (*GT*) grows towards the plant root (*R*). *Bar* 1 cm

gigantea is a common species in dunes of the Atlantic coast of the US (Koske and Halvorson 1981; Koske 1987), but it has also been found in other ecosystems (e.g., Gerdemann 1955). Newly formed spores (produced from August through December) are dormant, requiring exposure to cold temperatures to break dormancy (Gemma and Koske 1988a). Spores germinate in spring when soil temperatures reach ca. 20 °C, producing long germ tubes that grow up to 10 cm or more (Koske 1981b). The germ tubes are attracted to plant roots by volatile root exudates (Fig. 11.4; Koske 1982; Gemma and Koske 1988b). Germination occurs equally well in the presence or absence of host roots (Koske 1982). Spores are capable of producing up to ten consecutive germ tubes, so a single spore has numerous chances of contacting a plant root (Koske 1981a). Spores of *G. gigantea* retain high viability even after immersion in seawater for more than three weeks (Koske et al. 1996), suggesting that transport in longshore currents is possible.

11.3 AM Fungi in Coastal Dunes

Almost half of the ca. 150 described species of AM fungi have been found in dune ecosystems. However, many species remain undescribed, and accurate identification of species is difficult. Within the root zone of a single dune plant, 1–14 species of AM fungi have been found (Koske 1987), and the range typically is 3–7 (Rose 1988; Tadych and Blaszkowski 2000a).

Some studies report higher spore richness in stabilized dunes than in early successional or disturbed sites (Giovannetti and Nicolson 1983; Koske and Gemma 1997), but edaphic factors, temperature, and plant species may also influence the spore distribution (Koske 1987, 1988; Trufem 1995). As with plant species, temperature affects the distribution of AM fungi in dunes. In a 355-km-long transect in the Atlantic coast dunes of the US, some species of AM fungi (*Scutellospora weresubiae* and *S. fulgida*) were limited in their northern distribution by cooler temperatures, while others (e.g., *S. calospora*), were less common in the warmer, southern end of the transect (Koske 1987). However, some isolates of *S. calospora* are tolerant to warm conditions, occurring in Australian dunes (Koske 1975). Reports of the same species of AM fungi from diverse dune sites suggest that ecotypes may also be common (e.g., Koske 1975; Tadych and Blaszkowski 2000a).

11.4 Seasonality of AM Fungi in Coastal Dunes

The population of spores of AM fungi and the rate at which plant roots are colonized by AM fungi vary seasonally (Nicolson and Johnston 1979; Giovannetti 1985; Gemma and Koske 1988a; Corkidi and Rincón 1997a; Sigüenza et al. 1996). The mycorrhizal inoculum potential (MIP) of a soil is a measure of how rapidly roots growing in soil will be colonized by AM fungi. In soils with a high MIP, plant roots will be contacted by hyphae sooner and become colonized more rapidly than will roots in soil with a low MIP (Moorman and Reeves 1979). Early colonization of roots leads to early benefits to the plant (Fig. 11.3; Read and Birch 1988).

Seasonality of spore abundance is influenced by biotic and abiotic factors (e.g., the species and phenology of the host plant, temperature, and the abundance of other AM fungi present) (Koske 1981 c, 1987; Gemma et al. 1989; Sigüenza et al. 1996; Tadych and Blaszkowski 2000a,b). Spore populations increase as a result of sporulation and decline because spores germinate, emptying their contents into the resulting hyphae, and from losses due to predation, parasitism, and age (Lee and Koske 1994a,b). Fluctuations in spore abundance in 1 year can be dramatic, some species changing by 400–2000 % in a period of 3–4 months (Stürmer and Bellei 1994; Lee and Koske 1994a). No consistent relationship between sporulation and plant phenology has been found. Some studies showed higher spore abundance at the end of the growing season (Giovannetti 1985), while others report that individual species sporulate at different times (Sylvia 1986; Gemma et al. 1989; Stürmer and Bellei 1994; Sigüenza et al. 1996).

Less is known of the causes of declining spore abundance. A detailed study of parasitism of spores of *Gigaspora gigantea* in a Rhode Island sand dune showed that numerous species of Fungi Imperfecti and actinomycetes iso-

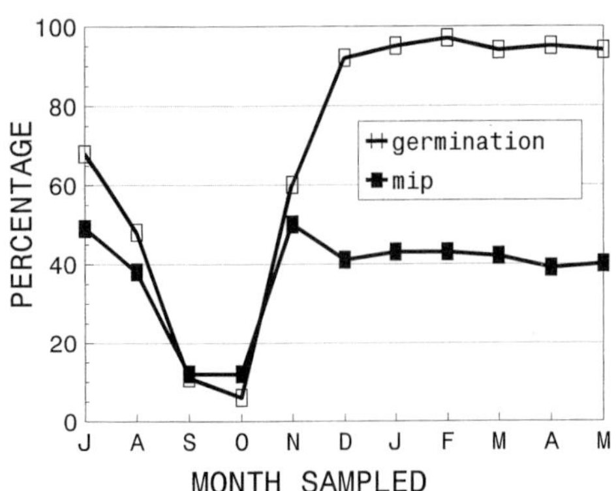

Fig. 11.5. Seasonal variation in mycorrhizal inoculum potential of dune soil from Rhode Island compared to germination rate of spores of *Gigaspora gigantea*, a dominant AM fungus in the dune. Spores are dormant in September and October when inoculum potential is lowest

lated from the dune soil were capable of penetrating and killing large numbers of healthy spores (Lee and Koske 1994a,b). Some pathogens attacked AM fungal hyphae as well as spores. Parasitism of AM fungi in non-dune systems has also been shown to reduce their ability to colonize roots (Ross and Ruttencutter 1977).

Seasonal differences of MIP in dunes reflect not only the population of spores, but their physiological state as well as the abundance of other types of inocula (hyphae, roots, and rhizome fragments) (Gemma and Koske 1988a; Koske and Gemma 1997). In dunes of the North Atlantic coast of the US, MIP was relatively consistent throughout the year except in the months of September and October when it declined by ca. 75 %. This reduction generally coincided with the very low germination rate of newly produced (but dormant) spores of *G. gigantea* (Fig. 11.5; Gemma and Koske 1988a).

11.5 Effects of Arbuscular Mycorrhizas on the Establishment and Growth of Coastal Dune Plants

Plant growth response to arbuscular mycorrhizas has been assessed in relatively few sand dune species. AM inoculation of European and American beachgrass (*Ammophila arenaria* and *A. breviligulata*), two premier dune-builders, increased survival (Koske and Polson 1984) and growth (Nicolson and Johnston 1979; Tadych and Blaszkowski 1999) in greenhouse studies. In field studies, addition of AM fungi improved establishment and tillering of *A. breviligulata*, greatly stimulated root formation (Fig. 11.6), and increased panicle production (Gemma and Koske 1989, 1997).

Similarly, mycorrhizal inoculation of the warm-season grass *Uniola paniculata* significantly increased the height, dry mass, and tissue phosphorus concentration (Sylvia 1989), and mycorrhizal plants of this species grew 219 %

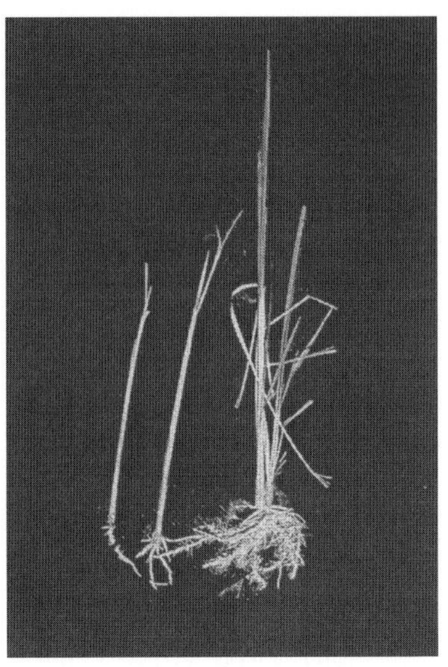

Fig. 11.6. Effect of AM fungi on growth of *Ammophila breviligulata* in a field study. Plant on *right* was inoculated 6 weeks before photo. Two plants on *left* were planted 40 cm away from first plant but were not inoculated. All plants were started as rootless tillers. Note profuse rooting and leaf elongation

larger (shoots) and had 53 % more tillers than nonmycorrhizal plants when they were transplanted from a nursery to a barren replenishment beach in Florida (Sylvia 1989). Addition of AM fungi increased growth of the temperate tree *Prunus maritima* (280 % increase in shoot weight after 20 weeks; Gemma and Koske 1997) and of the four tropical species *Chamaecrista chamaecristoides* (280 % increase in shoot weight, 500 % increase in root weight), *Palafoxia lindenii* (shoots 200 %, roots 50 %) and *Trachypogon plumosus* (shoots 350 %, roots 904 %) from Mexico (Corkidi and Rincón 1997b), and *Sesbania tomentosa*, an endangered Hawaiian species (shoots 116 %) (Gemma et al. 2002). AM fungi also improved the growth of five of six dune colonizers in Poland, some in excess of 5300 % (Tadych and Blaszkowski 1999). In the Polish study, the grass *Festuca rubra* ssp. *rubra* was unable to grow without AM fungi, a striking demonstration of its dependency on the symbiosis.

Species of AM fungi vary in their effectiveness in stimulating growth, and plant species vary in their response to inoculation (e.g., Sylvia and Burks 1988; Corkidi and Rincón 1997b). *Leymus arenarius* plants inoculated with indigenous AM fungi from a sand dune in Iceland had higher dry mass than plants inoculated with commercial inoculum (of non-dune origin) (Greipsson and El-Mayas 2000).

The presence of mycorrhizas does not necessarily indicate a beneficial association to plant growth. *Ipomoea pes-caprae* and *Sporobolus virginicus* are species reported as highly mycorrhizal in the sand dune natural conditions, but did not improve their growth response in presence of mycorrhizal inoculation (Corkidi and Rincón 1997b; Koske and Gemma 1990).

The degree to which plants are dependent on AM fungi is typically assessed by comparing the growth of inoculated and non-inoculated plants in pots (vs. field conditions) (e.g., Corkidi and Rincón 1997b). It should be noted that "dependency" in this sense refers only to growth, not to non-growth benefits. Because AM fungi are important to plants under stress conditions (for most plants in the wild, this is most of the time), it is necessary to measure dependency when plants are subjected to stress. In the artificial conditions that pot-grown plants experience (especially in maintaining levels of P and moisture comparable to those in the field), it is extremely difficult to determine dependency accurately, and too little or too much P can lead to large underestimations of dependency (Fig. 11.7; Miyasaka et al. 1993; Gemma et al. 2002). The concept of ecological mycorrhizal dependency has been suggested to emphasize the need for assessing mycorrhizal dependency under appropriate conditions (Gemma et al. 2002). This focus on growth in assessing dependency ignores the non-nutritional benefits of AM that may be more important, such as drought tolerance. For example, in greenhouse trails, AM fungi often have little effect on growth of *A. breviligulata* (Koske and Gemma, unpub. obser.), but when plants experienced moderate drought conditions, only 20 % of the controls survived in contrast to 78 % of the AM plants (Koske and Polson 1984). In addition, other studies on sand dune species have also shown that the mycorrhizal association can contribute to plant tolerance against salinity (Tsang and Maun 1999), burial (Perumal and Maun 1999), and infection with nematodes (Little and Maun 1996, 1997).

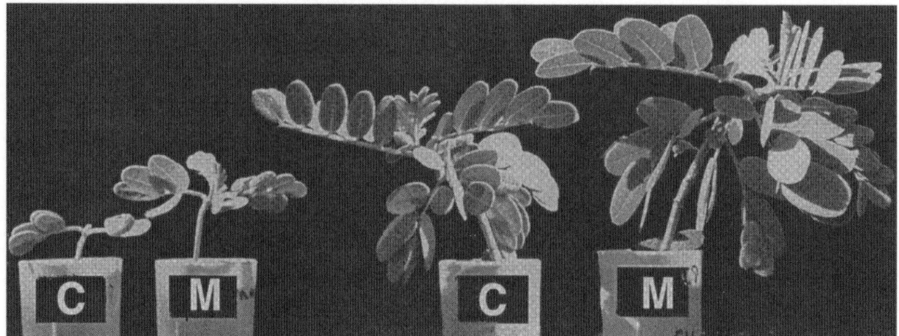

Fig. 11.7. Interaction of soil P and inoculation with AM fungi in *Sesbania tomentosa*. At low P (equivalent to field levels), plant inoculated with AM fungi (*M*) is 2.2 times larger than uninoculated plant (*C*). When soil P is similar to that of a productive agricultural field (plants on *right*), little difference between *M* and *C* plants is detected

11.6 Arbuscular Mycorrhizas and Coastal Dune Succession

Succession and AM fungi in sand dunes has been studied from two viewpoints: (1) plant succession as related to the activity of AM fungi in soil, and (2) fungal succession as related to the plant community and edaphic conditions. Several studies have demonstrated a strong positive correlation between the population of AM fungi (spore abundance, species richness, extent of root colonization, and inoculum potential) in dunes and the percent cover/successional stage of the vegetation (Nicolson 1959, 1960; Nicolson and Johnston 1979; Koske and Halvorson 1981; González et al. 1995; Trufem 1995; Corkidi and Rincón 1997a; Sigüenza et al. 1996; Koske and Gemma 1997; Tadych and Blaszkowski 2000b). In most cases, after increasing during early and mid-succession, AM fungal populations decline in the latest stage (Fig. 11.8), often as sites include ectomycorrhiza-forming species (Read 1989). As with plants, species of AM fungi have been identified as early-, mid-, and late colonizers in the dunes (Koske and Gemma 1997; Rose 1988; Tadych and Blaszkowski 2000b). Many of the species show a wide tolerance of conditions, occurring throughout much of the dune succession.

The means by which pioneer plant species that require AM fungi are able to become established in barren, primary successional dune sites is incompletely known. It was assumed that primary successional species could not be dependent upon AM (Read 1989). While this notion had intuitive appeal (pioneer sites lack plants [and therefore AM fungi], so only plants that do not require AM fungi can be pioneers), there has been little evidence to support this hypothesis. Nearly all of the major dune-building species have arbuscular mycorrhizas even in the earliest stages of dune formation (e.g., Nicolson 1960;

Fig. 11.8. Relationship between successional stage in a dune and abundance of spores of AM fungi and mean species richness of AM fungi. (Adapted from Koske and Gemma 1997)

Koske et al. 1975; Peterson et al. 1985; Koske and Gemma 1990; Gemma and Koske 1992; Corkidi and Rincón 1997a).

There are two ways by which plants and AM fungi become associated in the dunes. The first, which is the less efficient, depends upon the fungi being present in a site of potential plant establishment (by seed or vegetative fragment) before the plant arrives. Spores and fragments of roots and rhizomes could arrive at these sites in drift material (deposited by waves or blown in from vegetated areas of the dune) (Nicolson 1960; Sylvia 1986; Koske and Gemma 1990). However, in contrast to the prolific spore production of most wind-dispersed fungi, AM fungi form very few spores, limiting their capacity for dispersal. Spores and fragments in the dunes do not migrate downward into the sand with rainfall, so many of the propagules are too distant from the root zone to be able to form arbuscular mycorrhizas with seedlings (Friese 1984). While significant populations of AM fungi can occur in drift lines on the beach and berm (Nicolson 1960; Koske and Gemma 1990), in barren dune areas, soil collected from a few centimeters below the surface may have few or no AM fungi (Sylvia 1989; Koske and Gemma 1997).

A much more efficient mechanism for ensuring that AM fungi and plant arrive simultaneously at the same site (co-dispersal) has evolved in a number of primary colonizers (Gemma and Koske 1992). While the benefits of reproduction by vegetative fragmentation for primary colonizers of unstable habitats such as sand dunes is recognized (e.g., Grubb 1987), the co-dispersal of vegetative fragments with AM fungi may be more important (Koske and Gemma 1990). AM fungal spores and hyphae occur in the rhizomes (in addition to the roots) of several dune-building species including the grasses *A. breviligulata, Distichlis spicata, Ischaemum byrone, S. virginicus,* and *U. paniculata*, and in the tropical *Jacquemontia sandwicensis* (Convolvulaceae) (Koske and Gemma 1990; Koske and Gemma, unpubl. observ.). When rhizome fragments arrive in barren dune sites and sprout new roots and shoots, the resident AM fungi spread into the new roots, benefiting the recruit as well as establishing a hyphal network in the soil that facilitates the establishment of other plants that do not co-disperse (Gemma and Koske 1989, 1992, 1997). AM fungi in root fragments (like spores) retain their viability after immersion in seawater (Koske and Gemma 1990).

The extent of the hyphal network increases in the early stages of succession, as does the abundance of AM fungi spores and AM fungi species richness, increasing the MIP of the site (Corkidi and Rincón 1997a; Koske and Gemma 1997). Establishment of mid- and late-successional species in the dunes appears to be prevented until a hyphal network has developed in the dunes (Koske and Gemma 1997).

In addition to this important role, the hyphal network builds soil structure, binding sand grains into larger aggregates that resist wind erosion (e.g., Koske et al. 1975; Sutton and Sheppard 1976; Clough and Sutton 1978; Forster and Nicolson 1981; Jehne and Thompson 1981; Rose 1988). In sand dunes of Lake Huron in the US, up to 9 % of the sand in the mobile dunes is held in aggregates greater than 2 mm diameter by AM fungal hyphae (Koske et al. 1975). The network also links plants in the dune, and nutrients may be shunted from larger to smaller plants, even of different species (Smith and Read 1997).

Driftline vegetation is inconsistently associated with AM fungi (Nicolson 1960; Giovannetti and Nicolson 1983; Peterson et al. 1985; Koske and Halvorson 1989; Koske and Gemma 1990; Sigüenza, pers. observ.). The instability of this habitat prevents long-term colonization by plants, and the contribution of mycorrhizas to plant survival is of unknown significance there.

As a final note, the absence of arbuscular mycorrhizas from root samples does not always indicate that the plants lack mycorrhizas; unsampled roots may have had AM fungi. In some species, only the smaller roots had AM (Koske and Halvorson 1981). Also, rhizomatous species can be misleading because roots of young ramets may lack AM fungi, while older ramets (connected to the AM-free ramets) are highly mycorrhizal (Koske and Gemma, unpubl. observ.).

11.7 Arbuscular Mycorrhizas and Sand Dune Stabilization and Restoration

With the realization of the importance of AM fungi to the growth of dune species, field trials using AM inoculum in dune stabilization and restoration were initiated (Sylvia and Will 1988; Sylvia et al. 1993; Gemma and Koske 1989, 1997). These trials showed that inoculation resulted in more rapid establishment of transplants, stimulating the formation of AM hyphal networks that allow recruitment by mid- and late-successional species. Thus, addition of AM fungi has the potential to accelerate succession and stabilization of dunes (Gemma and Koske 1997).

One reason that arbuscular mycorrhizas were overlooked as important components of dune succession and restoration is that field plantings of dune-building species made without the intentional addition of AM fungi were successful. This apparent paradox was resolved when AM fungi spores and hyphae were found in the roots and rhizomes of planting stock used for dunes (*A. breviligulata*, *Prunus maritima*, *Rosa rugosa*, and *Spartina patens*) (Gemma and Koske 1997). When tillers of *A. breviligulata* were planted in barren sites in Cape Cod, MA (without addition of AM fungi), 78 % of the root samples from these plants were mycorrhizal 47 weeks after planting (Gemma and Koske 1997). Nevertheless, even though both "non-inoculated" and inoculated plants developed mycorrhizas, the inoculated ones grew and tillered faster, and the significant differences were noted for 81 weeks when the experiment was ended (Gemma and Koske 1997).

For species that establish from seed (e.g., *Uniola paniculata*), co-dispersal is not possible. Addition of mycorrhizal inoculum at the time of planting may be necessary if such species are planted in barren or sparsely vegetated areas (Sylvia 1989). Addition of particularly effective species or isolates of AM fungi at the time of sowing or outplanting can result in greater benefits to plants than are provided by the native species (Sylvia and Burks 1988; Sylvia 1989; Sylvia et al. 1993).

Future research will address methods to produce inoculum for use in dune restoration. In addition, it is necessary to identify the species and isolates of AM fungi that are most effective for particular conditions, possibly fitting particular isolates to specific seral stages. Much still is to be learned about the ecology of AM fungi in sand dunes ecosystems.

References

Al-Agely AK, Reeves FB (1995) Inland sand dune mycorrhizae: effects of soil depth, moisture, and pH on colonization of *Oryzopsis hymenoides*. Mycologia 87:54–60

Augé RM (2001) Water relations, drought and VA mycorrhizal symbiosis. Mycorrhiza 11:3–42

Blaszkowski J (1993) The occurrence of arbuscular fungi and mycorrhizae (Glomales) in plant communities of maritime dunes and shores of Poland. Bull Pol Ac Soc Biol 41:377–392

Blaszkowski J (1994) Arbuscular mycorrhizal fungi and mycorrhizae (Glomales) of the Hel Peninsula, Poland. Mycorrhiza 5:71–88

Clough KS, Sutton JC (1978) Direct observation of fungal aggregates in sand dune soil. Can J Microbiol 24:333–335

Corkidi L, Rincón E (1997a) Arbuscular mycorrhizae in a tropical sand dune ecosystem on the Gulf of Mexico. I. Mycorrhizal status and inoculum potential along a successional gradient. Mycorrhiza 7:9–15

Corkidi L, Rincón E (1997b) Arbuscular mycorrhizae in a tropical sand dune ecosystem on the Gulf of Mexico. II. Effects of arbuscular mycorrhizal fungi on the growth of species distributed in different early successional stages. Mycorrhiza 7:17–23

Cowles, HC (1899). The ecological relations of the vegetation on the sand dunes of Lake Michigan. Bot Gaz 29:95–117, 167–202, 281–308, 361–391

Dalpé Y (1989) Inventaire et repartition de la flore endomycorrhizienne de dunes et de rivages maritimes du Québec, du Nouveau-Brunswick et de la Nouvelle-Ecosse. Nat Can 116:219–236

Forster SM, Nicolson TH (1981) Aggregation of sand from a maritime embryo sand dune by microorganisms and higher plants. Soil Biol Biochem 13:199–203

Friese CR (1984) Spatial distribution of mycorrhizal fungi in sand dunes. M Sc thesis, Univesrity of Rhode Island, Kingston, RI

Gemma JN, Koske RE (1988a) Seasonal Variation in spore abundance and dormancy of *Gigaspora gigantea* and in mycorrhizal inoculum potential of a dune soil. Mycologia 80:211–216

Gemma JN, Koske RE (1988b) Pre-infection interactions between roots and the mycorrhizal fungus *Gigaspora gigantea*: chemotropism of germ-tubes and root growth response. Trans Br Mycol Soc 91:123–132

Gemma JN, Koske RE (1989) Field inoculation of American Beachgrass (*Ammophila breviligulata*) with V-A mycorrhizal fungi. J Environ Manage 29:173–182

Gemma JN, Koske RE (1992) Are mycorrhizal fungi present in early stages of primary succession? In: Read DJ, Lewis DH, Fitter AH, Alexander IJ (eds) Mycorrhizas in ecosystems, CABI, Wallingford, pp 183–189

Gemma JN, Koske RE (1997) Arbuscular mycorrhizae in sand dune plants of the North Atlantic coast of the U.S.: field and greenhouse studies. J Environ Manage 50:251–264

Gemma JN, Koske RE, Carreiro M (1989) Seasonal dynamics of selected species of V-A mycorrhizal fungi in a sand dune. Mycol Res 92:317–321

Gemma JN, Koske RE, Habte M (2002) Mycorrhizal dependency of some endemic and endangered Hawaiian plant species. Am J Bot 89:337–345

Gerdemann JW (1955) Relation of a large soil-borne spore to phycomycetous mycorrhizal infections. Mycologia 47:619–632

Giovannetti M (1985) Seasonal variations of vesicular-arbuscular mycorrhizas and Endogonaceous spores in a maritime sand dune. Trans Br Mycol Soc 84:679–684

Giovannetti M, Nicolson TH (1983) Vesicular-arbuscular mycorrhizas in Italian sand dunes. Trans Br Mycol Soc 80:552–557

Godoy R, González B (1994) Simbiosis micorrícica en la flora de los ecosistemas dunarios del Centro-Sur de Chile. Gayana Bot 51:69–80

González B, Godoy R, Figueroa H (1995) Dinámica estacional de los hongos micorrícicos vesículo-arbusculares en ecosistemas dunarios del Centro-Sur de Chile. Agric Tec 55:267–272

Greipsson S, El-Mayas H (2000) Arbuscular mycorrhizae of *Leymus arenarius* on coastal sands and reclamation sites in Iceland and response to inoculation. Restor Ecol 8:144–150

Grubb PJ (1987) Some generalizing ideas about colonization and succession in green plants and fungi. In: Gray AJ, Crawley MJ, Edwards PJ (eds) Colonization, succession and stability. Blackwell, Oxford, pp 31–56

Jasper DA, Abbott LK, Robson AD (1989) The loss of VA mycorrhizal infectivity during bauxite mining may limit the growth of *Acacia pulchella* R. Dr. Aust J Bot 37:33–42

Jehne W, Thompson CH (1981) Endomycorrhizae in plant colonization on coastal sand dunes at Cooloola, Queensland. Aust J Ecol 6:221–230

Koske RE (1975) *Endogone* spores in Australian sand dunes. Can J Bot 53:668–672

Koske RE (1981a) Multiple germination by spores of *Gigaspora gigantea*. Trans Br Mycol Soc 76:328–330

Koske RE (1981b) *Gigaspora gigantea*: observations on spore germination of a VA-mycorrhizal fungus. Mycologia 73:288–300

Koske RE (1981 c) A preliminary study of interactions between species of vesicular-arbuscular fungi in a sand dune. Trans Br Mycol Soc 76:411–416

Koske RE (1982) Evidence for a volatile attractant from plant roots affecting germ tubes of a VA mycorrhizal fungus. Trans Br Mycol Soc 79:305–310

Koske RE (1987) Distribution of VA mycorrhizal fungi along a latitudinal temperature gradient. Mycologia 79:55–68

Koske RE (1988) Vesicular-arbuscular mycorrhizae of some Hawaiian dune plants. Pac Sci 42:217–229

Koske RE, Gemma JN (1990) VA mycorrhizae in strand vegetation of Hawaii: evidence for long-distance codispersal of plants and fungi. Am J Bot 77:466–474

Koske RE, Gemma JN (1996) Arbuscular mycorrhizal fungi in Hawaiian sand dunes: Island of Kaua'i. Pac Sci 50:36–45

Koske RE, Gemma JN (1997) Mycorrhizae and succession in plantings of beachgrass in sand dunes. Am J Bot 84:118–130

Koske RE, Halvorson WL (1981) Ecological studies of vesicular-arbuscular mycorrhizae in a barrier sand dune. Can J Bot 59:1413–1422

Koske RE, Halvorson WL (1989) Mycorrhizal associations of selected plant species from San Miguel Island, Channel Islands National Park, California. Pac Sci 43:32–40

Koske RE, Polson WR (1984) Are VA mycorrhizae required for sand dune stabilization? Bioscience 34:420–424

Koske RE, Sutton JC, Sheppard BR (1975) Ecology of *Endogone* in Lake Huron sand dunes. Can J Bot 53:87–93

Koske RE, Bonin C, Kelly J, Martínez C (1996) Effects of sea water on spore germination of a sand dune-inhabiting arbuscular mycorrhizal fungus. Mycologia 88:947–950

Kulkarni SS, Raviraja NS, Sridhar KR (1997) Arbuscular mycorrhizal fungi of tropical sand dunes of West Coast of India. J Coastal Res 13:931–936

Lee P, Koske RE (1994a) *Gigaspora gigantea*: seasonal abundance and aging of spores in a sand dune. Mycol Res 98:453–457

Lee P, Koske RE (1994b) *Gigaspora gigantea*: parasitism of spores by fungi and actinomycetes. Mycol Res98:458–466

Little LR, Maun MA (1996) The "*Ammophila* problem" revisited: a role for mycorrhizal fungi. J Ecol 84:1–7

Little LR, Maun MA (1997) Relationships among plant-parasitic nematodes, mycorrhizal fungi and the dominant vegetation of a sand dune system. Ecoscience 4:67–74

Logan VS, Clarke PJ, Allaway WG (1989) Mycorrhizas and root attributes of plants of coastal sand dunes of New South Wales. Aust J Plant Physiol 16:141–146

Miyasaka SC, Habte M, Matsyama DT (1993) Mycorrhizal dependency of two Hawaiian endemic tree species: Koa and Mamane. J Plant Nutr 16:1339–1356

Moorman T, Reeves FB (1979) The role of endomycorrhizae in revegetation practices in the semi-arid west. II. A bioassay to determine the effect of land disturbance on endomycorrhizal populations. Am J Bot 66:14–18

Newsham KK, Fitter AH, Watkinson AR (1995) Multi-functionality and biodiversity in arbuscular mycorrhizas. Trends Ecol Evol 10:407–411

Nicolson TH (1959) Mycorrhiza in the Gramineae. I. Vesicular-arbuscular endophytes, with special reference to the external phase. Trans Br Mycol Soc 42:421–438

Nicolson TH (1960) Mycorrhiza in the Gramineae. II. Development in different habitats particularly sand dunes. Trans Br Mycol Soc 43:132–145

Nicolson TH, Johnston C (1979) Mycorrhiza in the Gramineae. III. *Glomus fasciculatus* as the endophyte of pioneer grasses in a maritime sand dune. Trans Br Mycol Soc 72:262–268

Perumal JV, Maun MA (1999) The role of mycorrhizal fungi in growth enhancement of dune plants following burial in sand. Funct Ecol 13:560–566

Peterson RL, Ashford AE, Allaway WG (1985) Vesicular-arbuscular mycorrhizal associations of vascular plants on Heron Island. A great barrier reef coral cay. Aust J Bot 33:669–676

Read DJ (1989) Mycorrhizas and nutrient cycling in sand dune ecosystems. Proc R Soc Edinb 96B:89–110

Read DJ, Birch CPD (1988) The effects and implications of disturbance of mycorhizal mycelial systems. Proc R Soc Edinb 94B:13–24

Rose SL (1988) Above and belowground community development in a maritime sand dune ecosystem. Plant Soil 109:215–226

Ross JP, Ruttencutter R (1977) Population dynamics of two vesicular-arbuscular mycorrhizal fungi and the role of hyperparasitic fungi. Phytopathology 67:490–496

Sigüenza C, Espejel I, Allen EB (1996) Seasonality of mycorrhizae in coastal sand dunes of Baja California. Mycorrhiza 6:151–157

Smith SE, Read DJ (1997) *Mycorrhizal symbiosis*. Academic Press, London, 605 pp

Stahl E (1900) Der Sinn der Mycorrhizenbildung. Eine vergleichend-biologische Studie. Jahrb Wiss Bot 34:539–668

Stürmer SL, Bellei MM (1994) Composition and seasonal variation of spore populations of arbuscular mycorrhizal fungi in dune soils on the island of Santa Catarina, Brazil. Can J Bot 72:359–363

Sutton JC, Sheppard BR (1976) Aggregation of sand-dune soil by endomycorrhizal fungi. Can J Bot 54:326–333

Sylvia DM (1986) Spatial and temporal distribution of vesicular-arbuscular mycorrhizal fungi associated with *Uniola paniculata* in Florida foredunes. Mycologia 78:728–734

Sylvia DM (1989) Nursery inoculation of sea oats with vesicular-arbuscular mycorrhizal fungi and outplanting performance on Florida beaches. J Coastal Res 5:747–754

Sylvia DM, Burks JN (1988) Selection of vesicular-arbuscular mycorrhizal fungus for practical inoculation of *Uniola paniculata*. Mycologia 80:565–568

Sylvia DM, Will ME (1988) Establishment of vesicular-arbuscular mycorrhizal fungi and other microorganisms on a beach replenishment site in Florida. Appl Environ Microbiol 54:348–352

Sylvia DM, Jarstfer AG, Vosátka M (1993) Comparisons of vesicular-arbuscular mycorrhizal species and inoculum formulations in a commercial nursery and on diverse Florida beaches. Biol Fertil Soils 16:139–144

Tadych M, Blaszkowski J (1999) Growth responses of maritime sand dune plant species to arbuscular mycorrhizal fungi. Acta Mycologica 34:115–123

Tadych M, Blaszkowski J (2000a) Arbuscular fungi and mycorrhizae (Glomales, Zygomycota) of the Sowiñski National Park, Poland. Mycotaxon 74:463–483

Tadych M, Blaszkowski J (2000b) Succession of arbuscular mycorrhizal fungi in a deflation hollow of the Sowiñski National Park, Poland. Acta Soc Bot Pol 69:223–236

Trappe JM (1987) Phylogenetic and ecologic aspects of mycotrophy in the angiosperms from an evolutionary standpoint. In: Safir GR (ed) Ecophysiology of VA mycorrhizal plants. CRC Press, Boca Raton, pp 5–26

Trufem S (1995) Ecological aspects of arbuscular mycorrhizal fungi from coastal sand dunes. Rev Bras Bot 18:51–60

Tsang A, Maun MA (1999) Mycorrhizal fungi increase salt tolerance of *Strophostyles helvola* in coastal foredunes. Plant Ecol 14:159–166

12 The Role of Algal Mats on Community Succession in Dunes and Dune Slacks

G. Vázquez

12.1 Introduction

Different kinds of algal communities can live in dune slacks that may become temporarily flooded or remain moist throughout the year due to fluctuations in the proximity of the water table (Brown and McLachlan 1990). This chapter focuses on changes in the composition of algal communities during periods of flooding and drought, with special emphasis on the hydrological characteristics of slacks as well as on morphological and physiological factors that allow algae to survive in these stressful environments. Also discussed is the role of algae in the first stages of dune succession, when they form part of a soil community of so-called microbial mats (Belnap and Gillete 1998) as well as part of the aquatic community. In both circumstances, algae participate actively in sand stabilization and facilitate the development of several pioneer plants. Finally, changes in algal community composition are reported for the case where slacks became flooded in a tropical mobile dune system.

12.2 Hydrological Dynamics in Slacks Within Coastal Dune Systems

Dune slacks were defined by Tansley (1949, in Boorman et al. 1997) as wet hollows left between dune ridges where groundwater reaches or approaches the surface of the sand. These systems are formed in different ways: (1) on coasts where the shoreline is advancing seawards, a new line of dunes can enclose an area of beach plain that, after desalinization, can become a dune slack or dune lake (Boorman et al. 1997), and (2) the action of the wind can form deep blowouts in already existing dune areas.

Ecological Studies, Vol. 171
M.L. Martínez, N.P. Psuty (Eds.)
Coastal Dunes, Ecology and Conservation

Hydrologically, the height of the water table is decisive in whether a dune slack is dry, moist, wet, or permanently inundated. Although slacks appear to be simple in shape, their hydrology is not, being determined by dune geomorphology and local conditions of climate and topography. For instance, in temperate zones such as the Wadden Islands and the Frisian Island of Texel in The Netherlands, it has been noted that some slacks become flow-through lakes during the wet season, in which groundwater discharges into one part of the slack and surface water infiltrates on another side (Grootjans et al. 1998; Adema et al. 2002). This process occurs mainly during the rainy season, when the water table is so high that the slack is inundated. A similar hydrological mechanism was observed in the dune systems of Mont Saint Frieux in France, an area characterized by parabolic dunes with wet valleys and several dune streams. There, slacks are fed by subsurface flow from infiltration areas, infiltration of surface water from dune streams, and direct infiltration by rainfall (Bakker and Nienhuis 1990).

Similar to the effect of local weather, geographical position and local topographical conditions influence seasonal changes in the water table. In temperate zones, slacks can become flooded during winter and early spring and dry out in summer (Grootjans et al. 1997). In tropical zones, in contrast, the rainy season – with the resulting increase in the water table – occurs in the summer, while the dry season occurs during the winter (Moreno-Casasola and Vázquez 1999). Long-term research of a tropical dune system in La Mancha, Veracruz, Mexico (Moreno-Casasola and Vázquez 1999) showed that in the early rainy season (June), the water table started to rise and flooding depended on the amount of surplus rain that fell in June and July (Fig. 12.1). The study by Moreno-Casasola and Vázquez (1999) showed that the amount of precipitation necessary for flooding to occur in La Mancha slacks was 500 mm, thus, in very rainy years, the probability of flooding therefore increases. During autumn and early winter (October-December), the water level started to decrease and the water table remained below ground level until the following rainy season.

12.3 Algal Communities in Slacks and Other Coastal Zones

The hydrological characteristics of slacks, blowouts, and other coastal zones - such as coastal tidal sand flats and sand dunes- determine the composition and abundance of the algal communities found in these systems (Pluis and De Winder 1990; Stal 2000). In dry conditions, algae are sometimes found in macroscopic structures similar to microbial mats that are made up of cyanobacteria, diatoms, green algae, and eubacteria (Simons 1987; Lange et al. 1992; Norris et al. 1993; Grootjans et al. 1998; Vázquez et al. 1998; Stal 2000). Species such as *Crinalium epipsammum* (Pluis and De Winder 1990), which

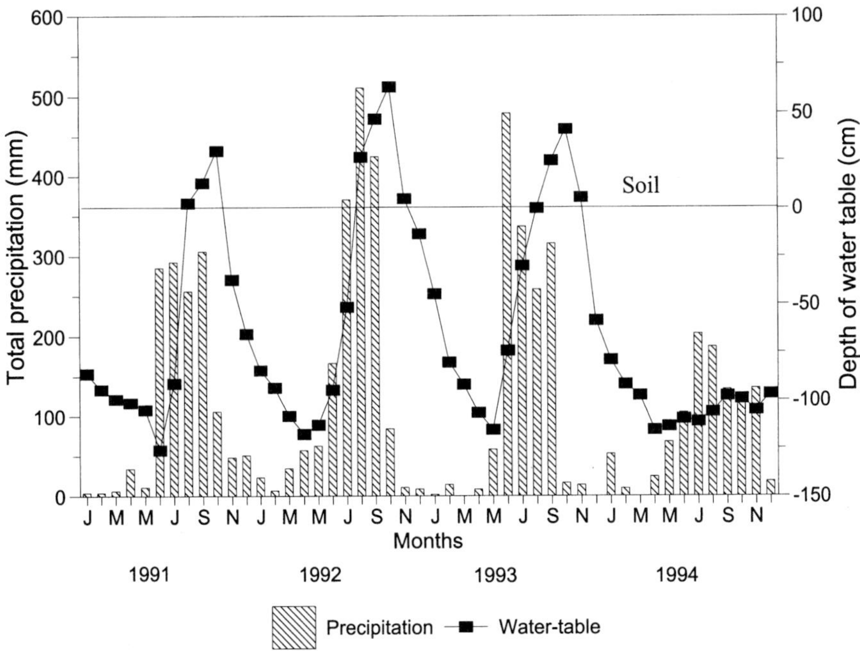

Fig. 12.1. Total monthly precipitation and fluctuation of water table from 1991 to 1994 in slacks from La Mancha, Ver. Depth of water table is established in relation to soil surface. Negative values denote depth beneath the sand surface

are unique band-shaped filamentous cyanobacteria, can also be found (Stal 2000). Another noteworthy species is the filamentous non-heterocystous *Microcoleus chthonoplastes,* which is observed in some desert crusts and intertidal marine zones (Brock 1975, Stal 2000). *Cylindrospermum, Oscillatoria, Lyngbya, Spirulina, Anabaena, Scytonema,* and *Nostoc*, along with unicellular species such as *Chroococcus, Chroococcidiopsis, Merismopedia,* and *Synechococcus*, also form mucilaginous colonies.

Microbial mats possess several characteristics that enable them to be successful in coastal environments. For instance, cyanobacteria form an extracellular polysaccharide sheath (EPS) that protects them against desiccation and grazing (Brock 1975 in Stal 2000). The EPS may retain large amounts of water, and organisms that produce it may thus tolerate long periods of drought. Due to this mechanism, in fact, they have been shown to be more successful than green algae (Stal 2000). When the moisture level increases, the sheath absorbs water and the organism resumes activity immediately, then particles of sand are trapped, thus stabilizing the soil. Although the mats are no more than a few millimeters thick, they play a decisive role in dune mechanics, as they influence rain interception as well as water infiltration into the soil, evaporation, soil stabilization, and moisture maintenance.

They also play a role in nutrient recycling, especially N, and carbon capture (Lange et al. 1992).

Cyanobacteria are the main primary producers in most microbial mats. They can photosynthesize with little oxygen availability and low light intensity, permitting them to make efficient use of light as an energy source, water as an electron donor, and CO_2 as a carbon source. This is very important because light is strongly attenuated in microbial mats, due both to sediment and absorption by the dense phototrophic community (Stal 2000). The dense biomass of cyanobacteria in the upper photic zones of microbial mats produces high rates of photosynthesis comparable to the productivity of rain forests, which are usually considered the most productive ecosystems on Earth (Guerrero and Mass 1989 in Stal 2000). Although their nutrient requirements are low, cyanobacteria can use dinitrogen as a nitrogen source, making it possible for them to grow regardless of whether or not there is nitrogen in the environment.

During the rainy season, slacks become small seasonal freshwater systems that are poor in nutrients and characterized by large fluctuations in volume due to the high evaporation rate. For this reason, the algal community changes dramatically. In dune pools and slacks from temperate zones such as The Netherlands (Simons 1987; 1994) and in tropical zones such as those discussed here, a diverse algal community can be found: filamentous green algae such as Zygnematales (*Spirogyra, Mougeotia, Zygnema, Zygogonium*), Oedogoniales (*Oedogonium*), and Klebsormidiales (*Klebsormidium*) as well as Desmidiales (*Cosmarium, Closterium, Staurastrum, Euastrum*) (Bowling et al. 1993), Charales (*Chara, Nitella*) (Simons 1987; Simons and Nat 1996), and diatoms (*Navicula, Cocconeis, Pinnularia, Melosira*) (Kling 1986).

Most of the dominant species are filamentous and macroscopic green algae in flooding conditions and can form large macroscopic clouds or are, like the filamentous periphyton, found on and around larger aquatic plants. When the slack border dries, these filamentous matrices stay on the border and maintain a relatively high humidity in the substrate, favoring germination of phanerogam species (Vázquez et al. 1998).

Some of the species found in such unstable environments have characteristics that allow them to survive drought conditions when the water table decreases and the slack dries out. For example, the thick walls of *Spirogyra* and *Oedogonium* zygospores contain sporopollenin, an inert material that protects the cell from drought (Simons 1987; Van den Hoeck et al. 1998). Thus, cells adopt a latent state that permits survival until the following rainy season (Simons 1987). Another adaptation to stressful environments is zygote dormancy and survival. Van den Hoeck et al. (1998) have shown that the length of the dormant period depends on temperature. At 4 °C, *Spirogyra maxima* zygotes remain dormant for 14 months, while at 18–20 °C they are dormant for only 3.5 months.

Also common in slacks during the rainy period are desmids such as *Cosmarium, Staurastrum*, and *Closterium*. Diverse desmid floras are characteristic of freshwater systems that are shallow, transparent, and stagnant, with low conductivity and little nutrients, as is the case in ponds lying in leached dunes. They also flourish in high sodium:potassium and calcium:magnesium ratios of less than 2. In this type of environment, they may live as phytoplankton, on the bottom as benthic dwellers, or on the submerged parts of plants. Coesel (1981, in Van den Hoek et al. 1998) found that a diverse desmid flora is favored by small-scale patchiness in nutrient concentration within natural habitats such as small ponds where mixing between ground water and rain water occurs. In desmids, sexual reproduction has been observed only sporadically in nature. Most desmids survive adverse conditions such as low temperature, low light intensities, or partial desiccation as vegetative cells rather than hypnozygotes.

12.4 The Role of Algae During Primary Succession in Coastal Dunes

Primary succession in dune slacks is considered to occur during four phases (Grootjans et al. 1997), and algae play a fundamental role during the first two. Algal and microbial mats are predominant during the first phase, when there is little organic matter in the soil. During the second, algae facilitate the colonization of phanerogams that can tolerate limited nutrient availability. Later, in the third phase, a layer of mosses and bryophytes develops and tall grasses and shrubs establish, increasing vegetation structure and composition. Finally, once there is more organic material, species replacement occurs, allowing trees to establish (Pluis and De Winder 1990; Grootjans et al. 1998; Vázquez et al. 1998).

One of the mechanisms for dune slack stabilization is the fixation of sand by algal crusts or mats. This diminishes the wind's impact by increasing the resistance surface: mucilage produced by algae remains adhered to grains of sand and acts as a fixing agent (Pluis 1994). In sandy soil, colonization starts just under the surface with the arrival of cyanobacteria such as *Microcoleus*, *Oscillatoria*, and *Tychonema*, which are considered primary colonizers and are sometimes followed by the green algae *Klebsormidium flaccidum*. It has been observed that cyanobacteria adapt rapidly to variations in water availability, while *K. flaccidum* appears to be associated with conditions of water retention caused previously by cyanobacteria. If unremoved by subsequent storms, algae can be followed by annual phanerogam communities. Slack size influences stabilization efficiency, as very large areas of small slacks are covered with algal crusts and these contribute to the establishment of colonizing species, especially in the rainy season, when moisture conditions favor both (Pluis and De Winder 1990).

During flooding, algal mats may also participate in succession by facilitating the germination of phanerogams. Water volume in slacks fluctuates considerably, decreasing due to high temperatures and evaporation and increasing after precipitation. When slack borders dry somewhat but remain moist, algal mats may stay on the soil, forming a crust that maintains moisture and thus facilitates the germination of the seeds found there. The mechanism that has been suggested under these conditions is that algae maintain the moisture necessary for germination even in dry conditions. This was demonstrated experimentally on three substrates (sand, algae, and cotton) with two irrigation treatments: the first one consisted of continuous watering to keep sand permanently moist (wet treatment) and the second involved watering once a week (dry treatment). Vázquez et al. (1998) found that tropical Cyperaceae (*Fuirena simplex, Fimbristilys cymosa*, and *Rhyncosphora colorata*) germinated successfully (>50 %) on moist algal mats of *Spirogyra, Mougeotia, Oedogonium*, and *Microspora*, which form dense clouds that can act as seed traps. In the wet treatment, the highest final humidity was maintained in the algal and cotton substrate, while in the dry treatment the algal substrate maintained more humidity than the sand and cotton substrate. *Cyperus articulatus* responded differently, as it showed a high germination percentage in every type of experimental substrate (algae, sand, and cotton), suggesting that humidity was not a determinant factor for the germinability of this species. The first three species are found principally on slack borders, so that algae appears to play a very important role in maintaining the moisture necessary for germination. *C. articulatus*, in contrast, can be found both in dry and wet parts of tropical dunes, indicating humidity requirements that are probably broader than those of the other three species (Fig. 12.2).

In a preliminary experimental study designed to determine the importance of algae for the growth of several Cyperaceae, Vázquez, Moreno-Casasola and Barrera (unpublished data) compared the growth of seedlings of *Fimbristylis cymosa* and *Cyperus articulatus* under different nutrient treatments: (1) wet sand with nutrients provided by filamentous algae decomposition lasting 3 months, (2) sand to which a diluted solution was added (1:10) that consisted of fertilizer with N-P-K (low nutrient values), (3) sand to which a concentrated fertilizer with N-P-K high-nutrient solution was added (1:1), and (4) sand with no nutrients added (control). The dry biomass that accumulated was measured every 5 to 6 weeks for each treatment. *Fimbristylis cymosa* showed the greatest biomass increment in both the algal and high-nutrient concentration treatments (Fig. 12.3a). Growth in sand and with low nutrient concentrations was, in contrast, significantly lower. *Cyperus articulatus* only underwent a significant biomass increment in the high nutrient concentration treatment, but no difference was noted between sand, algae, and low nutrient concentrations (Fig. 12.3b). These results suggest that algae maintain favorable growing conditions for *F. cymosa* seedlings, possibly supplying necessary nutrients, while the growth of *C. articulatus* does not seem to be affected by the presence of algae. This corroborates previous findings regard-

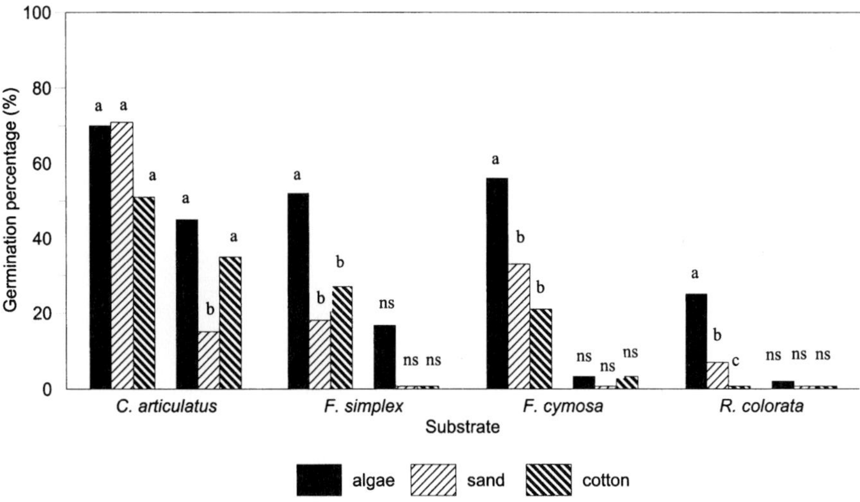

Fig. 12.2. Mean final germination percentages (%) of Cyperaceae species on different substrates in the wet and dry treatment for each species. Different letters indicate significant differences between treatments for each species. *ns* Non-significant differences between treatments

ing germination for the same species (Vázquez et al. 1998). Thus, algae facilitate the germination and growth of *F. cymosa*.

In temperate calcareous slacks in The Netherlands, Adema et al. (2002) found evidence of alternative stable states in the pattern of pioneer vegetation and in later successional stages, suggesting that positive-feedback mechanisms are responsible of these states, one of which relates to microbial mats. The combined metabolic activities of microbial mats (with cyanobacteria, colorless sulfur bacteria, purple sulfur bacteria, and sulfate-reducing bacteria) result in microgradients of oxygen and sulfide which are toxic for most higher plants as their growth is stunted. However, dune slack pioneer species can protect themselves against the toxic sulfide, as they release oxygen from their root system and favor colorless sulfur bacteria that detoxify free sulfide. This results in stable, open pioneer vegetation with a microbial mat that cannot be invaded by later species that have not adapted to anoxic soils containing free sulfide. Grootjans et al. (1997) also found that microbial mats prolonged the life of pioneer stage species such as *Samolus valerandi*, and inhibited the growth of later successional stage species (*Calamagrostis epigejos* and *Juncus alpinoarticulatus*). This is apparently due to the fact that the roots of the latter species cannot penetrate microbial mats and also to the limited accumulation of organic matter, a condition that favors pioneer species.

Another mechanism by which microbial mats may favor the growth of pioneer species is through precipitation of calcium carbonate, which prevents the

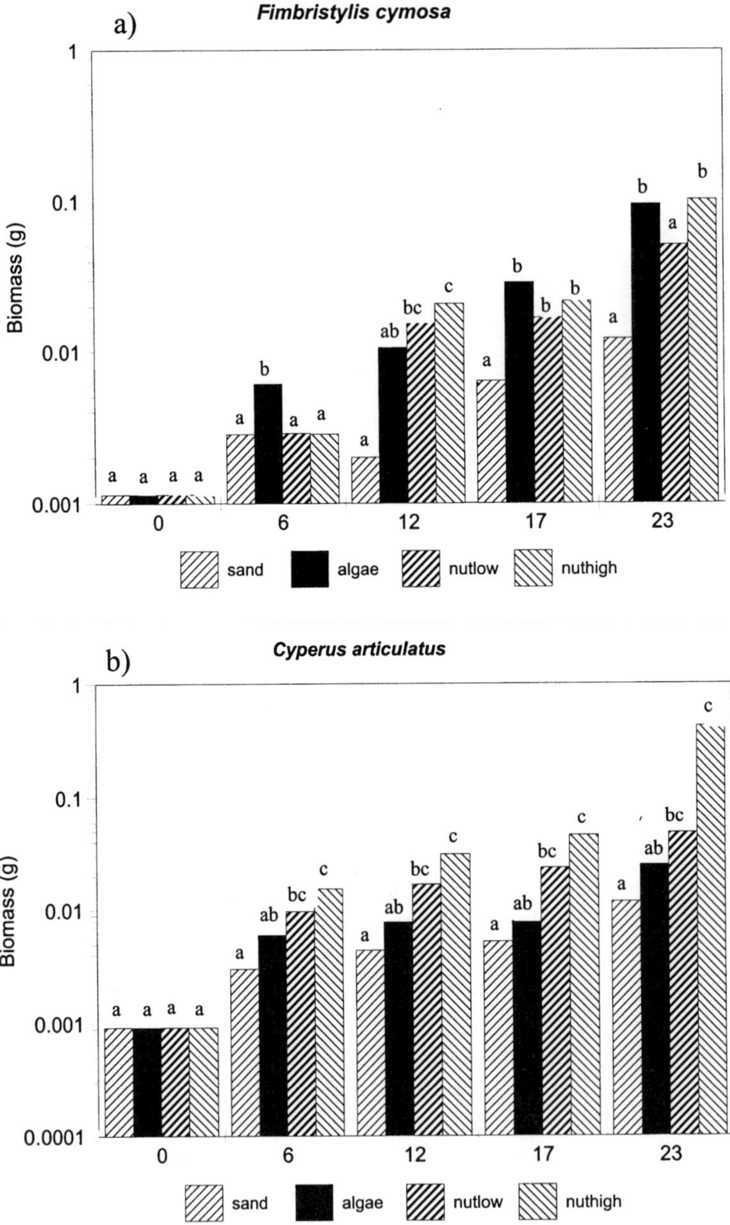

Fig. 12.3. Mean final biomass (g) of *Fimbristylis cymosa* (**a**) and *Cyperus articulatus* (**b**) over 23 weeks for different treatments with nutrient concentrations. Biomass was measured each 5 to 6 weeks. *Different letters* indicate significant differences between nutrient treatments

soil from acidifying rapidly. This generally occurs in infiltration areas characterized by surplus precipitation (Grootjans et al. 1996, 1997).

12.5 A Case Study on the Gulf of Mexico

Various systems of tropical dunes, both stabilized and mobile, exist along the Gulf of Mexico coast. In mobile systems, there are numerous slacks that follow the hydrological behavior described herein: in summer they flood for 2 or 3 months and then dry out in winter. Thus, during rainy years as well as rainy periods, very shallow aquatic systems form; they are nutrient-poor and facilitate the development of algae; both macro -and microscopic, they are of little-known taxonomic composition and undetermined variation over time. Considering the importance of these communities as primary producers, and particularly due to their participation in dune succession, research was conducted over a 2-year period (1990–1991) during the flooding months. The objective was to identify community structure, taxonomic composition, and variation over time, in the dune slacks of Doña Juana, located in the state of Veracruz, Mexico (96°20′W, 19°28′N). Dunes in this system are parabolic and reach up to 20–30 m in height; each has a slack at its lowest point. The climate in the region is warm and sub-humid. The rainy season normally runs from June to September, when approximately 81 % of total precipitation (1230 mm) is registered. Mean annual temperature is 25 °C. During the short flooding period (3 or 4 months), algae samples were obtained from the slacks every 15–20 days. Sample analysis was qualitative, and quantification was performed using an arbitrary scale: 1- scarce, 2- abundant, and 3- very abundant.

Results showed that the macroscopic and microscopic algal community at Doña Juana is composed principally by cyanobacteria (Chroococcales, Oscillatoriales and Nostocales), diatoms (Pennales), and green algae (Chlorococcales, Desmidiales, and Zygnematales) (Table 12.1). For the most part, macroscopic algae were represented by *Mougeotia, Oedogonium, Spirogyra*, and *Zygnema* (Zygnematales), with *Spirogyra* and *Oedogonium* (also reported in temperate zones) as the most important components of dune pools. Other microscopic algal groups were found immersed in the filamentous masses. Among them, the most important were desmids such as *Cosmarium, Euastrum*, and *Staurastrum*, followed by the green algae Chlorococcales (*Oocystis, Sphaerocystis, Tetraedron*), cyanobacteria (*Anabaena, Chroococcus, Oscillatoria)* and diatoms (*Amphora, Cocconeis placentula, Rhopalodia, Mastogloia smithii*). The desmids *Cosmarium* and *Staurastrum* were observed in asexual reproduction.

The two sampling years were different in terms of the proportions of each group (Fig. 12.4). The dominant groups forming the floating masses were the

Zygnematales, Desmidiales and Chlorococcales. In 1990, the dominant species during the initial flooding were filamentous green algae (*Spirogyra*) and desmids (*Cosmarium* and *Staurastrum*), diatoms (*Rhopalodia* sp., *Mastogloia smithii, Nitzschia* sp. and *Cocconeis placentula*), and cyanobacteria (*Oscillatoria limosa*). In late August (1990), when filamentous green algae and desmids were dominant, unicellular green algae (Chlorococcales) and cyanobacteria disappeared. The diminution of algal mats and microscopic species in late autumn (26 November 1990) may be related to lower temperatures and the end of the rainy period. In 1991, during initial flooding (September), Chlorococcales (*Oocystis lacustris, Tetraedron minimum*), and desmids (*Cosmarium* sp. and *Staurastrum gracile*) were noted again; later, in October, a larger number of cyanobacteria such as *Chroococcus minor, Gloeotrichia, Merismopedia*, and *Anabaena* appeared.

Table 12.1. List of algal species found in dune slacks of the Doña Juana dune system in the state of Veracruz, Mexico. The classification system was proposed by Van Den Hoek et al. (1998)

CLASS CYANOPHYCEAE (cyanobacteria)
Order Chroococcales
Chroococcus limneticus Lemmermann
Chroococcus minor (Kützing) Nageli
Merismopedia sp.
Order Oscillatoriales
Lyngbya sp.
Oscillatoria limosa (Dillwyn) C. Agardh
Order Nostocales
Anabaena sp.
Gloeotrichia sp.

CLASS BACILLARIOPHYCEAE (diatoms)
Order Pennales
Amphora coffeaeformis Agardh
Amphora sp.
Cocconeis placentula Ehrenberg
Mastogloia smithii Thwaites
Navicula sp.
Nitzschia sp.
Pinnularia viridis (Nitzsch) Her.
Rhopalodia sp.

CLASS CHLOROPHYCEAE (green algae)
Order Chlorococcales
Gloeocystis ampla (Kützing) Rabenhorst
Oocystis lacustris Chodat
Oocystis solitaria Wittrock
Sphaerocystis schroeteri Chodat
Tetraedron minimum (A. Braun) Hangs
Order Oedogoniales
Oedogonium sp.

CLASS ZYGNEMATOPHYCEAE (green algae)
Order Zygnematales
Mougeotia sp.
Spirogyra sp.
Zygnema sp.
Order Desmidiales
Cosmarium botrytis Meneghini
Cosmarium sp. 1
Cosmarium sp. 2
Cosmarium sp. 3
Cosmarium sp. 4
Cosmarium sp. 5
Cosmarium sp. 6
Euastrum sp. 1
Euastrum sp. 2
Staurastrum gracile Ralfs
Staurastrum sp.

Mat composition appeared to vary considerably during the two flooding periods (Fig. 12.5). In July and early August of 1990, *Spirogyra* dominated, although *Oedogonium*, *Zygnema*, and *Mougeotia* were also present in smaller quantities. In late August and October, the situation was reversed: *Oedogonium* and *Zygnema* increased considerably and the other species diminished, although they did not disappear. When slacks flooded the following year, in September of 1991, *Zygnema* dominated throughout and was always observed to be in a reproductive state. Large quantities of *Mougeotia* appeared only in mid-October, but without a decrease in the occurrence of *Zygnema*. *Spirogyra* and *Oedogonium* maintained a limited presence during this second cycle. During the study period, *Spirogyra, Oedogonium*, and *Zygnema*, had a high rate of sexual reproduction. *Spirogyra* is an especially common genus, with up to 70 species reported in The Netherlands (Simons 1987). Other studies have shown zygospores of *Spirogyra* and *Oedogonium* as decay-resistant resting spores, which are considered part of a life strategy that facilitates the tolerance of stressful conditions in ephemeral bodies of water (Simons 1987). As previously mentioned, these results are interpretable as a way that algae adapt to highly stressful slack conditions.

Although the two cycles studied represent a short period of time, the results indicate that important changes occur in the community from one

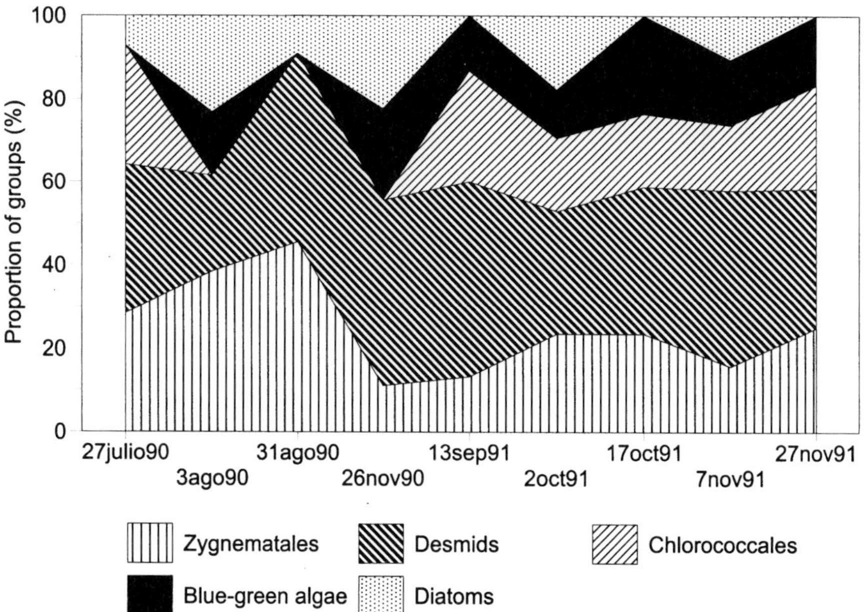

Fig. 12.4. Temporal variation in the number of species from the most important taxonomical groups of algae in slacks from a mobile dune system in Doña Juana, Ver.

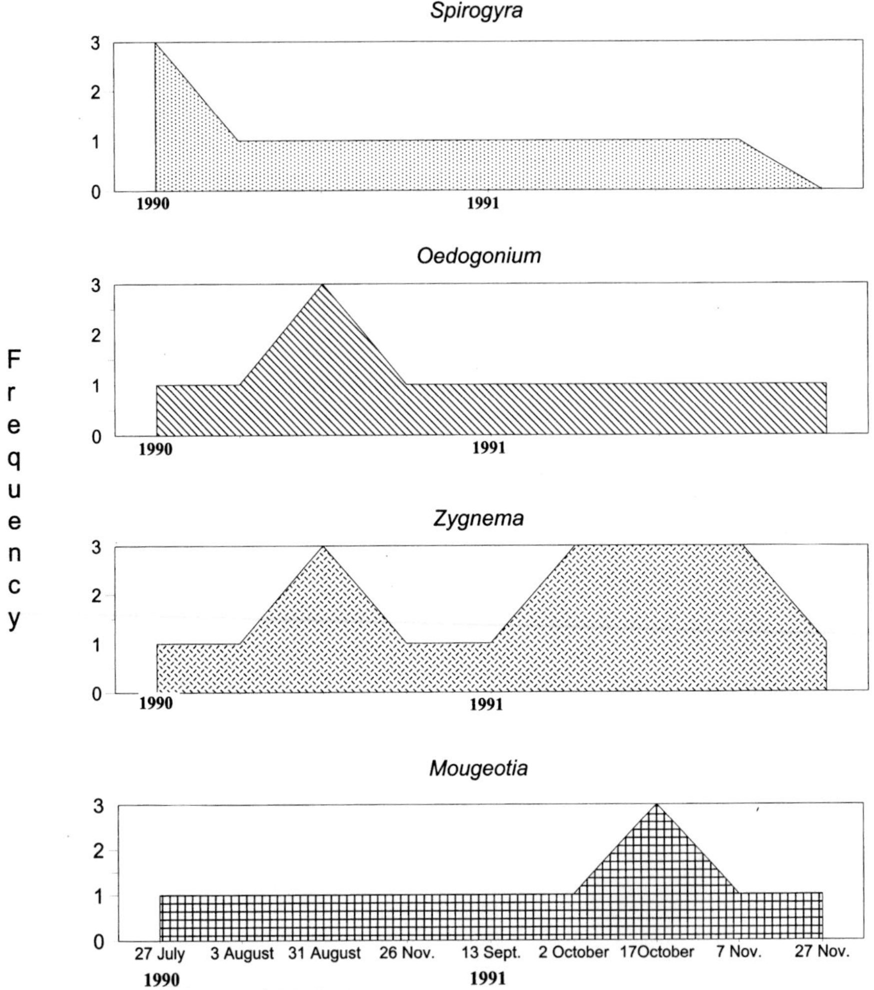

Fig. 12.5. Temporal variation in the abundance of the most important macroscopic green algae in a mobile dune system in Doña Juana, Ver.

year to the next. This can be considered a common phenomenon in tropical slacks due to the drastic environmental fluctuations that characterize them.

12.6 Importance of Algae for Slack Conservation

Dune slacks, which represent pioneer successional stages on mobile dunes, have great ecological importance due to their diversity in plants and animals (rare, endemic, and protected), which is the result of environmental conditions that are unique to slacks (Grootjans et al. 1998). Interannual variation of the water table and sand movement favor an internal cyclical succession in which algal mats plays different roles. As analyzed in this chapter, these roles in both temperate and tropical zones have been studied by different authors (Van den Ancker et al. 1985; Pluis and De Winder 1990; Grootjans et al. 1997; Vázquez et al. 1998; Adema et al. 2002) who have proved experimentally the existence of different stabilization mechanisms within the slacks studied: sand stabilization, the establishment of phanerogams through facilitation mechanisms, and in particular, the inhibition of late-successional stage species so that the presence of pioneer stage species is favored. Thus, some of these systems' conservation measures have led to the use of algae to maintain pioneer-stage succession in dunes. This prevents the loss of diversity within these systems that would occur if succession occurred in only one direction, so that pioneer-stage species would be replaced by shrubs and trees, eventually becoming a forest.

12.7 Conclusions

Algal communities play a very important role in dune slack dynamics and other systems such as blowouts. The importance of algae lies in their participation on ecological processes that occur in highly stressful systems. Algae favor the establishment and colonization of sand during early succession; they facilitate or detain the germination and establishment of different phanerogam species during various successional stages. As primary producers, their taxonomic composition and abundance influence nutrient availability in an aquatic system.

The taxonomic composition of these communities depends on the hydrological characteristics of the system in which they are found. Due to morphological and physiological characteristics, cyanobacteria (Cyanophyceae) can survive despite the highly stressful condition of limited water present in soil during periods of drought. During flooding, moreover, succession occurs and filamentous species such as *Spirogyra, Zygnema, Oedogonium,* and *Microspora* (green algae) become quite common.

During this study it became clear that while vast information exists on algae, most of it is limited to a soil setting; little research has addressed the composition and function of algal communities in dune systems. Further long-term studies are needed in order to better understand algal dynamics in slacks and the role that they play during dune succession.

Acknowledgements. I am grateful to A.P. Grotjans, Mario Favila, and M.L. Martínez for their comments on the manuscript and Ingrid Márquez for translating it. Data for this study were obtained from a project financed by CONACYT (225260-5-3465N) and the Instituto de Ecología, A.C. (902-17).

References

Adema EB, Grootjans AP, Petersen J, Grijpstra J (2002) Alternative stable states in a wet calcareous dune slack in The Netherlands. J Veg Sci 13:107–114

Bakker TWM, Nienhuis PR (1990) Geohydrology of les dunes de Mont Saint Frieux, Boulonnais, France. Catena Suppl 18:133–143

Belnap J, Gillete DA (1998) Vulnerability of desert biological crusts to wind erosion: the influences of crust development, soil texture, and disturbance. J Arid Environ 39:133–142

Boorman LA, Londo G, Van der Maarel E (1997) Communities of dune slacks. In: Van der Maarel E (ed) Dry coastal ecosystems, Part C. Elsevier, Amsterdam, pp 275–293

Bowling LC, Banks MR, Crome RL, Tyler PA (1993) Reconnaissance of Tasmania II. Limnological features of Tasmanian freshwater coastal lagoons. Arch Hydrobiol 126:385–403

Brown AC, McLachlan A (1990) Ecology of sand dunes. Elsevier, Amsterdam

Grootjans AP, Stuyfzand PJ, Sival FP (1996) Hydrogeochemical analysis of a degraded dune slack. Vegetatio 126:27–38

Grootjans AP, Van den Ende FP, Walsweer AF (1997) The role of microbial mats during primary succession in calcareous dune slacks: an experimental approach. J Coastal Conserv 3:95–102

Grootjans AP, Ernst WHO, Stuyfzand PJ (1998) European dune slacks: strong interactions of biology, pedogenesis and hydrology. Tree 13:96–100

Kling GW (1986) The physicochemistry of some dune ponds on the Outer Banks, North Carolina. Hydrobiologia 134:3–10

Lange OL, Kidron GJ, Budel B, Meyer A, Kilian E, Abeliovich A (1992) Taxonomic composition and photosynthetic characteristics of the abiological soil crusts covering sand dunes in the western Negev Desert. Funct Ecol 6:519–527

Moreno-Casasola P, Vázquez G (1999) Succession in tropical dune slack after disturbance by water-table dynamics. J Veg Sci 10:515–524

Norris RH, Moore JL, Mather WA, Wensing LP (1993) Limnological characteristics of two coastal dune lakes. Aust J Mar Freshwater Res 44:437–458

Pluis JLA (1994) Algal crust formation in the inland dune area, Laarder Wasmeer, The Netherlands. Vegetatio 113:41–51

Pluis JLA, De Winder B (1990) Natural stabilization. Catena Suppl 18:195–208

Simons J (1987) *Spirogyra* species and accompanying algae from dune waters in The Netherlands. Acta Bot Neerl 36:13–31

Simons J (1994) Field ecology of freshwater macroalgae in pools and ditches, with special attention to eutrophication. Neth J Aquat Ecol 28:25–33

Simons J, Nat E (1996) Past and present distribution of stoneworts (Characeae) in The Netherlands. Hydrobiologia 340:127–135

Stal LJ (2000) Cyanobacterial mats and stromatolites. In: Whitton BA, Potts M (eds) The ecology of cyanobacteria. Their diversity in time and space. Kluwer, Dordrecht, pp 61–120

Van den Ancker J, Jungerius PD, Mur LR (1985) The role of algae in the stabilization of coastal dune blowouts. Earth Surf Process Landforms 10:189–192

Van den Hoek C, Mann DG, Jahns HM (1998) Algae. An introduction to phycology. Cambridge University Press, Cambridge

Vázquez G, Moreno-Casasola P, Barrera O (1998) Interaction between algae and seed germination in tropical dune slack species: a facilitation process. Aquat Bot 60 4:409–416

13 Plant–Plant Interactions in Coastal Dunes

M.L. Martínez and J.G. García-Franco

13.1 Introduction

Biotic interactions among plants have long been recognized as important in determining community structure and dynamics (Pickett 1980). Interactions range from negative to positive and function simultaneously. The result is a balance among mechanisms that cause a gradient of potential effects. In general, it is accepted that as the abiotic stress increases, so will the importance of positive interactions, and the opposite occurs when physical stress is reduced, and the importance of competition is expected to increase (Bertness and Callaway 1994). In coastal dunes, in particular, high environmental heterogeneity generates the possibility of the occurrence of a wide variety of interactions among plants. The number of studies on plant–plant interactions in these environments has increased during the last decades (see Table 13.1) and has shown that both positive and negative interactions are important, although the abiotic environment exerts an important control as well. However, although the successional gradients and dynamics have been widely studied in these environments (for example, Cowles 1899; Yarranton and Morrison 1974; Lichter 2000), there have been few studies of the relative impact of the different interactions during different successional stages. In this chapter we will focus on the most studied plant–plant interactions of the literature on coastal dunes: facilitation, competition, and epiphytism. We will review the information reported in the literature and will include our own data, to test the prediction that positive interactions are more frequent during the environmentally harsh early successional stages, when the sand is highly mobile, while competition and epiphytism become more frequent as the system stabilizes and the plant cover increments.

Ecological Studies, Vol. 171
M.L. Martínez, N.P. Psuty (Eds.)
Coastal Dunes, Ecology and Conservation

Table 13.1. Chronological survey of the literature on plant–plant interactions studied in dune communities throughout the world, including successional stage, location and nature of the study. (The list is not exhaustive but merely suggestive.) Unlike field experiments, field data refer to observations without previous manipulation

Authors	Interaction	Successional stage	Location	Type of study
Barbour (1970)	Competition	Early	USA	Greenhouse
Yarranton and Morrison (1974)	Facilitation	Stable	Canada	Field data
Pemadasa and Lovell (1974)	Competition	Early to stable	United Kingdom	Greenhouse
Mack and Harper (1977)	Competition	Stable	United Kingdom	Greenhouse
Silander and Antonovics (1982)	Competition, facilitation	Early to stable	USA	Field experiments
De Jong and Klinkhamer (1988)	Facilitation	Early	The Netherlands	Field experiments
Ehrenfeld (1990)	Competition	Early to stable	Worldwide	Literature review
Barros-Henriques and Hay (1992)	Competition	Early to stable	Brazil	Field experiments
Kellman and Kading (1992)	Facilitation	Stable	Canada	Field experiments
Olff et al. (1993)	Competition	Early to stable	The Netherlands	Field experiments
Jungerius et al. (1995)	Competition	Early	The Netherlands	Field data
Alpert and Mooney (1996)	Facilitation	Early	USA	Field data
García-Franco (1996)	Epiphytism	Stable	Mexico	Field data
García-Franco and Rico-Gray (1996)	Parasitism	Stable	Mexico	Field data
García-Franco and Rico-Gray (1997)	Parasitism	Stable	Mexico	Field data
Veer (1997)	Competition	Early to stable	The Netherlands	Field experiments
Kooijman et al. (1998)	Competition	Early to stable	The Netherlands	Field experiments
Vázquez et al. (1998)	Facilitation	Early	Mexico	Greenhouse
Lammerts et al. (1999)	Competition, facilitation	early to stable	The Netherlands	Field experiments
Lichter (2000)	Competition	Early to stable	USA	Field experiments
Shumway (2000)	Facilitation	Early	USA	Field experiments
Martínez et al. (2001)	Facilitation	Early	Mexico	Field data

13.2 Facilitation

Facilitation occurs when the establishment of a plant is enhanced by the amelioration of the environmental extremes provided by previously established plants. Because of the frequent disturbance events in the form of sand deposition, and because of the environmental harshness (drought, low nutrients, temperature extremes, intense solar radiation) experienced by dune plants, such positive interactions are expected to occur in these communities. Studies that examine facilitation in coastal dunes are scarce and have shown these interactions to be frequent but not always present. This interaction has been demonstrated for dunes from different latitudes (Canada, The Netherlands, USA and Mexico) (Table 13.1). In general, facilitation seems to occur mostly during the early stages of succession, although it has also been observed during late successional stages, when the dunes are covered by forests (Table 13.1). Descriptive field data and field experiments have demonstrated that beneath the canopy of established plants, late colonizers show better establishment of seedlings (Yarranton and Morrison 1974; Kellman and Kading 1992). In general, the result of these facilitative interactions is a spatial aggregation of the interacting species (Yarranton and Morrison 1974; Kellman and Kading 1992; Callaway 1995).

In South Wellfleet, Massachusetts (USA), Shumway (2000) observed that the two herbaceous sand dune species (*Solidago sempervirens* and *Ammophila breviligulata*) grew larger, were more likely to flower, produced a larger number of seeds and were in better physiological conditions (higher midday water potentials, tissue nitrogen and photosynthetic efficiency) beneath the nitrogen-fixing shrubs of *Myrica pensylvanica* than beyond its shade. Similarly, Martinez (2003) found that in the tropical coastal dunes of La Mancha (central Gulf of Mexico), the shade (with significant lower temperature and sand accretion) of the also nitrogen-fixing shrub *Chamaecrista chamaecristoides* was beneficial for seedling establishment of two tall perennial grasses, *Schizachyrium scoparium* and *Trachypogon plumosus* (Table 13.2). Although survival in the shade only was significantly different for *T. plumosus*, in both species the only individuals that reached a mature state and reproduced were those growing beneath *C. chamaecristoides*. Furthermore, adult long-term survival was greater beneath the shade of the shrub than beyond it and reproduction was exclusive to the shade (Table 13.2). As a result of these positive interactions, the studied grasses plus an additional annual (*Triplasis purpurea*) and a herbaceous perennial composite (*Pectis saturejoides*) were significantly closer to *Chamaecrista* than expected by chance (Table 13.3). This spatial aggregation was maintained throughout the seasons of the year: winter (with strong winds), dry and rainy seasons, except for the annual *Triplasis purpurea*, in which the facilitative interaction was not observed when this species was only present in the seed bank during the rainy

Table 13.2. Environmental conditions and seedling survival in two contrasting locations: exposed sand and protected sites beneath the facilitative shrub, *Chamaecrista chamaecristoides* (*P* significant differences between the two conditions after a t-test; – not tested statistically)

	Exposed	Protected	*P*
Environment			
Maximum midday photosynthetic active radiation (μmol s^{-1} m^{-2})	1964	996	0.05
Maximum temperature on the sand surface (°C)	64	53	0.05
Maximum temperature in the air (°C)	39	39	n.s.
Nitrates in the sand (mg/l)	0.14	0.16	n.s.
Phosphates in the sand (ppt)	2	2.6	n.s.
Organic matter (%)	0.16	0.12	n.s.
Water content in the sand (mg/g)	70	66	n.s.
Maximum gusts of wind (km/h)	120	90	–
Maximum sand accretion (cm)	22	7	0.05
Seedling survival			
Schizachyrium scoparium (%)	1	6	n.s.
Trachypogon plumosus (%)	0	6	0.01
Long-term adult survival			
Schizachyrium scoparium (years)	1	3	–
Trachypogon plumosus (years)	0	3	–
Reproductive effort (at age 1 year)			
Schizachyrium scoparium (spikes per individual)	0	4	0.05
Trachypogon plumosus (spikes per individual)	0	3	0.05

season. In both examples, the areas beneath benefactors were environmentally improved, resulting in beneficial effects on the surrogate species. The areas beneath the canopies of the shrub were more shaded, had lower temperatures on the surface of the sand, and registered a decreased impact of disturbances by substrate mobility. Additional studies focused on facilitation in coastal dunes have demonstrated that the sand in the vicinity of early mobile dune colonizers had higher nutrient and moisture levels (McLeod and Murphy 1977; De Jong and Klinkhamer 1988; Alpert and Mooney 1996), which were probably also beneficial for late colonizer species.

Facilitation has also been demonstrated in forests that have developed on dunes during late successional stages. In the Wasaga Beach dunes, Ontario (Canada), Kellman and Kading (1992) found that the establishment of pine seedlings of *Pinus strobus* and *P. resinosa* was enhanced beneath the pre-established oak trees *Quercus rubra*. Similarly, at La Mancha, Muñiz (2001) registered seedlings of 18 tropical rain forest tree species growing in the shade of isolated *Diphysa robinoides* leguminous trees that were separated (by at least 20–30 m) from the nearest tropical rain forest patch. No seedlings were

Table 13.3. Comparison of observed and expected mean distances (cm) (± standard error) from different species to the nearest shrub of *Chamaecrista chamaecristoides* in a mobile area in coastal dunes located on the Gulf of Mexico. Distances were measured during three seasons: nortes (see text), dry and rainy; – no data; *P* represents significant differences between observed and expected distances; t-test)

	N	Observed	Expected	*P*
Winter with strong winds				
Schizachyrium scoparius	58	50.0±4.5	87.7±10.6	<0.001
Trachypogon plumosus	65	52.5±2.6	87.7±10.6	<0.01
Triplasis purpurea	200	51.7±2.4	87.7±10.6	<0.001
Pectis saturejoides	113	43.1±2.9	87.7±10.6	<0.001
Dry				
Schizachyrium scoparius	54	58.1±4.1	110.1±10.12	<0.001
Trachypogon plumosus	31	63.8±7.8	110.1±10.12	<0.001
Triplasis purpurea	8	44.0±8.1	110.1±10.12	<0.01
Pectis saturejoides	135	51.7±3.6	110.1±10.12	<0.001
Rainy				
Schizachyrium scoparius	36	38.97±4.8	100.3±9.8	<0.001
Trachypogon plumosus	51	52.5±4.01	100.3±9.8	<0.001
Triplasis purpurea	–	–	–	–
Pectis saturejoides	129	46.6±2.8	100.3±9.8	<0.001

found in the surrounding grassland. The above suggests a facilitative phenomenon although other factors such as predominant seed dispersal beneath these leguminous trees could be playing an important role rather than establishment.

In summary, the evidence described above indicates that facilitation is a frequent (but not necessarily obligate) (see Houle 1997) interaction in the dune environments. The literature focused on facilitation in dunes has shown that they occur during early and late successional stages, as well as between different growth forms: annuals and perennials, and even between different tree species.

13.3 Competition

Competition is the mutually adverse effect of organisms on one another, because they are striving for a common resource. Although the literature on competition between plants is vast, relatively few studies have focused on dune environments. In general, competition increases as the system becomes naturally covered by vegetation (Table 13.1) and also as a result of human management. Not all authors agree with this statement. For instance, Barros-Henriques and Hay (1992) have argued that because resources are limited during early succession, competition is expected to occur during these stages, as niche overlap is high, while Lichter (2000) concluded that dune succession is best explained by differential colonization and competitive abilities of plants.

Some general trends observed as a result of competition are: decreased plant weight, cover and growth of individual species (Mack and Harper 1970; Silander and Antonovics 1982; Pemadasa and Lovell 1974); and restrictions in the distribution of species such as *Cakile* spp. due to competitive exclusion by *Ammophila arenaria* (Barbour 1970). It has also been observed that competition for light occurs in late successional stages (Olff et al. 1993). Furthermore, in early successional stages, the initial colonizers are gradually replaced by alleged competitive interactions with late colonizers (Silander and Antonovics 1982; Martínez et al. 2001). Probably, competition for light is also important in this case.

Specific and well-known examples of competition between plants growing on coastal dunes are grass encroachment and the invasion of exotic plants. In both cases it is argued that species are replaced by competition (Kooijman et al. 1998).

13.3.1 Grass Encroachment

The encroachment by grasses occurs when aggressive and competitive grasses (e.g. *Elytrichia atherica*, *Calamagrostis epigejos*, *Ammophila arenaria*, *Schizachyrium scoparium*) (Kooijman and Haan 1995; Martínez et al. 2001) spread on dune areas, leading to the dominance of a few tall grass species over formerly species-rich communities. This process has been observed under varying circumstances: atmospheric Nitrogen deposition (beneficial for the grasses) or a reduced disturbance regime due to a decreased recreational use or a few or no herbivores (such as rabbits) that initiate small-scale substrate mobility (Kooijman and Haan 1995; Martínez et al. 2001). In particular, grass encroachment is considered problematic on the coasts of The Netherlands, Belgium, Germany and certain parts of France, because of drastic reductions in diversity due to the dominance of grasses (Fig. 13.1; Veer 1997). In The Netherlands, habitat management techniques such as mowing or grazing have been used to re-initiate substrate mobility, which stimulates the growth and development of locally lost pioneer vegetation, and combats grass encroachment in the dune areas (Jungerius et al. 1995). However, caution must be taken before destabilizing dunes for management purposes, since this can lead to a mobile dune field with serious erosive problems.

The paucity of evidence for grass encroachment in tropical dunes is due to the lack of studies in these latitudes. In an ongoing 10-year-long study at La

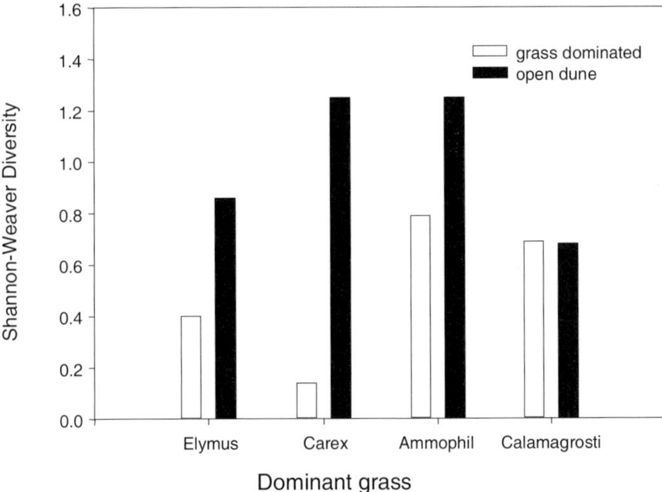

Fig. 13.1. Decreased species diversity in sites dominated by different grass species in The Netherlands. (Modified from Veer 1997)

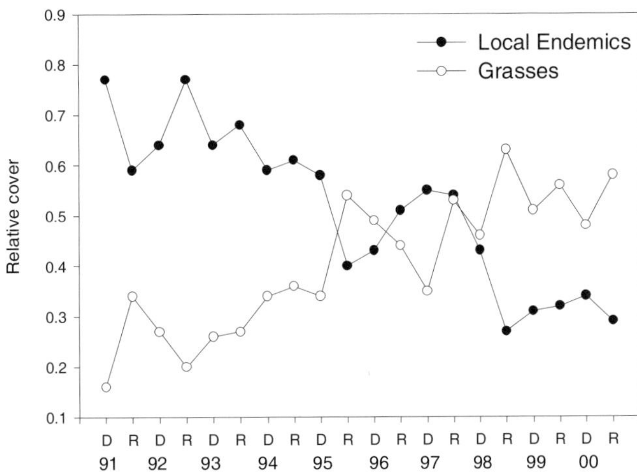

Fig. 13.2. Decreased relative cover of local endemics as grass cover increases in Mexican sand dunes during a ten year period (1991–2000). Plant cover was registered every 6 months during the dry (*D*) and rainy (*R*) periods. (Modified from Martínez et al. 2001)

Mancha, tall grasses (*Schizachyrium scoparium* and *Trachypogon plumosus*) have been gradually increasing their cover, while the Mexican endemics (*Chamaecrista chamaecristoides* and *Palafoxia lindenii*) are becoming locally extinct (Fig. 13.2; Martínez et al. 2001). This extremely rapid stabilization and species replacement has probably been enhanced by the removal of cattle since 1988 (M.L. Martínez, unpubl. data) when the dune system became a nature reserve under protection. Interestingly, the rabbit population has also started to increase, and appears to gradually reverse this situation by reducing the density of the dominant grasses. It has been quantified that in the 480-m^2 dune area monitored during 10 years the maximum plant cover of the dominant grass *Schizachyrium scoparium* was 60 % before rabbits were observed, and decreased to 45 % 2 years after these herbivores had increased their population noticeably (M.L. Martínez, unpubl. data), resulting in larger areas of exposed sand and rendering the sampled area more mobile. Slowly, local endemics started to increase their cover since the dominant grasses became less abundant and apparently their competitive impact had decreased. In this sense, herbivores may greatly affect the end result of plant–plant interactions.

13.3.2 Invasive Plants

The introduction of exotic species to stabilize mobile sand was a common practice in the past. For instance, because of its high tolerance to burial, the European marram grass (*Ammophila arenaria*) was widely used for this purpose. In many occasions, the grass became naturalized, and spread widely, replacing the native sand binders, such as the endemic *Desmoschoenus spiralis* (New Zealand) and the native Tasmanian species *Festuca littoralis*, *Spinifex hirsutus* and *Acaena novae-zelandiae*. In the west coast of North America it replaced the native *Ammophila breviligulata* (van der Maarel 1993a, b).

Conifers have also been introduced to stabilize dune systems in Europe. The large-scale introduction of pines to dune systems has resulted in the shading out of most of the dune flora, including rare species, and the lowering of the water table through a combination of an increased transpiration and artificial drainage. Some additional woody invasive species that eliminate native and endangered species through competition are: *Hippophae rhamnoides*, *Acer pseudoplatanus*, *Lupinus arboreus*, *Pinus* spp., *Rhododendron ponticum* (van der Maarel 1993a). In The Netherlands, common Reed (*Phragmites australis*) overgrows native and characteristic slack species such as the tussock forming sedge (*Schoenus nigricans*) and the grass (*Parnassia palustris*) resulting in the local extinction of such species and thus, in biodiversity loss (Grootjans et al. 1991).

Recent studies have demonstrated that the introduction of alien and exotic species is not necessarily needed for stabilization, since indigenous dune species can be used successfully to stabilize the shifting substrate (Carter 1991). Thus, the current new trends in management policies prevent further unwanted species invasions and employ an ecological approach using the available indigenous species.

Because of the frequent disturbance events, coastal dunes are also prone to natural invasions by species from other communities. Castillo and Moreno-Casasola (1996) found that throughout the coastal dunes located along the Gulf of Mexico, 71 species (10.83 %) were considered coastal, while the remaining 89.8 % (573 plant species) were either ruderal/secondary species or from other nearby communities. These natural invasions potentially lead to an increased biological diversity of the dune communities. However, when the invaders exert a negative impact on the community they displace native species by competing for resources, interfering with successional processes, altering disturbance regimes and disrupting food chains. Thus, dune conservation should aim at preserving the natural dynamics of these systems, where both coastal and non-coastal species coexist, maintaining species diversity.

13.4 Epiphytes

13.4.1 Non-Parasites

Approximately 23,000 of all plant species are epiphytic. They are taxonomically widespread among vascular plants (Kress 1989), although more than half of them are orchids, ferns, bromeliads, or cacti (Benzing 1990). Typically, the ecological interaction between non-parasitic epiphytes (named epiphytes hereafter) and their host woody plants has been described as commensalism. Epiphytes obtain support and adequate environmental conditions without a metabolic relationship, while host plants (phorophytes) are neither benefited nor damaged (Lütte 1989). Nevertheless, in some cases a high epiphyte infestation results in competition for light and nutrients. It also generates high mechanical pressure that breaks the branches and stems, creating microenvironmental conditions that could promote the establishment of host pathogens (Benzing 1990).

Epiphytic plants are distributed worldwide, although vascular epiphytes are more abundant in tropical or sub-tropical subtropical latitudes. There are only a few reports of epiphytes for coastal dune systems. Australia has only one epiphytic species recorded, while six vascular epiphytes and one non-vascular epiphyte have been reported for Europe (Table 13.4). The American Continent is the best-known region, with 32 vascular epiphyte species recognized from different coastal ecosystems (Table 13.4). In terms of epiphytes, La Mancha, Mexico, is the best-studied place.

Because epiphytes are dependent on phorophytes, an advanced successional stage with woody plants established is critical for epiphyte colonization in the dune systems. According to García-Franco (1996), these host trees and shrubs should first have growth of heavy branches, a clumped distribution, and a relatively long time of exposure so that the seeds from the epiphytes can be dispersed onto these branches. Secondly, epiphyte seed dispersal should occur during the season when seeds can be better dispersed by wind (winter season for bromeliads, orchids, and ferns; García-Franco 1996) or by birds (dry season for aroids and hemiparasites; López de Buen and Ornelas 2001).

And thirdly, the distance between the sources of epiphyte seeds and the dune system should be such (ca. 50 m) that successful establishment of phorophytes is favored while individuals are dispersed (García-Franco 1996).

In addition, environmental harshness, (drought, light intensity, strong wind speed, nutrient limitation and constant salt spray) can limit epiphyte distribution and diversity (García-Franco 1996). For example, vascular epiphytes, such as *Tillandsia concolor, Tillandsia ionantha,* and *Tillandsia paucifolia,* considered as full or nearly full sun-plants and xerophytes (sensu Benzing 1990), endure drought and are able to establish in the young forests and shrubs at La Mancha, Mexico. In contrast, other species such as *Brassavola nodosa, T. usneoides,* and *T. schiedeana* resemble full sun and xerophytic plants, but they are less tolerant to the dune environment and grow in the tropical rain forest established on stabilized dunes at the same dune system (García-Franco 1996).

Table 13.4. Epiphytes and parasites species recorded on coastal dune systems of different regions of the World. *p* Parasite, *e* epiphyte. (Survey from Davy and Costa 1992; Dillenburg et al. 1992; van der Maarel 1993a, b; Rhoades 1995; García-Franco 1996)

Geographic location	Family	Species
Africa (Kenya, W. Africa, Senegal)	Scrophulariaceae	*Striga gesneroides* (p)
America (Brazil, Cuba, Galapagos,	Araceae	*Anthurium scandens* (e)
Mexico, USA		*Philodendron bipinnatifidum* (e)
	Bromeliaceae	*Aechmea bracteata* (e)[a]
		A. recurvata (e)
		Hohembergia peduncularis (e)
		Tillandsia aëranthos (e)
		T. concolor (e)[a,b]
		T. ionantha (e)[a,b]
		T. paucifolia (e)[a,b]
		T. recurvata (e)[a]
		T. schiedeana (e)[a,b]
		T. strepthophylla (e)[a]
		T. usneoides (e)[a,b]
		T. utricularia (e)[a]
		T. xerographica (e)
	Cactaceae	*Hylocereus undatus* (e)[a]
		Lepismium cruciforme (e)
		Rhipsalis baccifera (e)
		Selenicereus testudo (e)[a]
	Cuscutaceae	*Cuscuta acuta* (p)
	Cytinaceae	*Bdallophyton americanum* (p)[a]
	Lauraceae	*Cassytha filiformis* (p)
	Moraceae	*Ficus aurea* (e)
	Loranthaceae	*Psittacanthus caliculatus* (p)[a]
		P. shiedeanus (p)[a]
		Struthanthus venetus (p)

Epiphytes living on relatively ephemeral supports (small branches and stems) frequently fall to the ground due to the wind action (Matelson et al. 1993). This is particularly important in coastal zones, where constant seashore winds play a key role in the community dynamics. At La Mancha, Mexico, strong winds occurring during hurricanes and winter storms (both with wind speeds potentially higher than 80 km/h) affect epiphytes. Frequent dramatic population changes could occur when reproductive individuals and juvenile stages growing on the exterior upper branches fall to the ground by the wind action. For example, studies by García-Franco and Martínez (unpubl. data) showed that when strong northerly winds occur, the smallest individuals of *Tillandsia ionantha, T. concolor* and *T. balbisiana* are the ones that most frequently fall to the ground (Fig. 13.3). Probably, the loss of the smallest individuals has a demographic impact since recruitment possibilities might be decreased as well.

Geographic location	Family	Species
	Orchidaceae	*Brassavola cucullata* (e)[a,b]
		B. nodosa (e)[a,b]
		B. tuberculata (e)
		Cattleya intermedia (e)
		Myrmecophyla tibicinis (e)[a,b]
		Oncidium cebolleta (e)[a,b]
		Oncidium pumilium (e)[b]
	Polypodiaceae	*Polypodium aureum* (e)
		Polypodium polypodioides (e)[a]
		Microgramma vacciniifolia (e)
		Microgramma sp. (e)[a]
		Vittaria lineata (e)
	Viscaceae	*Phoradendron tamaulipensis* (p)[a]
Asia (Japan, New Guinea)	Cuscutaceae	*Cistanche helipaea* (p)
	Lauraceae	*Cassytha fifliformis* (p)
Europe (Croatia, Yugoslavia)	Moraceae	*Ficus casica* spp. *caprificus* (e)
		Parietaria lusitanica (e)
	Polypodiaceae	*Polypodium australe* (e)
		Sedum telephium spp. *maximum* (e)
		Selaginella helvetica (e)
		Umbilicus pendulinus (e)
	Non-vascular (liverworts)	*Riccia* spec. *Plur.* (e)
Oceania (Australia, New Zealand)	Lauraceae	*Cassytha filiformis* (p)
	Polypodiaceae	*Pyrrosia serpens* (e)
		Amysema preissii

[a] Species recorded for CICOLMA
[b] Species full sun and xerophyte

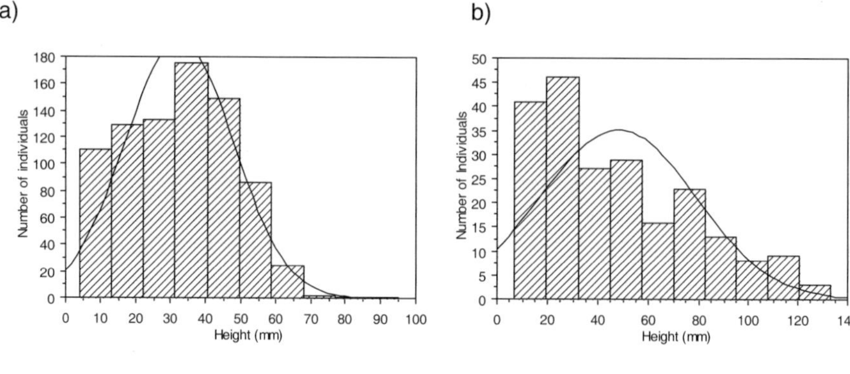

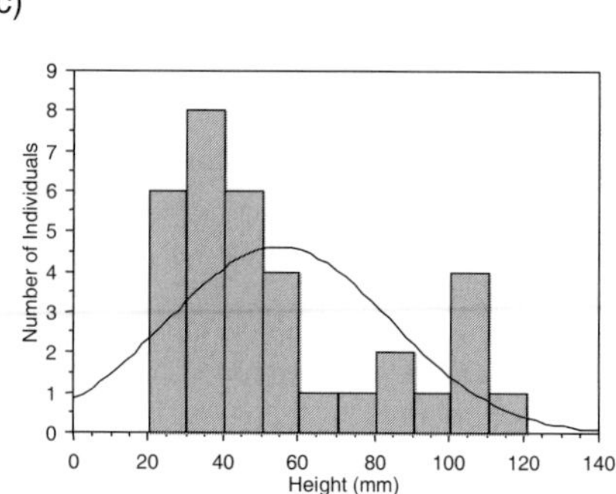

Fig. 13.3. Frequency of bromeliads fallen to the ground as a consequence of strong winds (80 km/h) occurring at La Mancha in 2000. Data for: **a** *Tillandsia ionantha*, **b** *Tillandsia concolor*, **c** *Tillandsia balbisiana*

Finally, epiphytes may have an indirect impact on the coastal dune dynamics. Potentially, a medium to heavy epiphyte infestation on the long-lived shrubs established since earlier successional stages (e.g. *Randia laetevirens* at La Mancha) could help reduce wind speed and then sand carry-over as well, promoting dune stabilization and plant colonization.

13.4.2 Parasites

A special case of epiphytism are the parasitic interactions (Benzing 1990), where there is physiological communication between the plants by means of the haustorium, resulting in a negative impact on the host while the parasite benefits from the interaction by drawing water and nutrients (Kuijt 1969). According to the level of host dependence, parasitic angiosperms have been divided into hemiparasites (photosynthetic obligate and facultative parasites), and holoparasites (non-photosynthetic obligate parasites), and they are further divided into stem and root parasites (Stewart and Press 1990).

Approximately 1 % of all known plant species are parasites (Norton and Carpenter 1998). Parasitic angiosperms are distributed worldwide, and all plant communities contain at least one parasitic species interacting with one or several hosts (Stewart and Press 1990). Despite their wide distribution, there are few parasite records on the coastal dune ecosystems and the ecological studies are fewer. One parasite species has been reported for Africa and Oceania, and two more for Asia (Table 13.2). The American continent has seven parasitic species recorded, most of them from the coastal dunes of La Mancha, Mexico (Table 13.2).

The distribution of parasites is obviously host dependent, but the environmental conditions and the parasite's environmental range also limit their distribution (García-Franco and Rico-Gray 1996). Therefore, parasite establishment on coastal dune systems appears to be restricted to advanced successional stages, and, probably, depends on the presence of some specific susceptible woody plant species. In general, stem hemiparasites as mistletoes are bird dispersed (Howe and Smallwood 1982; van der Pijl 1982; López de Buen and Ornelas 2001) and root holoparasites are primarily mammal dispersed (Kuijt 1969). For example, in the coastal dune system of La Mancha, Mexico, *Phorandendron tamaulipensis*, stem hemiparasite, shows strong preference for *Randia laetevirens* growing on semi-mobile dunes. The overlapping fruiting periods between the host and the parasite coupled with the behavior of *Minus polyglottos* (Passeriforme: Mimidae) which perches and eats fruits from both plant species, enables the seeds of the parasites to be dispersed and deposited on the branches of new hosts (García-Franco 1996). In contrast, in the tropical rain forest developing on stabilized sand dunes at La Mancha, *Bdallophyton americanum* (root holoparasite) is dispersed by mice and ants (García-Franco and Rico-Gray 1997). This parasite always grows on the superficial fine roots (first 10 cm depth and 0.55 cm diameter) of *Bursera simaruba* that develop during the rainy season (García-Franco and Rico-Gray 1996). Dispersers such as mice and ants deposit seeds on the ground close to the host's roots. In general, the seeds from parasitic plants are chemotropic, which means that germination occurs in response to a chemical gradient through the contact with the host's epidermis (Stewart and Press 1990; Kuijt 1969). The environmental clues that trigger the germination of the parasite seeds still remain largely unknown.

13.5 Conclusions

Different plant–plant interactions occur throughout the different successional stages of the dune environment. It was generally accepted that early succession was controlled by environmental factors (i.e. soil nutrient status) and positive interactions, while late succession was controlled by vegetation processes such as competition (Lammerts et al. 1999). Contrary to this hypothesis, our observations and the literature indicate that the interactions mentioned in this review are not restricted to a certain successional stage. For example, facilitation and competition occur during both early and late stages. Epiphytes and parasites are more abundant during advanced successional stages, when woody vegetation is present, although they can also be found in earlier stages and relatively more mobile sites of the dune environment, as long as hosts (woody vegetation) are available. In all cases, plant–plant interactions play a key role in community dynamics, in terms of substrate stabilization and species successional replacement. In many occasions, plant–plant interactions are greatly affected by the activities of animals. Herbivory can reduce plant–plant competition while epiphytes and parasites frequently depend on their dispersal vectors (often animals) in order to colonize a new host.

The importance of plant–plant interactions adds a new dimension to the proper approach to coastal dune conservation and restoration. Frequently, restoration projects aim at the reintroduction of previously existing species. Research to date indicates that interactions and processes should also be restored.

Acknowledgements. The work that generated the information presented in this revision received financial support to MLM from CONACyT (1841P-N) and Instituto de Ecología, A.C. (902-16 and 902-17-516). We are thankful to Dr. R. Callaway and Dr. P. Hietz for their comments which greatly improved the manuscript. The text was elaborated during a sabbatical stay of JGGF at Florida International University (USA), (CONACyT C000/C300/0308).

References

Alpert P, Mooney HA (1996) Resource heterogeneity generated by shrubs and topography on coastal sand dunes. Vegetatio 122:83–93

Barbour MG (1970) Seedling ecology of *Cakile maritima* along the California coast. Bull Torrey Bot Club 97:280–289

Barros-Henriques RP, Hay JD (1992) Nutrient content and the structure of a plant community on a tropical beach-dune system in Brazil. Acta Oecol 13:101–117

Benzing DH (1990) Vascular epiphytes. Cambridge University Press, Cambridge

Bertness MD, Callaway RM (1994) Positive interactions in communities. Tree 9:191–193

Callaway RM (1995) Positive interactions among plants. Bot Rev 61:306–349

Carter RWG (1991) Coastal environments. Academic Press, London

Castillo SA, Moreno-Casasola P (1996) Coastal sand dune vegetation: an extreme case of species invasion. J Coastal Conserv 2:13–22

Cowles HC (1899) The ecological relations of the vegetation on the sand dunes of Lake Michigan. Bot Gaz 27:95–117. In: Real LA, Brown JH (eds) Foundations of ecology (1991). University of Chicago Press, Chicago, pp 28–58

Davy AJ, Costa CSB (1992) Development and organization of Saltmarsh communities. In: Seeliger U (ed) Coastal plant communities of Latin America, Academic Press, San Diego, pp 157–178

De Jong TJ, Klinkhamer PGL (1988) Seedling establishment of the biennials *Cirsium vulgare* and *Cynoglossum officinale* in a sand dune area: the importance of water for differential survival and growth. J Ecol 76:393–402

Dillenburg LR, Waechter JL, Porto ML (1992) Species composition and structure of a sandy coastal plain forest in northern Rio Grande do Sul, Brazil. In: Seeliger U (ed) Coastal plant communities of Latin America. Academic Press, San Diego, pp 349–366

Ehrenfeld JG (1990) Dynamics and processes of Barrier Island vegetation. Aquat Sci 2:437–480

García-Franco JG (1996) Distribución de especies epífitas en matorrales costeros de Veracruz, México. Acta Bot Mex 37:1–10

García-Franco JG, Rico-Gray V (1996) Distribution and host specificity in the holoparasite *Bdallophyton bambusarum* (Rafflesiaceae) in a tropical deciduous forest in Veracruz, Mexico. Biotropica 28:759–762

García-Franco JG, Rico-Gray V (1997) Dispersión, viabilidad, germinación y banco de semillas de *Bdallophyton bambusarum* (Rafflesiaceae) en la costa de Veracruz, México. Rev Biol Trop 44–45:87–94

Grootjans AP, Hartog PS, Fresco LFM, Esslink H (1991) Succession and fluctuation in a wet dune slack in relation to hydrological changes. J Veg Sci 2:545–554

Houle G, (1997) No evidence for interspecific interactions between plants in the first stage of succession on coastal dunes in subartic Quebec, Canada. Can J Bot 75:902–915

Howe HF, Smallwood J (1982) Ecology of seed dispersal. Annu Rev Ecol Evol 13:201–228

Jungerius PD, Koehler H, Kooijman AM, Mücher HJ, Graefe U (1995) Response of vegetation and soil ecosystem to mowing and sod removal in the coastal dunes "Zwanenwater", The Netherlands. J Coastal Conserv 1:3–16

Kellman M, Kading M (1992) Facilitation of tree seedling establishment in a sand dune succession. J Veg Sci 3:679–688

Kooijman AM, Haan MWA de (1995) Grazing as a measure against grass encroachment in Dutch dry dune grasslands: effects on vegetation and soil. J Coastal Conserv 1:127–134

Kooijman AM, Dopheide JCR, Sevink J, Takken I, Verstraten JM (1998) Nutrient limitations and their implications on the effects of atmospheric deposition in coastal dunes; lime-poor and lime-rich sites in The Netherlands. J Ecol 86:511–526

Kress WJ (1989) The systematic distribution of vascular Epiphytes. In: Lüttge U (ed) Vascular plants as epiphytes. Evolution and ecophysiology. Ecological Studies 76. Springer, Berlin Heidelberg New York, pp 234–261

Kuijt J (1969) The biology of parasitic flowering plants. University of California Press, Berkeley

Lammerts EJ, Pegtel DM, Grootjans AP, van der Veen A (1999) Nutrient limitation and vegetation changes in a coastal dune slack. J Veg Sci 10:111–122

Lichter J (2000) Colonization constraints during primary succession on coastal Lake Michigan sand dunes. J Ecol 88:825–839

López de Buen L, Ornelas JF (2001) Seed dispersal of the mistletoe *Psittacanthus schiedeanus* by birds in Central Veracruz, Mexico. Biotropica 33:487–494

Lüttge U (1989) Vascular epiphytes: Setting the scene. The evolution of Epiphytism. In: Lüttge U (ed) Vascular plants as epiphytes. Evolution and ecophysiology. Ecological Studies 76. Springer, Berlin Heidelberg New York, pp 1–14

Mack RN, Harper JL (1977) Interference in dune annuals: spatial pattern and neighbourhood effects. J Ecol 65:345–363

Martínez ML (2003) Facilitation of seedling establishment by an endemic shrub in tropical coastal sand dunes. Plant Ecol 168(2):333-345

Martínez ML, Vázquez G, Sánchez-Colón S (2001) Spatial and temporal dynamics during primary succession on tropical coastal sand dunes. J Veg Sci 12:361–372

Matelson TJ, Nadkarni NM, Longino JT (1993) Longevity of fallen epiphytes in a neotropical montane forest. Ecology 74:265–269

McLeod KW, Murphy PG (1983) Factors affecting growth of *Ptelea trifoliata* seedlings. Can J Bot 61:2410–2415

Muñiz MA (2001) Reclutamiento de especies de árboles y arbustos característicos de selva bajo la copa de *Diphysa robinoides* (Leguminosae) en dunas costeras estabilizadas en La Mancha, Ver. In: Martínez ML, López-Portillo J, Cervantes L, García-Franco JG (eds) Memorias del Curso de Ecología de Campo 2001. Posgrado en Ecologia y Manejo de Recursos, Instituto de Ecología, AC Xalapa

Norton DA, Carpenter A (1998) Mistletoes as parasites: host specificity and speciation. Tree 13(3):101–105

Olff H, Huisman J, van Tooren BF (1993) Species dynamics and nutrient accumulation during early primary succession in coastal sand dunes. J Ecol 81:693–706

Pemadasa MA, Lovell PH (1974) Interference in populations of some dune annuals. J Ecol 62:855–868

Picket STA (1980) Non-equilibrium coexistence of plants. Bull Torrey Bot Club 107:238–248

Rhoades FM (1995) Nonvascular epiphytes in forest canopies: Worldwide distribution, abundance, and ecological roles. In: Lowman M, Nadkarni NM (eds) Forest canopies. Academic Press, San Diego, pp 353–408

Shumway SW (2000) Facilitative effects of a sand dune shrub on species growing beneath the shrub canopy. Oecologia 124:138–148

Silander JA, Antonovics J (1982) Analysis of interspecific interactions in a coastal plant community – a perturbation approach. Nature 298:557–260

Stewart GR, Press MC (1990) The physiology and biochemistry of parasitic angiosperms. Annu Rev Plant Physiol Mol Biol 41:127–151

van der Maarel E (1993a) Dry coastal ecosystems: polar regions and Europe. Elsevier, Amsterdam

van der Maarel E (1993b) Dry coastal ecosystems: Africa, America, Asia and Oceania. Elsevier, Amsterdam

van der Pjil L (1982) Principles of dispersal in higher plants. Springer, Berlin Heidelberg New York

Vázquez G, Moreno-Casasola P, Barrera O (1998) Interaction between algae and seed germination in tropical dune slack species: a facilitation process. Aquat Bot 60:409–416

Veer MAC (1997) Nitrogen availability in relation to vegetation changes resulting from grass encroachment in Dutch dry dunes. J Coastal Conserv 3:41–48

Yarranton GA, Morrison RG (1974) Spatial dynamics of a primary succession: nucleation. J Ecol 62:417–428

14 Ant–Plant Interactions: Their Seasonal Variation and Effects on Plant Fitness

V. Rico-Gray, P.S. Oliveira, V. Parra-Tabla, M. Cuautle, and C. Díaz-Castelazo

14.1 Importance of Interspecific Interactions

Interactions between species have the potential to influence many evolutionary processes including patterns of adaptation, genetic variation, community organization, and the stability of species (Bondini and Giavelli 1989; Rico-Gray 2001). Similar to species, interspecific interactions can evolve and multiply, forming links between species that affect their evolutionary trajectories through time (Thompson 1999). Organisms in nature are not isolated, and to survive and reproduce, have adapted combinations of their own genetic information and that of other species in the process of coevolution (Thompson 1999). Furthermore, the effect of interspecific interactions may encompass more than two species, e.g., via top-down and bottom-up forces in a community (Dyer and Letourneau 1999).

Interspecific interactions change in space and time and are based on cost/benefit systems, so a continuum from antagonism to mutualism should be expected (Thompson 1994; Bronstein 2001). A species may be antagonistic in one stage of its life cycle while mutualistic in another; whereas a population or species may be antagonistic in one portion of its distribution or habitat while another population of the same species may be mutualistic in another portion of its distribution (Puterbaugh 1998). Interspecific interactions can be defined on the basis of whether the net effect or outcome of the interaction is an increase or decrease in fitness, or no effect (neutral) for each interacting species; thus, basically two types of interactions can be considered: antagonistic and mutualistic (Rico-Gray 2001).

This chapter discusses the importance of nectar to ants, the effect of ants on plant fitness (either mediated by nectaries or Homoptera), plus seasonal variation and diversity of interactions in the tropical coastal regions of the Yucatan Peninsula and Veracruz, Mexico. Tropical coastal dunes are rich envi-

Ecological Studies, Vol. 171
M.L. Martínez, N.P. Psuty (Eds.)
Coastal Dunes, Ecology and Conservation

ronments for ant–plant interactions, yet diversity of interactions, seasonality and the effect of ants on plant fitness have rarely been studied.

14.2 Richness and Seasonal Variation of Ant–Plant Interactions

Interactions of species vary spatially and seasonally, and should be analyzed using a landscape approach (Bronstein 1995; Ortiz-Pulido and Rico-Gray 2000). Ant–plant interactions vary in their probability of occurrence along environmental gradients and under different disturbance regimes (Koptur 1992; Rico-Gray et al. 1998). The pattern of interactions in different ecological conditions (Cushman and Addicott 1991) and between habitats (Barton 1986), also exhibits significant temporal variation (Alonso 1998). The structure of ant communities and of ant–plant interactions has been studied in a variety of habitats, assessing that neither the spatial nor the temporal dimensions can be ignored (Herbers 1989). Ant assemblages are very dynamic and extrapolating superficially similar characteristics from one ant community to another may lead to erroneous inferences, precluding broad generalizations (Herbers 1989; Feener and Schupp 1998).

Ant–plant associations in tropical sand dunes are abundant (e.g., over 350 specific associations have been recorded in the coastal vegetation of central Veracruz, Mexico (Rico-Gray 1993; C. Díaz-Castelazo and V. Rico-Gray, unpubl. data) relative to temperate semiarid or humid mountain sites (Rico-Gray et al. 1998)). Studies at two coastal sites in Mexico [La Mancha in Veracruz (Rico-Gray 1993; Rico-Gray and Castro 1996; Rico-Gray et al. 1998;

Table 14.1. Results from the generalized linear models fitted to the number of ant–plant interactions, minimum temperature and precipitation per month data curves. (Modified from Rico-Gray et al. 1998)

Source	χ^2	χ^2	df[a]	Explanation of variation (%)		Probability	
	LM[b]	SB		LM	SB	LM	SB
Temperature	9.713	12.33	1	32.38	17.98	<0.005	<0.001
Precipitation	0.0034	2.605	1	-	0	NS	NS
Interaction	13.91	0.328	1	46.38	0	<0.001	NS
Residual	6.3678	53.31	8				
Total	29.991	68.57	11	78.76	17.98		

[a] Degrees of freedom

[b] LM, La Mancha, Veracruz (sand dune scrub, tropical dry and deciduous forests, mangroves); SB, San Benito, Yucatan (sand dune scrub, mangroves

Oliveira et al. 1999), and San Benito in Yucatan (Rico-Gray 1989; Rico-Gray and Thien 1989a, b; Rico-Gray et al. 1989, 1998)] have shown significant within-habitat seasonal variation, as well as considerable variation among habitats in the number, diversity, and seasonal distribution of ant–plant interactions. The observations suggest that inter-habitat variation of ant–plant interactions is the effect of variation in environmental parameters (Table 14.1), e.g., the richness of plants with nectaries in the vegetation (C. Díaz-Castelazo and V. Rico-Gray, unpubl. data) and the richness in habitat heterogeneity. Thus, the diversity of the vegetation determines the nature of the ant community to a certain extent, and the diversity of the vegetation is driven by the abiotic environment (Rico-Gray et al. 1998).

Seasonal variation of ant–plant interactions is illustrated using research conducted at La Mancha, Veracruz, Mexico (Fig. 14.1). The number of ant associations with extrafloral nectaries (efns) and nectaries located in the reproductive structures (floral or circum-floral) of plants exhibited seasonal changes throughout the year (Rico-Gray 1993; Rico-Gray et al. 1998; Oliveira et al. 1999). Ant/efn associations increased significantly during the wet season (Spearman, r=0.58, $P<0.001$), and were constant throughout the year in comparison with other plant resources. Most plant species in tropical dry forests produce a flush of new leaves at the onset of the rainy season (Bullock and

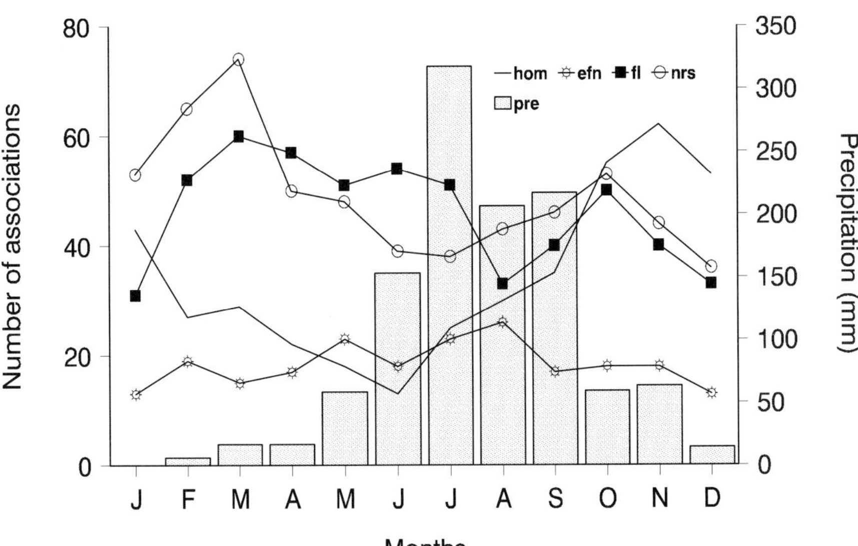

Fig. 14.1. Number of ant–plant associations registered per month per food resource for the tropical the coastal vegetation at La Mancha, Veracruz, Mexico. *hom* Homopteran honeydew; *efn* nectar from extrafloral nectaries; *fl* floral nectar; *nrs* nectar from reproductive structures; *pre* monthly precipitation. (Modified from Rico-Gray 1993)

Solís-Magallanes 1990). Extrafloral nectaries are associated with leaves and many studies have shown that secretion of extrafloral nectar is greatest during periods of rapid vegetative growth, e.g., the expansion of new leaves. Ant presence is highly correlated with peaks of nectar flow, and an increase of ant/efn associations would therefore be expected as new leaves appear. Ant associations with reproductive structures (Spearman, $r=-0.59$, $P<0.001$) and flowers (Spearman, $r=-0.76$, $P<0.001$) peaked during the dry season, and decreased during the wet season.

There are two main flowering peaks in the dry tropical lowlands of Middle America, one in mid-dry season and one at the start or during the wet season (Bullock and Solís-Magallanes 1990; Castillo and Carabias 1982). These major periods of flowering are supplemented by erratic flowering of many species year-around, presenting ants with year-round sources of liquids and energy. Ant associations with honeydew-producing Homoptera increased significantly after the start of the rainy season (Spearman, $r=0.70$, $P<0.001$) and decreased abruptly once the dry season began. During the warm-humid months plants produce new soft vegetative tissues, creating ideal feeding conditions for Homoptera (Cuautle et al. 1999). Interestingly, there was a significant negative association (Spearman, $r=-0.66$, $P<0.001$) between the number of ant visits to flowers and to Homoptera. Ant/Homoptera associations decrease sharply during the dry season, while ant–flower associations peak at this time. This complementary pattern may reflect the use of alternative resources with similar nutritional value, as 62.5 % of the ant species using floral nectar also foraged for honeydew. Ants have been shown to prefer and select sugar solutions containing a complex mixture of amino acids to sugar-only solutions (Lanza 1988; Smith et al. 1990; Völkl et al. 1999), and they also discriminate between poor and rich homopteran honeydew, preferring honeydew rich in trisaccharides and with higher total sugar concentration (Völkl et al. 1999).

Finally, the simultaneous increase in the number of ant/efn and ant/Homoptera associations, and the decrease in the number of ant/circum-floral and ant/flower associations during the wet season, may reflect a decrease in nectar production. It is more likely, however, that ants are switching to alternative food sources (efns, honeydew-producing Homoptera). Ants could also be feeding on a variety of insect prey during the wet season, when insects exhibit their peak activity in lowland tropical seasonal vegetation (Smythe 1982; Rico-Gray 1989; Rico-Gray and Sternberg 1991). In summary, because more food resources are available during the wet season, ants are able to diversify their foraging activity at that time. During the dry season ants concentrate on available limited resources.

14.3 Importance of Nectar to Ants in Tropical Seasonal Environments

Ants frequently visit flowers (Fig. 14.2) and other reproductive parts (e.g., buds, inflorescence spikes, fruits) (Fig. 14.3) of plants in a variety of environments (Rico-Gray and Thien 1989a; Puterbaugh 1998), especially in the lowland dry tropics. Ants are typically considered to be robbers of floral nectar that decrease plant fitness (e.g., McDade and Kinsman 1980; Norment 1988). Some evolutionary trends in floral morphology associated with a decrease in the range of effective pollinators, have also been thought to increase plant adaptedness by excluding non-pollinating nectarivores, such as ants (e.g., Herrera et al. 1984). Nevertheless, a high number (up to 40 %) of plant species in tropical coastal habitats possess flowers visited by ants foraging for nectar (Rico-Gray 1980, 1989, 1993; García-Franco and Rico-Gray 1997; Rico-Gray et

Fig. 14.2. Ants foraging for floral nectar, clockwise from top left. *Crematogaster brevispinosa* and *Avicennia germinans*, *Camponotus* sp. and *Passiflora foetida*, *Camponotus sereceiventris* and *Bdallophyton bambusorum*, and *Camponotus planatus* and *Coccoloba uvifera*

Fig. 14.3. Examples of ants foraging for circum-floral nectar, clockwise from top. *Pseudomyrmex* sp. foraging on the calyx and petals of *Iresine celosia*, *Pseudomyrmex* sp. foraging on the calyx and floral peduncle of *Canavalia rosea*, and *Ectatomma tuberculatum* foraging on the fruit of *Myrmecophyla christinae*

al. 1998; Oliveira et al. 1999), which suggests ants play an important role in nectar consumption in these coastal habitats. If ants were merely robbing nectar, there should be a considerable decrease in fitness of many individual plants. Alternatively, a yet undiscovered mutualistic interaction may be occurring, besides possible pollination. Here we analyze the importance of nectar to ants in tropical coastal seasonal habitats.

Plants are defended against herbivory in many ways, e.g., by covering themselves with tough, spiny or inedible surfaces, sclerophylly, suffusing their tissues with chemical deterrents, toxins or digestibility-reducing compounds or by employing the services of animals like ants to ward off herbivores (Koptur 1991, 1992). There is a strong relationship between some of these characteristics and low plant nitrogen concentrations (Mattson 1980). Moreover, nitrogen deficiencies are usually accompanied by increased tissue toughness, reducing the digestibility of the plant material by increasing indigestible bulk and hydrogen bonding with carbohydrates and proteins (Chauvin and

Gueguen 1978: Mattson 1980; Coley 1983). Several of these attributes are common in plants living in dry environments and function as adaptations to cope with low moisture (Mattson 1980), making plants less palatable to herbivores, and reducing food sources during the dry season (Smythe 1982) resulting in few food alternatives for insects that prey on herbivores. Plant organs with high turnover rates (flowers, fruits, seeds) invariably contain higher nitrogen concentrations than more quiescent tissues (Mattson 1980; Koptur 1984). Thus, plant reproductive structures and associated nectar may be the most 'attractive' plant parts available as food during the dry season. When the wet season begins, the increasing moisture triggers prolific new vegetative growth that is both nitrogen-rich and succulent; consequently, there is an abundance of herbivores and other insects that prey on them, and diets may change with the availability of new food resources (Mattson 1980; Smythe 1982; Rico-Gray and Sternberg 1991).

Many plant species in the tropical dry seasonal vegetation along the Gulf of México (central portion of the state of Veracruz) and the Caribbean (Yucatan Peninsula) flower during the dry season. Flowering is not simultaneous, but there is always at least one species in flower from December through June (Rico-Gray 1989, 1993; Rico-Gray et al. 1998). Throughout the dry season ants forage for the nectar produced on the buds, flowers, and fruits of many species (Rico-Gray 1980, 1989, 1993; Rico-Gray et al. 1998), representing at least a third of the flora. The floral and circum-floral nectar produced by these structures is probably the major liquid-energy source for ants during the dry season, since there are few alternative food sources, such as insects or new vegetative growth (e.g., soft plant tissues are sometimes chewed by ants to extract the sap). Owing to food shortage during the dry season in tropical seasonal habitats, ants will rely on the nectar produced by flowers and other reproductive structures. Thus, nectar may play an important role not only in insect nutrition but in water balance as well. And, since plant reproductive organs are thus particularly vulnerable in the dry season, there should be high selection for defense by ants.

14.4 Effect of Ants on Plant Fitness

Insect herbivores may consume nearly all types of plant tissue, and the damage may occur at any stage of a plant's life cycle. However, since herbivore damage includes both vegetative and reproductive tissue, the impact of herbivory on plant fitness may vary with the type of tissue being consumed (Marquis 1992). Numerous plant traits are hypothesized to have evolved as a response to selection exerted by herbivores, including structural, chemical, physiological, and life history traits (Marquis 1992). One of such defense strategies involves mutualistic associations with ants, and many plant species produce domatia and/or food rewards to attract ants which in turn provide the plant with some protection against herbivores (Del Claro et al. 1996; De la Fuente and Marquis 1999).

Extrafloral nectaries are nectar-secreting organs found on virtually all above-ground plant parts not directly involved in pollination (Elias 1983; Koptur 1992) (Fig. 14.4). Plants bearing efns are distributed worldwide, and available evidence suggests that these glands are more common in tropical than in temperate environments (Coley and Aide 1991). Although efns attract a variety of nectar feeders (Koptur 1992; Pemberton and Lee 1996; Cuautle et al. 1999), ants are by far the most frequent visitors to efn-bearing plants both in temperate and tropical habitats (Oliveira and Brandão 1991). Many field

Fig. 14.4. Ants foraging for extrafloral nectar, clockwise from top. *Crematogaster brevispinosa* foraging for the nectar produced by nectaries located at the base of the leaf petiols of *Turnera ulmifolia* and a nectary located on the rachis of the foliole of a Leguminosae, and *Camponotus abdominalis* foraging for nectar produced by the foliar nectaries of *Inga vera*

experiments have demonstrated that ants visiting efn may increase plant fitness by deterring leaf herbivores (e.g., Koptur et al. 1998), bud or flower herbivores (e.g., Rico-Gray and Thien 1989a; Oliveira et al. 1999), and seed predators (e.g., Inouye and Taylor 1979; Keeler 1981). Some plants, however, receive no apparent benefit from ant visitation (O'Dowd and Catchpole 1983).

Ant/plant mutualisms mediated by efns are facultative and non-specialized, as indicated by the wide variety of associated ant visitors (Bronstein 1998). In fact, ant-derived benefits to efn-bearing plants can vary with factors such as time (Tilman 1978), habitat type (Barton 1986), aggressiveness of ant visitors (Oliveira et al. 1987; Rico-Gray and Thien 1989a), as well as the capacity of herbivores to circumvent ant predation (Heads and Lawton 1985; Koptur 1984; Freitas and Oliveira 1996). Research on plant defense by ants in tropical sand dunes is scarce and mainly restricted to four systems [*Myrmecophyla (Schomburgkia tibicinis) christinae*, *Paullinia fuscescens*, *Opuntia stricta*, and *Turnera ulmifolia*], which, however, represent a wide range of mutualistic interactions between ants and plants.

14.4.1 *Myrmecophyla* (*Schomburgkia tibicinis*) *christinae* (Orchidaceae)

Myrmecophyla christinae is a large epiphyte inhabiting the sand dune scrub in the state of Yucatan, Mexico (Fig. 14.5). Its large, hollow pseudo-bulbs, each with an opening at the base, are inhabited by ants. At least 13 ant species live in the hollow pseudo-bulbs, in old inflorescence spikes or in the soil directly beneath the plants. Only five ant species are common foragers on the inflorescence of the orchid (*Camponotus abdominalis*, *C. planatus*, *C. rectangularis*, *Crematogaster brevispinosa* and *Ectatomma tuberculatum*). Several ant species may occupy the same plant, but strong territoriality separates the species; two species never occur or nest in the same pseudo-bulb. Only one ant species will dominate all the inflorescence spikes of a given orchid plant, and its workers will forage day and night, throughout the reproductive season of the orchid (December–June) for nectar produced at the tip of the developing inflorescence spikes, the apex of the floral buds, the base of the floral pedicels and on the fruits. The main herbivore is *Stethobaris* sp. (Coleoptera: Curculionidae). This snout beetle (size ca. 2.5 mm) is present in relatively large numbers during early inflorescence development, decreasing through flowering. They are very active and bore holes at the tip of the growing inflorescence spike, on the buds and flowers. The beetle's most damaging effect is on the inflorescence spike. Beetle attack prior to bud differentiation kills the spike; if attack is during bud differentiation, it decreases flower number. Attack to the fruits is not as important, because beetle numbers decrease towards the end of April. The rare presence of larger beetles (size ca. 30 mm) my cause severe damage to a few fruits.

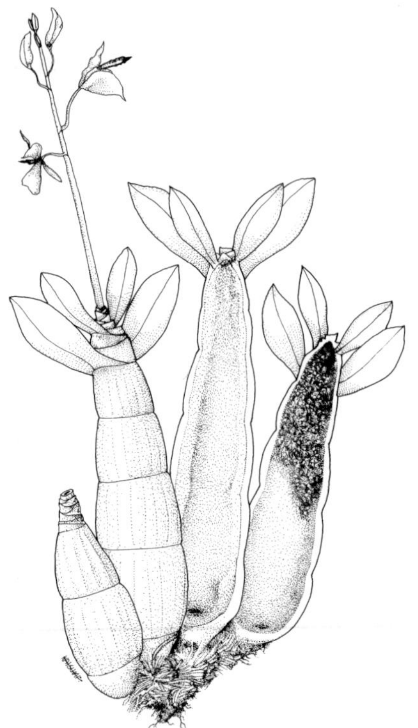

Fig. 14.5. A typical individual *Myrmecophyla (Schomburgkia tibicinis) christinae*. From *left* to *right*: external view of pseudo-bulb, pseudo-bulb with inflorescence, inside of pseudo-bulb without organic matter, inside of pseudo-bulb with organic matter (the last two show at the *bottom* the entrance hole for ants)

An ant exclusion experiment showed that orchids associated with the smallest ant species (*C. brevispinosa*) exhibited the highest number of dead spikes and produced the least number of fruits (Table 14.2). The possible protection offered by this ant species is negated because it herds the common citrus mealybug (*Planococcus citri*, Homoptera: Pseudococcidae) and by its small size relative to *Stethobaris* sp. (Rico-Gray and Thien 1989b). On the other hand, orchids associated with the largest ant species (*C. abdominalis, E. tuberculatum*) yielded significantly less dead inflorescence spikes, and produced significantly more flowers and fruits than control plants (Rico-Gray and Thien 1989a; Table 14.2). Furthermore, *C. brevispinosa* and the three *Camponotus* species pack some pseudo-bulbs with organic debris (dead ants and other insects, plant material, seeds) which are decomposed by bacteria and fungi, absorbed by the orchid, and utilized for growth and reproduction (Rico-Gray et al. 1989). It is clear that in this ant/plant interaction, some ant species are antagonistic in one stage of the life cycle of the orchid while mutualistic in another.

Table 14.2. Results from the ant exclusion experiment. (Modified from Rico-Gray and Thien 1989a, b)

Parameter/treatment	CTL[a]	CB	CP	CR	CA	ET
No. of inflorescences used	62	62	62	62	62	62
No. of dead inflorescences	25	40	16	12	0[b]	9[b]
Dead inflorescences (%)	40.3	64.5	25.3	19.3	0	15.4
No. of flowers produced	441	198	642[b]	547[b]	824[b]	742[b]
No. of fruits produced	9	2[b]	8	21	32[b]	33[b]
Ant size (mm)	-	2.5	3.5	5.0	8.0	10.5

[a] CTL, control; CB, *Crematogaster brevispinosa*; CP, *Camponotus planatus*; CR, *C. rectangularis*; CA, *C. abdominalis*; ET, *E. tuberculatum*

[b] Significantly different from control (χ^2 Yates corrected, $P<0.01$)

14.4.2 *Paullinia fuscescens* (Sapindaceae)

Paullinia fuscescens is a deciduous, nectarless, woody vine, often found in association with the shrub *Randia laetevirens*, whose foliage is renewed with the onset of rains in June. They inhabit the sand dune scrub at La Mancha, Veracruz, Mexico. The interaction between *P. fuscescens*, the ant *Camponotus planatus* (Hymenoptera: Formicidae), and an unidentified aphid species (Homoptera: Aphididae) occurs during bud and flower development. Inflorescence growth usually starts in September; flowers are present during October (a few flowers may be present as late as March). Fruiting begins in October, and seeds are present until February or March.

In 1990, we examined 4312 inflorescences. Flower production averaged 134.6 flowers per inflorescence (range 88 to 344); fruit production ranged from 0 to 42 fruits per inflorescence. *C. planatus*, a generalist ant using a wide range of food resources (Rico-Gray, 1993), is present in the area throughout the year. *C. planatus* foraged for nectar produced by extrafloral nectaries of neighboring and sometimes intertwined individuals of *Passiflora* sp. (Passifloraceae). When the latter stops flowering in late September, their efns ceased to produce nectar, and the ants forage on the honeydew of aphids feeding on the inflorescences of *P. fuscescens*. Aphids feed at the beginning of the most stressful time of the year (dry season and winter) on plant tissues (young inflorescences) that are rich in energy and nutrients.

A three-year ant–aphid exclusion experiment showed that the outcome of this interaction varied between years (Table 14.3; Rico-Gray and Castro 1996). The ant–aphid association significantly reduced average seed production per inflorescence in the first two years, whereas it had no effect in the third year. These results assess that full benefits to plants harboring ant-tended Homoptera are rarely demonstrated. The effect of treatment (with or without

Table 14.3. Mean number of seeds (±SE) per inflorescence and percent of inflorescences with seeds in a plot with ants and aphids (control) and a plot with ants excluded (experimental) in 1990, 1991, and 1993 (*n*number of inflorescences per treatment). (Rico-Gray and Castro 1996)

Response (RM-ANOVA)	Treatment	1990 (n=344)	1991 (n=344)	1993 (n=344)
No. seeds per inflorescence	Control	4.79aA[a] (±0.34)	2.99aB (±0.21)	2.98aB (±0.18)
	Experimental	12.01bA(±0.52)	3.08aB (±0.26)	5.55bC (±0.31)
Inflorescences (%) with seeds	Control	69.8 %aA	58.9 %aB	71.2 %aA
	Experimental	93.1 %bA	60.5 %aB	78.2 %bC

[a] Treatment means or percentages within a year followed by the same lowercase letter are not significantly different (P>0.05). Year means or percentages within a treatment followed by the same uppercase letter are not significantly different (P>0.05)

ants and aphids) on percentage of flowers producing seeds changed across years (G=48.92, df=2, P<0.0001). Furthermore, between-year variation in precipitation and temperature had an equal effect on the study plots (control, experimental), so within-year variation per plot was attributed to the effect of the ant–aphid association. Because of the importance of the conditional nature of interactions (Cushman and Addicott 1991; Bronstein 2001), the results suggest that (1) the presence of two potentially mutualistic species is not enough to generate mutualism; (2) habitat fragmentation, patch size and distribution, and location in space are vital to the outcome of an interaction; and (3) it is difficult to classify interactions as antagonistic or mutualistic, because geographically or seasonally, interactions can shift in their outcome.

14.4.3 *Opuntia stricta* (Cactaceae)

Opuntia stricta is a succulent cactus that commonly occurs along the coastal dunes of the Gulf of Mexico and the Caribbean. Its flowers can be pollinated by bees and birds, and the fruits are consumed by several birds, rodents, and other mammals (Oliveira et al. 1999 and references therein). *O. stricta*'s efns are located in the areoles of the developing tissue of emerging cladodes and flower buds. Ants actively visit the efns on a round-the-clock basis (Fig. 14.6). The main herbivores of *O. stricta* in the sand dune scrub at La Mancha, Veracruz are (Oliveira et al. 1999): (1) *Narnia* sp. (Hemiptera: Coreidae), whose adults mate on the plant and egg batches (8–14) are laid on the spines, and nymphs and adults suck plant juice from cladodes and produce typical white rings around punctures; (2) *Hesperolabops* sp. (Hemiptera: Miridae): egg batches were not seen, and nymphs and adults suck plant juice from cladodes

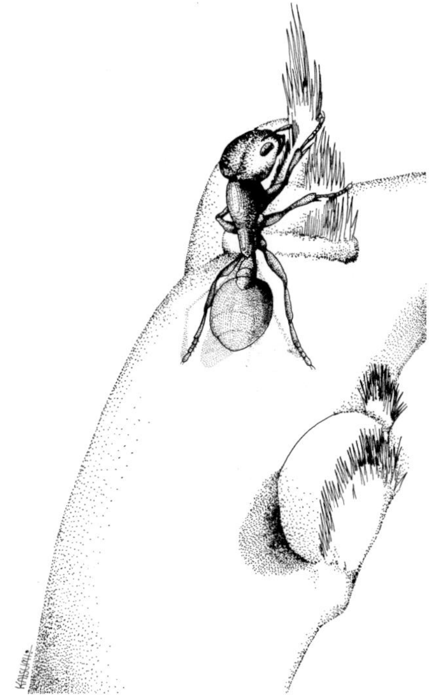

Fig. 14.6. *Opuntia stricta* with *Cremato-gaster brevispinosa* ant foraging nectar at an areole

and punctures are detectable by white dots; (3) Mining insects (Diptera): mining/feeding activity by developing larvae leave easily detectable tunnels within infested cladodes; and (4) a bud-destroying moth (Lepidoptera: Pyralidae, Phycitinae), which lays eggs on floral buds and developing cladodes, and larval burrowing/feeding activity within the plant organs leaves characteristic external marks.

The associated ant assemblage was formed by nine species distributed in four subfamilies, however, the dominant ant visitor changed markedly from day (*Camponotus planatus*) to night (*C. abdominalis*). Cladodes of control (ants present) and treated (ants excluded) plants of *Opuntia* were equally infested by sucking bugs (*Narnia* sp., *Hesperolabops* sp.: Hemiptera) (RM-ANOVA, $F_{1,36}$=0.067, P=0.797), and mining Diptera, both before (χ^2=1.279, df=2, P=0.734) and after (χ^2=0.973, df=2, P=0.807) ant treatment. Damage to buds by a pyralid moth (Pyralidae: Lepidoptera), however, was significantly higher on treatment than on control plants (X±SD, 0.84±1.92 vs 0.10±0.3, respectively; Mann-Whitney *U*-test, U=893.0, P<0.0001, N=19). Ant visitation to *Opuntia*'s efns translated into a 50 % increase in the plant's reproductive output, as expressed by the number of fruits produced (X±1 SD) by experimental control (3.62±1.80) and treatment (2.40±0.34) branches (paired *t*-test, t=2.564, df=18, P=0.0195). Moreover, fruit production by ant-visited branches was positively and significantly associated with the mean monthly rate of ant visitation to efns. Although the consequences of damage by sucking and mining insects remain unclear for *Opuntia*, the results show how the association of efns with vulnerable reproductive plant organs can result in a direct ant-derived benefit to plant fitness.

14.4.4 *Turnera ulmifolia* (Turneraceae)

Turnera ulmifolia is a polymorphic polyploid complex of herbaceous, perennial weeds, bearing extrafloral nectaries, and native throughout much of the neotropics (Torres-Hernández et al. 2000, and references therein). *T. ulmifolia* inhabits a variety of vegetation associations, exhibiting two contrasting patterns of floral morphology, where populations are either dimorphic or monomorphic for a range of floral traits (e.g., style length, stamen height, pollen size). *T. ulmifolia* grows on the semi-stabilized and stabilized sand dunes, is monomorphic, self-compatible with long styles and a range of stamen heights, they flower and fruit year-around, with a peak during the summer (rainy season) at La Mancha, Veracruz, Mexico. Branches grow continuously from an apical meristem, producing leaves regularly, flowers are axillar and one to three flowers are in anthesis per day; not all leaves are associated with flowers. Flowers remain in anthesis less than a day, and the associated leaf remains throughout fruit development. Extrafloral nectaries are located at both sides of the petiole, close to the insertion of the floral pedicel in leaves with flowers (Fig. 14.4); the nectar produced is a balanced solution of sucrose, glucose, and fructose. Ants (*Camponotus planatus*, *C. abdominalis*, *Conomyrma* sp., *Crematogaster brevispinosa*, *Forelius* sp., *Pseudomyrmex* sp.), wasps (*Polistes* sp. and an undetermined species), and honey bees (*Apis mellifera*) forage for nectar produced by the efns. The main leaf herbivore is a caterpillar (*Euptoieta hegesia*, Lepidoptera: Nymphalidae), which is highly active between June and August. A previous survey determined that the experimental removal of ≥50 % of leaf area significantly reduces fruit production.

The results of an allelochemical survey (Torres-Hernández et al. 2000) showed that *T. ulmifolia* does not exhibit a significant chemical arsenal to deter herbivores. Since there is usually a trade-off in plant defenses, i.e., a lack of redundancy of defenses that act over the same temporal, spatial, and/or herbivore scales, it was hypothesized that ants visiting extrafloral nectaries were responsible for plant defense against herbivores. The effect of different ant species on the reproductive fitness (estimated as end-of-season fruit set per treatment) of *T. ulmifolia* has been studied for several years (Torres-Hernández et al. 2000; M. Cuautle and V. Rico-Gray, unpubl. data). The results show that (1) plants associated with the larger ant species (*C. abdominalis*) produced more fruits than plants associated with the smaller ant species or those without ants (Kruskal-Wallis, H=22.158, df=4, P<0.001; Student-Newman-Keuls, P<0.05), and (2) the percent of leaf tissue removed by caterpillars of *Euptoieta hegesia* was significantly lower in plants with ants than in plants with ants excluded (Kruskal-Wallis H=37.272, df=1, P<0.001). Similar to results obtained for *Myrmecophyla christinae* (see above), ant presence is not synonymous with plant protection, and the level of protection by ants will depend on the size of the worker ants in a guild of ant visitors. Moreover, wasps visited efns when ants were excluded, exerting a higher level of protection than that offered by the smaller ant species.

Recent work has demonstrated that individuals of *T. ulmifolia* present less herbivorous damage and more unripe fruits when either ants (*Camponotus abdominalis*, *C. planatus*) or wasps (*Polybia occidentalis*, *Polistes instabilis*) were present, relative to plants in which both of these insects were excluded (M. Cuautle and V. Rico-Gray, unpubl. data). However, when both ants and wasps were present there was no increase in fruit production or decrease in herbivorous damage, relative to plants with either ants or wasps; apparently a competitive ant–wasp interaction does not allow for both these insects to simultaneously participate in plant protection (i.e., the protection exerted by ants and wasps is not additive) (M. Cuautle and V. Rico-Gray, unpubl. data). Finally, the differential effect of wasps and the dispersal by ants of *T. ulmifolia* elaiosome-bearing seeds are currently being studied in detail (M. Cuautle and V. Rico-Gray, unpubl. data).

14.5 Conclusion

Interspecific interactions are one of the most important processes influencing patterns of adaptation, variation of species, and community organization and stability. Ant–plant interactions vary in their probability of occurrence along environmental gradients and under different disturbance regimes, their outcome varies in different ecological conditions or between habitats, and they exhibit significant temporal variation. The structure of ant communities and of ant–plant interactions has been studied in a variety of habitats, assessing that neither the spatial nor the temporal dimensions can be ignored. Ant assemblages are very dynamic and extrapolating results from one ant community to another can lead to erroneous inferences, precluding broad generalizations.

Ant–plant associations in tropical sand dunes are abundant, relative to temperate semiarid or humid mountain sites. Studies at two coastal sites in Mexico, La Mancha in Veracruz and San Benito in Yucatan, show significant within-habitat seasonal variation, as well as considerable variation among habitats in the number, diversity and seasonal distribution of ant–plant interactions. They suggest that inter-habitat variation of ant–plant interactions is the effect of variation in environmental parameters, richness of plants with nectaries in the vegetation, and richness in habitat heterogeneity.

Many components in the tropical seasonal vegetation along the Gulf of Mexico (state of Veracruz) and the Caribbean (Yucatan Peninsula) flower during the dry season. Flowering is not simultaneous, but there is at least one species in flower from December through June. Throughout the dry season ants forage for nectar produced by buds, flowers, and fruits of many species. The floral and circum-floral nectar produced by these structures is probably the major liquid-energy source for ants during the dry season, since there are not many alternative food sources, such as insects or new vegetative growth. Owing to food shortage during the dry season in these habitats, ants rely on the nectar produced by the reproductive structures as their main liquid-energy source. As plant reproductive organs are thus particularly vulnerable in the dry season, there should be high selection for defense by ants.

Research on plant defense by ants in tropical sand dunes is scarce and mainly restricted to four systems [*Myrmecophyla christinae* (Orchidaceae), *Paullinia fuscescens* (Sapindaceae), *Opuntia stricta* (Cactaceae) and *Turnera ulmifolia* (Turneraceae)], which, however, represent a wide range of mutualistic interactions between ants and plants. They demonstrate that: (1) ant presence is not synonymous to defense, protection from herbivores being related to ant size; (2) the outcome of the interactions varies between seasons; (3) ant–Homoptera associations can be more harmful than beneficial to the associated plant; (4) wasps may play a significant role in plant defense; and (5) these interactions are more complex than solely defense, for example, the feeding of plants by ants (*M. christinae*) or seed dispersal by ants (*T. ulmifolia*).

Acknowledgements. Financial support was provided by CONACYT (VRG, VPT, MC, CDC), by Instituto de Ecología, A.C. (902–16) (VRG, PSO, MC, CDC), and by the Conselho Nacional de Desenvolvimento Científico e Tecnológico and the Fundação de Amparo à Pesquisa do Estado de São Paulo (PSO).

References

Alonso LE (1998) Spatial and temporal variation in the ant occupants of a facultative ant–plant. Biotropica 30:201–213

Barton AM (1986) Spatial variation in the effect of ants on an extrafloral nectary plant. Ecology 67:495–504

Bondini A Giavelli G (1989) The qualitative approach in investigating the role of species interactions on stability of natural communities. BioSyst 22:289–299

Bronstein JL (1995) The plant–pollinator landscape. In: Hansson L, Fahrig L, Merriam G (eds) Mosaic landscapes and ecological processes. Chapman and Hall, London, pp 256–288

Bronstein JL (1998) The contribution of ant–plant protection studies to our understanding of mutualism. Biotropica 30:150–161

Bronstein JL (2001) The exploitation of mutualisms. Ecol Lett 4:277–287

Bullock SH, Solís-Magallanes A (1990) Phenology of canopy trees of a tropical deciduous forest in México. Biotropica 22:22–35

Castillo S Carabias J (1982) Ecología de la vegetación de dunas costeras: fenología. Biotica 7:551–568

Chauvin G, Gueguen A (1978) Dévelopment larvaire et bilan d'utilisation d'energie en fonction de l'hygrométrie chez *Tinea pellionella* L. (Lepidoptera: Tineidae). Can J Zool 56:2176–2185

Coley PD (1983) Herbivory and defensive characteristics of tree species in a lowland tropical forest. Ecol Mon 53:209–233

Coley PD, Aide TM (1991) Comparison of herbivory and plant defenses in temperate and tropical broad-leaved forests. In: Price PW, Lewinsohn TM, Fernandes GW, Benson WW (eds) Plant–animal interactions: evolutionary ecology in tropical and temperate regions. Wiley, New York, pp 25–49

Cuautle M, Rico-Gray V, García-Franco JG, López-Portillo J, Thien LB (1999) Description and seasonality a plant–ant–Homoptera interaction in the semiarid Zapotitlán Valley, México. Acta Zool Mex (N.S.) 78:73–83

Cushman JH, Addicott JF (1991) Conditional interactions in ant–plant–herbivore mutualisms. In: Huxley CR, Cutler DF (eds) Ant–plant interactions. Oxford University Press, Oxford, pp 92–103

De la Fuente MAS, Marquis RJ (1999) The role of ant-tended extrafloral nectaries in the protection and benefit of a neotropical rainforest tree. Oecologia 118:192–202

Del-Claro K, Berto V, Réu W (1996) Effect of herbivore deterrence by ants on the fruit set of an extrafloral nectary plant, *Qualea multiflora* (Vochysiaceae). J Trop Ecol 12:887–892

Dyer LA, Letourneau DK (1999) Relative strengths of top-down and bottom-up forces in a tropical forest community. Oecologia 119:265–274

Elias TS (1983) Extrafloral nectaries: their structure and functions. In: Bentley BL, Elias TS (eds) The biology of nectaries. Columbia University Press, New York, pp 174–203

Feener DH, Jr., Schupp EW (1998) Effect of treefall gaps on the patchiness and species richness of neotropical ant assemblages. Oecologia 116:191–201

Freitas AVL, Oliveira PS (1996) Ants as selective agents on herbivore biology: effects on the behaviour of a non-myrmecophilous butterfly. J Anim Ecol 65:205–210

García-Franco JG, Rico-Gray V (1997) Reproductive biology of the holoparasite *Bdallophyton bambusarum* (Rafflesiaceae). Bot J Linn Soc 123:237–247

Heads PA, Lawton JH (1984) Bracken, ants, and extrafloral nectaries. II. The effect of ants on the insect herbivores of bracken. J Anim Ecol 53:1015–1032

Herbers JM (1989) Community structure in north temperate ants: temporal and spatial variation. Oecologia 81:201–211

Herrera CM, Herrera J, Espadaler X (1984) Nectary thievery by ants from southern Spanish insect-pollinated flowers. Insect Soc 31:142–154

Inouye DW, Taylor OR (1979) A temperate region plant–ant–seed predator system: consequences of extrafloral nectar secretion by *Helianthella quinquenervis*. Ecology 60:1–7

Keeler KH (1981) Function of *Mentzelia nuda* (Loasaceae) postfloral nectaries in seed defense. Am J Bot 68:295–299

Koptur S (1984) Experimental evidence for defense of *Inga* saplings (Mimosoideae) by ants. Ecology 65:1787–1793

Koptur S (1991) Extrafloral nectaries of herbs and trees: modeling the interaction with ants and parasitoids. In: Huxley CR, Cutler DF (eds) Ant–plant interactions. Oxford University Press, Oxford, pp 213–230

Koptur S (1992). Extrafloral nectary-mediated interactions between insects and plants. In: Bernays E (ed) Insect-plant interactions, Volume IV. CRC Press, Boca Raton, pp 81–129

Koptur S, Rico-Gray V, Palacios-Rios M (1998) Ant protection of the nectaried fern *Polypodium plebeium* in central México. Am J Bot 85:736–739

Lanza J (1988) Ant preferences for *Passiflora* nectar mimics that contain amino acids. Biotropica 20:341–344

Marquis RJ (1992) Selective impact of herbivores. In Fritz RS, Simms EL (eds) Plant resistance to herbivores and pathogens: ecology, evolution, and genetics. University of Chicago Press, Chicago, pp 301–325

Mattson WJ Jr (1980) Herbivory in relation to plant nitrogen content. Annu Rev Ecol Syst 11:119–161

McDade LA, Kinsman S (1980) The impact of floral parasitism in two neotropical hummingbird-pollinated plant species. Evolution 34:944–958

Norment CJ (1988) The effect of nectar-thieving ants on the reproductive success of *Frasera speciosa* (Gentianaceae). Am Midl Nat 120:331–336

O'Dowd DJ, Catchpole EA (1983) Ants and extrafloral nectaries: no evidence for plant protection in *Helichrysum* spp.-ant interactions. Oecologia 59:191–200

Oliveira PS, Brandão CRS (1991) The ant community associated with extrafloral nectaries in the Brazilian cerrados. In: Huxley CR, Cutler DF (eds) Ant–plant interactions. Oxford University Press, Oxford, pp 198–212

Oliveira PS, da Silva AF, Martins AB (1987) Ant foraging on extrafloral nectaries of *Qualea grandiflora* (Vochysiaceae) in cerrado vegetation: ants as potential antiherbivore agents. Oecologia 74:228–230

Oliveira PS, Rico-Gray V, Díaz-Castelazo C, Castillo-Guevara C (1999) Interaction between ants, extrafloral nectaries, and insect herbivores in neotropical coastal sand dunes: herbivore deterrence by visiting ants increases fruit set in *Opuntia stricta* (Cactaceae). Fun Ecol 13:623–631

Ortiz-Pulido R, Rico-Gray V (2000) The effect of spatio-temporal variation in understanding the fruit crop size hypothesis. Oikos 91:523–527

Pemberton RW, Lee JH (1996) The influence of extrafloral nectaries of parasitism of an insect herbivore. Am J Bot 83:1187–1194

Puterbaugh MN (1998) The roles of ants as flower visitors: experimental analysis in three alpine plant species. Oikos 83:36–46

Rico-Gray V (1980) Ants and tropical flowers. Biotropica 12:223–224

Rico-Gray V (1989) The importance of floral and circum-floral nectar to ants inhabiting dry tropical lowlands. Biol J Linn Soc 38:173–181

Rico-Gray V (1993) Use of plant-derived food resources by ants in the dry tropical lowland of coastal Veracruz, Mexico. Biotropica 25:301–315

Rico-Gray V (2001) Interspecific interaction. Encyclopedia of life sciences. Macmillan, Nature Publ Group/www.els.net

Rico-Gray V, Castro G (1996) Effect of an ant–aphid–plant interaction on the reproductive fitness of *Paullinia fuscecens* (Sapindaceae). Southwest Nat 41:434–440

Rico-Gray V, Sternberg LSL (1991) Carbon isotopic evidence for seasonal change in feeding habits of *Camponotus planatus* Roger (Formicidae) in Yucatan, Mexico. Biotropica 23:93–95

Rico-Gray V, Thien LB (1989a) Effect of different ant species on the reproductive fitness of *Schomburgkia tibicinis* (Orchidaceae). Oecologia 81:487–489

Rico-Gray V, Thien LB (1989b) Ant–mealybug interaction decreases reproductive fitness of *Schomburgkia tibicinis* Bateman (Orchidaceae) in Mexico. J Trop Ecol 5:109–112

Rico-Gray V, Barber JT, Thien LB, Ellgaard EG, Toney JJ (1989) An unusual animal-plant interaction: feeding of *Schomburgkia tibicinis* by ants. Am J Bot 76:603–608

Rico-Gray V, García-Franco JG, Palacios-Rios M, Díaz-Castelazo C, Parra-Tabla V, Navarro JA (1998) Geographical and seasonal variation in the richness of ant–plant interactions in Mexico. Biotropica 30:190–200

Smith LL, Lanza J, Smith GC (1990) Amino acid concentrations in extrafloral nectar of *Impatiens sultanii* increase after simulated herbivory. Ecology 71:107–115

Smythe N (1982) The seasonal abundance of night-flying insects in a neotropical forest. In Leigh Jr EG, Rand AS, Windsor DM (eds) The ecology of a tropical forest. Smithsonian Institution Press, Washington, DC, pp 309–318

Thompson JN (1994) The coevolutionary process. University of Chicago Press, Chicago

Thompson JN (1999) The evolution of species interactions. Science 284:2116–2118

Tilman D (1978) Cherries, ants and tent caterpillars: timing of nectar production in relation to susceptibility of caterpillars to ant predation. Ecology 59:686–692

Torres-Hernández L, Rico-Gray V, Castillo-Guevara C, Vergara A (2000) Effect of nectar-foraging ants and wasps on the reproductive fitness of *Turnera ulmifolia* (Turneraceae) in a coastal sand dune, México. Acta Zool Mex (N.S.) 81:13–21

Völkl W, Woodring J, Fischer M, Lorenz MW, Hoffmann KH (1999) Ant–aphid mutualisms: the impact of honeydew production and honeydew sugar composition on ant preferences. Oecologia 118:483–491

V Environmental Problems and Conservation

15 Environmental Problems and Restoration Measures in Coastal Dunes in The Netherlands

A.M. Kooijman

15.1 Introduction

The more than 350-km-long coastal dune zone forms one of the last large semi-natural areas in The Netherlands and is home to some 70 % of the plant species occurring in this country, of which many are almost exclusive. The vegetation consists to a large part of open, species-rich dune grasslands belonging to the plant communities *Phleo-Tortuletum, Anthyllido-Silenetum, Taraxaco-Galietum, Festuco-Galietum* and *Violo-Corynephoretum* (Schaminee et al. 1996). However, during the last decades grass-encroachment has transformed the species-rich dune grasslands into monospecific stands of tall grasses (Weeda et al. 1994; Kooijman and de Haan 1995; Fig. 15.1). While increases in biomass and changes in species richness naturally occur in the course of succession, the loss of species diversity seemed unnaturally high. This was attributed, similar to other Dutch ecosystems (Aerts 1989; Bobbink 1989), to increased atmospheric deposition, which may amount to 30 kg N/ha annually (Dopheide and Verstraten 1995). However, since the effect of high N availability also depends on the availability of P, the ecosystem responses appeared to differ between dune districts and successional stages with different soil chemistry. The goal of this chapter is to give an overview of how grass-encroachment could have developed in the lime-rich and lime-poor dune areas of the Netherlands and which restoration measures are effective against it. Both aspects were studied as part of the larger Dutch restoration program 'Restoration plan forest and nature (OBN)'. Detailed methods and results are given in Kooijman and de Haan (1995), van der Meulen et al. (1996), Veer (1997, 1998), Veer and Kooijman (1997), Kooijman et al. (1998, 2000) and Kooijman and Besse (2002).

Ecological Studies, Vol. 171
M.L. Martínez, N.P. Psuty (Eds.)
Coastal Dunes, Ecology and Conservation

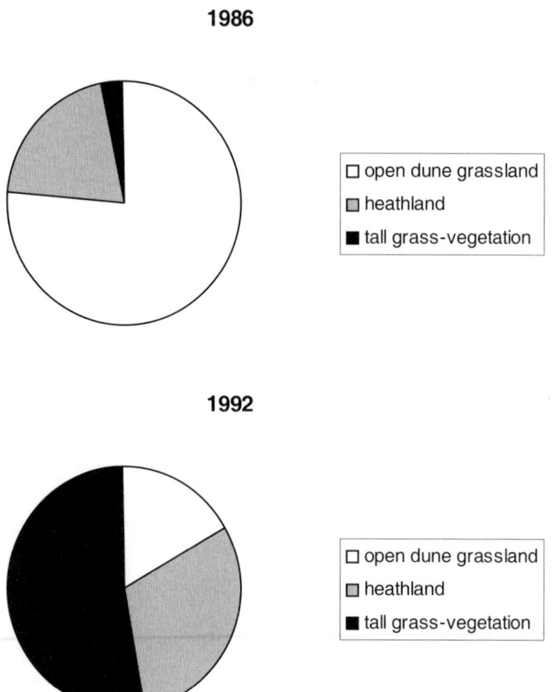

Fig. 15.1. The decrease in species-rich dune grassland from 1986 to 1992 in a 50 ha ungrazed part of the Zwanenwater area, Wadden district. Data from Kooijman and de Haan (1995)

15.2 Differences Between Renodunaal and Wadden Districts

The Dutch dunes are separated into two different districts: the lime-poor Wadden district in the north and the lime-rich Renodunaal district in the south. Both districts are representative of European coastal dune areas north and south of the Netherlands. Whereas in Chapter 6 (Grootjans et al.) the wet dune slacks are treated, the present study concentrates on the dry dunes, which comprise approximately 65 % of the total dune area.

The Renodunaal and Wadden district are distinguished on the basis of their differences in initial lime content, but also because of the mineral composition of the sand. In the Renodunaal district initial lime content ranges from 2–10 %, while in the Wadden district these values rarely exceed 1 % (Eisma 1968). The Renodunaal district has approximately 15-fold higher amounts of amorphic iron and aluminium (hydr)oxides than the Wadden district (Eisma 1968; Kooijman et al. 1998).

Apart from differences between districts, successional stages are important as well. In the lime-rich Renodunaal district three main soil types can be distinguished: (1) calcareous soils with high pH, (2) soils in which the topsoil has become decalcified, but part of the root zone is still calcareous and (3) soils decalcified to more than 1 m depth which have become acid. Because of the low initial lime content in the Wadden district the calcareous and partly decalcified soils are limited to a very small zone near the sea.

Several tall grass species are involved in grass-encroachment, each dominating under particular conditions (Weeda et al. 1994). In the Renodunaal district, *Elytrichia atherica* (Link) Carreras Mart. is a dominant species in calcareous soils and *Calamagrostis epigejos* (L.) Roth in decalcified soils. In the Wadden district, *Ammophila arenaria* (L.) Link is the main invading species.

15.3 Impact of Availability of P on Biomass Production and Successional Trends

15.3.1 Renodunaal District

In the lime- and iron-rich Renodunaal district plant biomass production seems to be primarily regulated by the availability of P (Kooijman et al. 1998; Kooijman and Besse 2002). Aboveground productivity, N-mineralization and P-mineralization values showed a peak around pH 5 (Fig. 15.2). The correspondence between N-mineralization and biomass production is not surprising, since these two factors can be closely coupled (e.g., Veer 1997; Neitzke 1998), but the peak at a particular pH is more difficult to explain. At high pH mineral N mainly occurs as nitrate and at low pH as ammonium, but both are highly soluble at all pH values.

The productivity peak at pH 5 corresponds very well, however, with the chemical behaviour of phosphate (Lindsay and Moreno 1966). In calcareous soils with pH>6.5 P-availability is low due to fixation in calcium phosphates, which is shown by the negative 'mineralization' values. In partly decalcified soils with pH 5 calcium phosphates have dissolved and become available to plant roots. This is indicated by the decrease in mineral-P from calcareous to partly decalcified soils (Kooijman et al. 1998), but especially by the high P-'mineralization' (Fig. 15.2). In decalcified soils P-availability is low again due to the chemical fixation in iron and aluminium phosphates.

This suggests that the biomass production in the Renodunaal district is primarily regulated by the P-availability. The peak in N-mineralization at pH 5 may be explained as a response to the increase in biomass production (and litter production) allowed by the higher P-availability. This increase in natural fertility corresponds with the large-scale formation of shrubland especially in this dune zone (e.g., Westhoff et al. 1970; Doing 1988). It also suggests that in this zone species-rich dune grasslands may not be a permanent stage, but be maintained only when the vegetation is kept short by grazers.

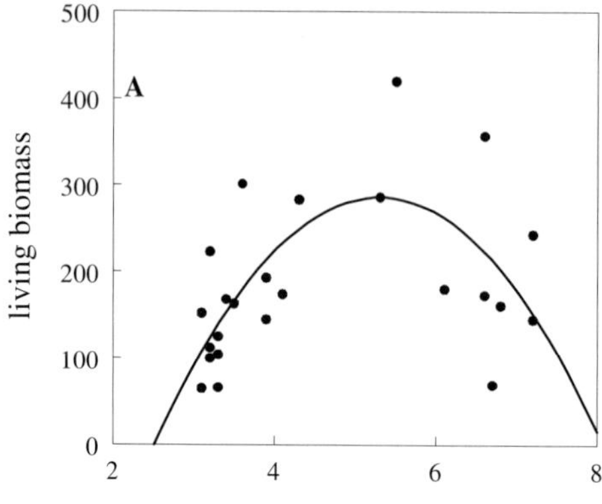

Fig. 15.2. Biomass production and nutrient availability in coastal dune grasslands over a pH-gradient in the Renodunaal district. **A** Aboveground biomass production (g m^{-2}) in July measured in exclosures (r=0.61), **B** in situ N-mineralization (g m^{-2}) measured over the period April–October(r=0.61) and **C** in situ P-mineralization (mg m^{-2}) over the same period (r=0.56). Data are derived from Kooijman and Besse (2002). All three correlations are significant (p<0.05)

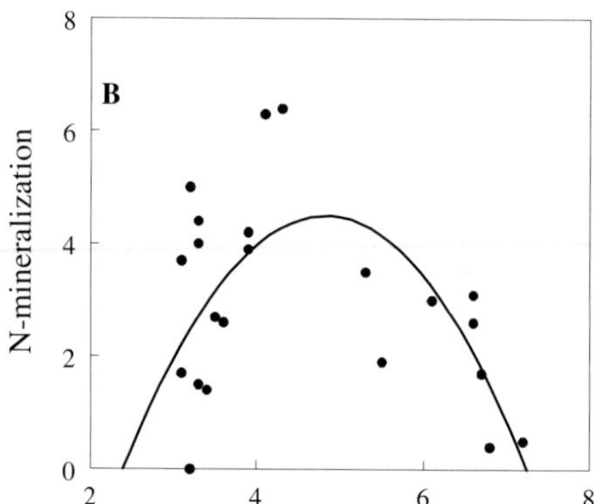

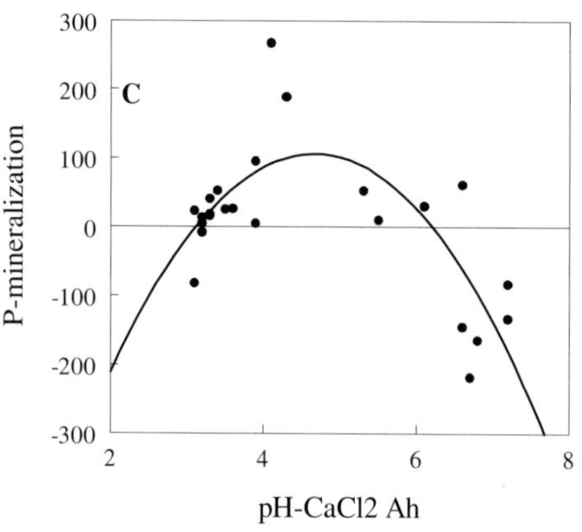

Regulation of plant biomass production by the availability of P suggests that the Renodunaal district is not very sensitive to atmospheric deposition of N. However, increased atmospheric deposition is probably a very important factor when both acid and nitrogen deposition are taken into account. Acid deposition leads to increased decalcification and dissolution of calcium phosphates, whereas nitrogen deposition simultaneously increases N-availability. The higher availability of both nutrients stimulates biomass production. This in turn not only increases internal acidification of the soil through root exchange processes and higher litter decomposition, but also increases N-mineralization, thus further stimulating plant productivity. This implies that increased atmospheric deposition may not change the direction, but nevertheless accelerates succession from calcareous to decalcified dunes. Because many characteristic plant and animal species prefer open calcareous dune grasslands, this is a problem for nature conservation.

15.3.2 Wadden District

Although in a different way, the availability of P is also a key-factor in regulating biomass production and response to atmospheric deposition in the Wadden district (Kooijman et al. 1998; Kooijman and Besse 2002). Foliar N/P ratios, which are 11 for both open dune grassland and tall grass vegetation, suggest that the vegetation is limited by N instead of P (Koerselman and Meuleman 1996). This is also illustrated by the P-mineralization rates, which are about ten times higher than the values in the (equally acid) decalcified soils of the Renodunaal district. This can be partly ascribed to the very low levels of inorganic iron and aluminium (hydr)oxides. As a result, chemical P-fixation in iron and aluminium phosphates is limited. Instead, phosphate is basically bound to iron and aluminium-organic matter complexes, which are much more reversible bindings (Scheffer 1982). The high P-availability implies that the ecosystem is relatively N-limited and sensitive to atmospheric N-deposition.

15.4 Mineralization of Nitrogen

15.4.1 Impact of Litter Production

Cycling of N seems to be more rapid in the Wadden district: N-mineralization rates were two to four times higher than in the Renodunaal district. In theory, this can be attributed to a number of factors. N-mineralization may be regulated by the input of litter. In the Wadden district the relationship between N

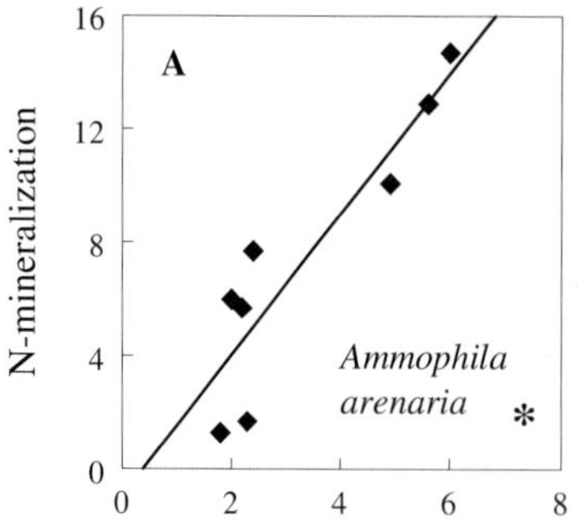

Fig. 15.3. Relationship between N in the living aboveground vegetation (g m^{-2}) in exclosures as indicator for plant performance and in situ N-mineralization over the period April–October (g m^{-2}) in coastal dune grasslands. **A** Wadden district with *Ammophila arenaria* (r=0.91); **B** calcareous soils in Renodunaal district with *Elytrichia atherica* (r=0.45). Data derived from Kooijman and Besse (2002). * Significant correlation (p<0.05)

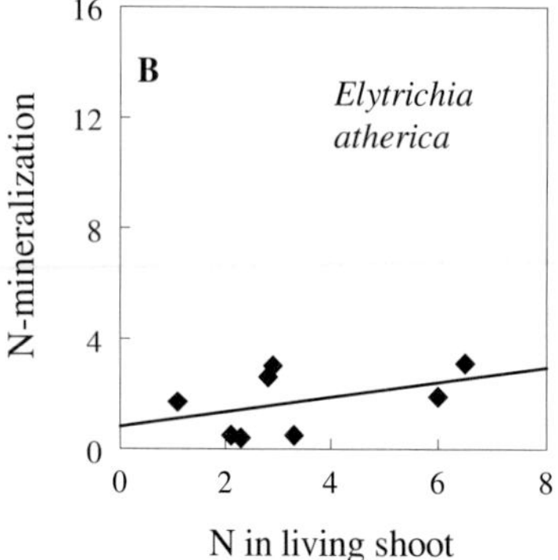

in biomass and N-mineralization appeared to be very strong indeed (Fig. 15.3). However, the amount of N in litter cannot explain the higher N-mineralization compared to the Renodunaal district, since this was about the same in both districts.

Rates of N-mineralization have been reported to increase over a successional dune gradient, due to the accumulation of soil organic matter and the development of a N-cycle in the soil (Gerlach et al. 1994). However, this cannot explain the higher N-mineralization values in the Wadden district, since soil organic matter contents were about the same in both districts. N-mineralization may also increase with increased atmospheric deposition (Sjöberg and Persson 1998). However, this cannot explain the higher N-mineralization values in the Wadden district, because atmospheric N-deposition is lower instead of higher than in the Renodunaal district (Dopheide and Verstraten 1995).

15.4.2 Impact of Litter Decomposition

The key to the high N-mineralization rates in the Wadden district seems to lie in the mechanism of litter decomposition, although in a different way than usually assumed. In the 'common wisdom' low decomposition leads to low N-mineralization rates and vice versa (Swift et al. 1979, Lambers et al. 1998, Aerts and Chapin 2000). The reasoning behind this is that a high biological activity leads to a rapid turnover of carbon, and as such of nutrients. Also, differences in C/N ratio between litter and micro-organisms are smaller in highly degradable litter with high N-concentrations, which supposedly means that microbial N-need is satisfied earlier in the breakdown process and net N-mineralization rates are higher.

However, in the coastal dunes the relationship between litter decomposition and N-mineralization seems to be negative instead of, as expected, positive (Kooijman and Besse 2002). The lowest values for litter decomposition and highest for N-mineralization were found in the Wadden district, while the Renodunaal district showed the opposite (Fig. 15.4). The combination of low decomposition and high N-mineralization and vice versa does not seem to be unique for the Dutch coastal dunes, but has been reported from calcareous-acid fens as well (Verhoeven et al. 1988, 1990). Also, N-mineralization appeared to be four times higher in acid than in calcareous beech forests (Davy and Taylor 1974).

The above suggests that acid soils with low-degradable litter are characterized by a high instead of low N-availability to the vegetation. This may be due to a lower microbial N-demand because of the lower overall biological activity, but also because in low-degradable litter the C-limitation stage is reached

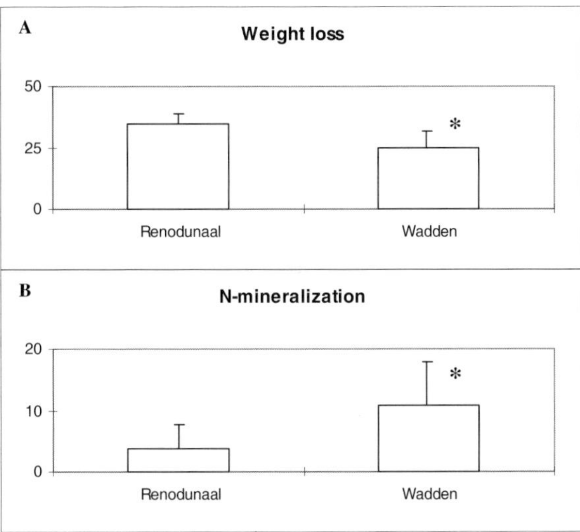

Fig. 15.4. Litter breakdown and N-mineralization in coastal dune grasslands dominated by tall grasses. **A** Weight loss after 1 year (%); **B** N-mineralization in the period April–October (g m^{-2}). Mean values (n=12 for Renodunaal and n=4 for Wadden district) and standard deviations. Data derived from Kooijman and Besse (2002). * Significant differences between dune districts

at higher C/N ratios (Berg and Ekbohm 1983; Berg and McClaugherty 1987), i.e., earlier in the decomposition process. In addition, microbial C/N ratios are higher (and N-demand lower) in soils unfavourable for decomposition (Hassink 1994; Hassink et al. 1993). A relatively large attribution of N to the vegetation instead of soil microbes may also explain the strong relationship between litter input and N-mineralization for *A. arenaria*, as opposed to the weak one for *E. atherica* (Fig. 15.3).

15.5 Role of *Ammophila arenaria* in the Wadden District

Grass encroachment, once started, follows its course driven by positive feedback mechanisms (Veer and Kooijman 1997). The large aboveground biomass of tall grasses increases root biomass and nutrient uptake capacity. The large biomass also increases litter input and therefore N-mineralization and nutrient availability. Both factors lead to enhanced nutrient uptake and even higher biomass production. The effect on smaller species in the vegetation is obvious: strongly reduced light availability and poor survival.

In the Wadden district these feedback mechanisms are aggravated by specific characteristics of *A. arenaria*. This species has low foliar N-levels and a high Nutrient Use Efficiency (Pavlik 1983), which means a relatively low N-need per unit biomass and thus a high biomass production. Its low N and high C content also lead to low-degradable leaves. As a result dead material stays on the plant for a relatively long time, thus further reducing light availability to smaller species. The low-degradable litter also seems to lead to a low microbial N-demand and a relatively high N-availability to plants. While in the Renodunaal district annual N-mineralization was estimated to be 1–5 % of total soil N, in the *Ammophila*-dominated plots in the Wadden district this amounted to 18 % (Kooijman et al. 2000).

In the Wadden district biomass production could already be high due to the relatively high P-availability and the response to increased atmospheric N-deposition. The low N-demand per unit biomass and the efficient recycling of N in the soil further contribute to the explosion of *A. arenaria* throughout most of the Wadden district.

15.6 Restoration

The effects of atmospheric deposition, grass-encroachment and enhanced succession may be counteracted by management practices. One way to accomplish this is directed towards removing aboveground biomass, e.g., by annual mowing and grazing by cattle. The second way is to develop new substrates, in order to set back succession of vegetation and soil, e.g. by sod-cutting and the stimulation of aeolian activity. Methods and results are briefly discussed below; the details are given in van der Meulen et al. (1996), Veer (1997, 1998) and Kooijman et al. (2000).

15.6.1 Effect of Grazing and Annual Mowing

A comparison of four Dutch coastal dune areas where mowing was applied for seven years and nine which were grazed by cattle for five years or more suggests that both measures are effective to counteract grass-encroachment (Fig. 15.5). Mean aboveground biomass was significantly reduced at all sites where mowing or grazing was applied. Available light at the soil surface was 3–5 % of full sunlight in the untreated tall-grass vegetation and increased significantly after mowing or grazing. Both measures increased species richness per square metre.

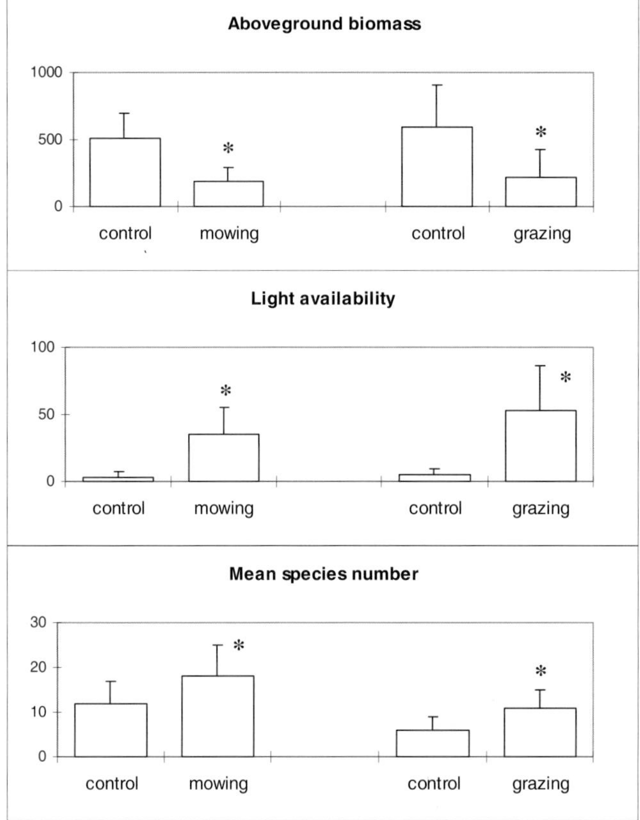

Fig. 15.5. Effect of annual mowing and grazing by cattle on aboveground biomass (g m^{-2}), light availability (% full daylight) and species number (mean number of plant species m^{-2}) in coastal dune grasslands dominated by tall grasses. Data derived from Kooijman et al. (2000). Mean values (n=4 for mowing and n=9 grazing) and standard deviations. * Significantly different from the associated control treatment (p<0.05)

The reduction in aboveground biomass thus clearly leads to improved conditions for small species in the understorey. However, for long-term ecosystem functioning and the development of management plans it is important to know whether this is only due to the temporary removal of biomass, or also due to a reduction in productivity as a result of reduced litter input. The latter would especially harm tall grasses, which have a higher nutrient demand (Veer and Kooijman 1997). Reduced input of litter and as such of nutrients to the soil would theoretically lead to a reduction in N-mineralization and biomass production. In this way competition between tall grasses and smaller herbs would change from light to nutrients (Olff and Ritchie 1998).

A reduction in N-mineralization and aboveground biomass production by grazing was indeed detected in the Wadden district (Fig. 15.6). Also, the area of tall-grass vegetation decreased in favour of species-rich dune grassland (ten Haaf 1999a). In the Renodunaal district, however, N-mineralization was only affected in decalcified soil and biomass production not at all. Also, the area of tall-grass vegetation did not decrease and development of shrubland continued despite the grazing regime in calcareous and partly decalcified

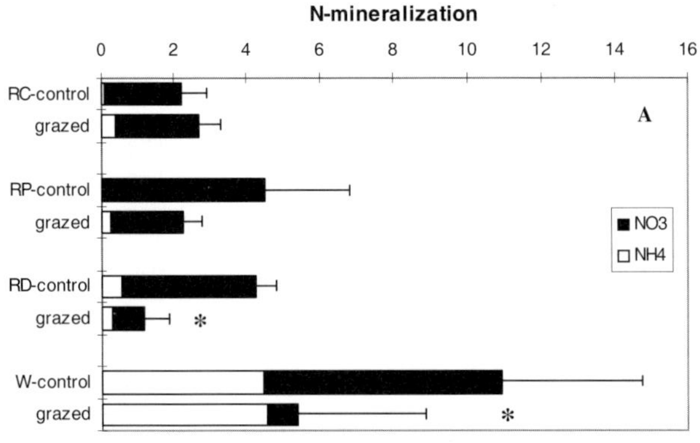

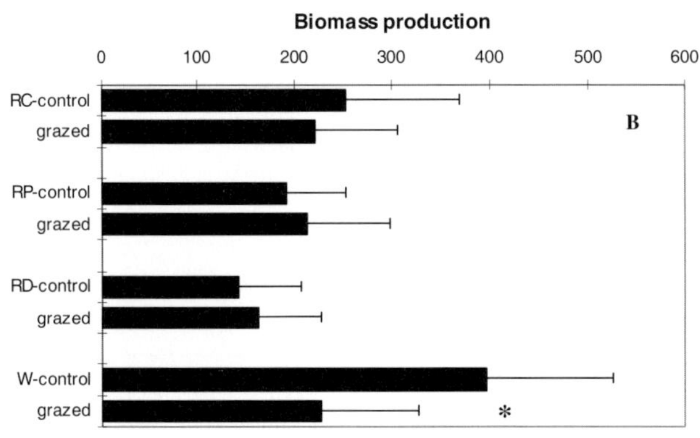

Fig. 15.6. The effect of grazing on nutrient availability and productivity in coastal dune grasslands dominated by tall grasses. **A** N-mineralization (NO_3+NH_4) from April–October (g m^{-2}) and **B** biomass production in July (g m^{-2}). *RC* Renodunaal district, calcareous soils; *RP* Renodunaal district, partially decalcified soils; *RD* Renodunaal district, soils decalcified to more than 1 m; *W* Wadden district. Data derived from Kooijman et al. (2000). Mean values (n=4) and standard deviations. * Significantly lower values in grazed plots (p<0.05)

dune zones (ten Haaf 1999b; Everts et al. 2000). Annual mowing, which was only studied in the Renodunaal district, did also not lead to changes in nutrient availability in calcareous and partly decalcified soils. In a decalcified site with high P-availability due to former agricultural practices, however, the mown treatment strongly indicated N-limitation. The above suggests that acid soils with low-degradable litter respond more strongly to management practices than calcareous soils.

While annual mowing and grazing by cattle are primarily applied to counteract grass-encroachment, they may reduce succession in calcareous and partly decalcified soils, because soil acidification is higher in highly productive systems. Over a seven year-period soil pH values had decreased in the tall-grass control treatments with 0.5–1 in calcareous and partially decalcified soisl respectively, but were about the same as before in the mown treatments. This suggests that removal of aboveground biomass retards soil acidification and further succession at least to some degree.

15.6.2 Effect of Sod-Cutting

Sod-cutting naturally leads to a decrease in soil organic matter and nutrient stocks and an increase in soil pH. It also led to a decrease in nutrient availability and biomass production. After seven years vegetation cover was still below 100 %, mean light availability at the soil surface (41 %) was higher than in the tall-grass control treatments (3 %) and the total number of species had increased from 27 to 42 (Fig. 15.7). However, not all results were positive. In the calcareous soils the establishment of the shrub *Hippophae rhamnoides* L. may give rise to the development of shrubland instead of dune grasslands. In the partly decalcified soil calcicole species, which were still present in the grass-dominated control treatments, did not re-establish. Thus, because of these potentially unfavourable side effects, sod-cutting should be applied with some care.

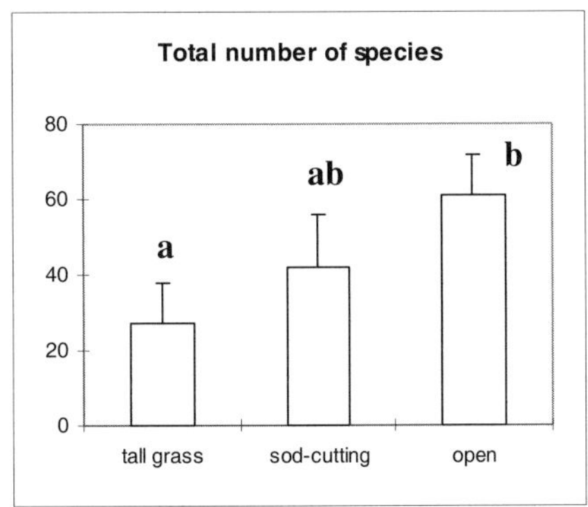

Fig. 15.7. Effect of sod-cutting on total species number in dune grasslands. *Tall grass* Untreated control plots in tall-grass vegetation; *sod-cutting* plots where sod-cutting was applied seven years earlier; *open* open, species-rich dune grassland used as reference vegetation. Data derived from Kooijman et al. (2000). Mean values (n=4) and standard deviations. *Different letters* indicate significant differences (p<0.05)

15.6.3 Effect of Increased Aeolian Activity

Until very recently, blow-outs in coastal dunes were stabilized as much as possible and since the 1950s the area of aeolian dynamic dunes have strongly decreased. During the last decades nature management organisations, however, began to realize that the loss of aeolian dynamics and lack of rejuvenation may be a loss for the dune ecosystem as a whole.

Monitoring of blow-outs over the past twenty years suggests that they do not expand very rapidly, which is in line with van der Meulen and Jungerius (1989). Two dune areas in the Renodunaal district where fixation practices were abandoned around 1980 showed an increase in aeolian activity from 1980 to1990, indicated by the increase in bare sand area from 3–6 % in one and from 7–12 % in the other location (Kooijman et al. 2000). However, in the following decade aeolian activity decreased again. These changes were ascribed to some large storms in the first period and wet years in the second. Blow-outs reactivated by buldozers showed some expansion in surface area over a seven year period in the Renodunaal district, but in the Wadden district the area of bare sand also strongly decreased.

Thus, on a landscape scale expansion of blow-outs seems to be limited. This may be partly due to increased atmospheric N-deposition. Natural stabilization of blow-outs occurs through algae (Pluis 1993), which need water and nutrients to grow. Early successional stages appeared to be N-limited in both the Wadden and Renodunaal district, as indicated by the foliar N/P ratios ranging from 9–11. This suggests that increased N-input results in enhanced growth of algae and blow-out stabilization.

Although aeolian activity clearly leads to an increase in pioneer vegetation, it does not seem to arrest grass-encroachment. Nineteen years after succession started in a Renodunaal district site the vegetation still consisted of (tall) pioneer vegetation. It is not known how long it takes for dune grassland to develop, but these results suggest that it requires much more than twenty years. Also, in a second site the area of species-rich dune grassland decreased from close to 50 % in 1979 to less than 20 % in 1997, due to burial by sand or continuing grass (and shrub) encroachment. A particular problem in the Wadden district may be that aeolian activity especially favours *Ammophila arenaria*, which is the main species in the grass-encroachment process.

Despite the (short-term) negative effects on the areas of species-rich dune grasslands, aeolian activity seems to be a very effective measure against enhanced acidification of the soil and thus vegetation succession (Table 15.1). In all areas, values in pioneer vegetation were higher than pH 7. Even in the control vegetation further away, which was supposedly outside the direct reach of the blow-outs, pH values were (much) higher than in comparable dune zones. In dune zones decalcified to some depth, blow-out development may bring calcareous sand to the surface, which is hardly possible with sod-cutting.

Table 15.1. Differences in pH between coastal dune grasslands with different aeolian activity in Renodunaal district (R) and Wadden district. Data derived from Kooijman et al. (2000). Mean values (n=4) and standard deviations. Different letters indicate significant differences within a column (p<0.05)

	R-calcareous soils	R-partly decalcified soils	R-decalcified soils	Wadden district
Pioneer vegetation in blow outs	8.3 (0.3) b	7.9 (0.2) c	8.4 (0.3) c	7.6 (0.5) c
Reference vegetation in surroundings of blow outs	7.8 (0.1) a	7.5 (0.8) b	7.8 (0.3) b	6.6 (1.2) b
Reference vegetation in the same dune zone	7.4 (0.3) a	5.0 (0.5) a	4.1 (0.1) a	4.2 (0.3) a

Aeolian activity also leads to changes in nutrient availability. As said, the pioneer stages are clearly N-limited. This may be due to the low organic matter content and low litter input, as well as the high pH, which all contribute to low N-mineralization. However, the high pH also leads to low P-availability, which suggests that biomass productivity will be limited for some time, probably until values have dropped below pH 6.5. This may mean that, in spite of the negative short-term effects on the areas of species-rich dune grasslands, the long-term prospects are much better.

15.7 Concluding Remarks

The results are based on dunes in a small country, but they suggest that the availability of nutrients and sensitivity to increased N-deposition is regulated by more general factors such as lime and iron contents in the soil. This implies that the mechanisms behind grass-encroachment have a wider application. There are unfortunately no data on N and P-availability and soil chemistry in other European dune areas to confirm this, but some indirect indications and personal observations are in line. Grass-encroachment has not (yet) been reported from the calcareous dunes in England, possibly due to a low availability of N and P at high pH. The northern part of the dunes in Denmark is low-productive, possibly due to a combination of low pH and high iron levels and thus low P-availability. However, on the Wadden isles of Germany and Denmark with soils low in lime and iron grass-encroachment is becoming a serious problem.

This study suggests that in (initially) lime- and iron-rich soils plant productivity can be limited by P-fixation at high and low pH. The ecosystem seems less responsive to N, which may mainly be a secondary factor responding to the productivity allowed by P. Management practices can affect light availability and increase chances for survival of small species, but hardly seem to alter nutrient availability and biomass production. Increased atmospheric deposition seems to be a problem mainly because of the simultaneous impact on both N and P, leading to more rapid succession of soil and vegetation. Stimulation of aeolian activity may be the best way to counteract this.

In lime- and iron-poor soils P does not seem to be a limiting factor, because of the absence of P-fixation. These ecosystems may instead strongly respond to increases in N-availability. This is aggravated by the strong relationship between litter input, N-mineralization and plant productivity, due to low rates of decomposition and thus low microbial N-demand and high N-mineralization per unit litter input. However, the strong relationship between litter input and N-mineralization also means that the ecosystem is sensitive to management practices. Removal of biomass not only leads to improved light conditions to small species, but also to lower nutrient availability and productivity, which are important responses in the longer term.

References

Aerts MAPA (1989) Plant strategies and nutrient cycling in heathland ecosystems. PhD Thesis, Univ of Utrecht

Aerts MAPA, Chapin III FS (2000) The mineral nutrition of wild plants revisited: a re-evaluation of process and patterns. Adv Ecol Res 30:1–67

Berg B, Ekbohm G (1983) Nitrogen immobilisation in decomposing needle litter at variable carbon:nitrogen ratios. Ecology 64:63–67

Berg B, McClaugherty C (1987) Nitrogen release from litter in relation to the disappearance of lignin. Biogeochemistry 4:219–224

Bobbink R (1989) *Brachypodium pinnatum* and the species diversity in chalk grasslands. PhD Thesis, Univ of Utrecht

Davy AJ, Taylor K (1974) Seasonal patterns of nitrogen availability in contrasting soils in the chiltern hills. J Ecol 62:793–807

Doing H (1988) Landschapsecologie van de Nederlandse kust. Stichting Duinbehoud en Stichting Publikatiefonds Duinen, Leiden

Dopheide JCR, Verstraten JM (1995) The impact of atmospheric deposition on the soil and soil water composition of the coastal dry dunes. Rep Lab Phys Geogr and Soil Sci 54, Univ of Amsterdam

Eisma D (1968) Composition, origin and distribution of Dutch coastal sands between Hoek van Holland and the island of Vlieland. PhD Thesis, Rijksuniversiteit Groningen

Everts FH, Fresco LMF, Pranger DP, Berg GJ, Til M van (2000) Beweiding op het Eiland van Rolvers; analyse permanente kwadraten 1983–1999. Rapport Everts & de Vries e.a., ecologisch advies en onderzoek. EV 00/17

Gerlach A, Albers EA, Broedlin W (1994) Development of the nitrogen cycle in the soils of a coastal dune succession. Acta Bot Neerl 43:189–203

Haaf C ten (1999a) Zwanenwater Slahoek 1998–1999; Monitoring van effectgerichte maatregelen tegen verzuring in open droge duinen. Referentie projekt begrazing. Ten Haaf en Bakker ecologisch en hydrologisch adviesbureau, Alkmaar

Haaf C ten (1999b) Duin en Kruidberg Zuidervlak 1998–1999; Monitoring van effectgerichte maatregelen tegen verzuring in open droge duinen. Referentie projekt begrazing. Ten Haaf en Bakker ecologisch en hydrologisch adviesbureau, Alkmaar

Hassink J (1994) Effects of soil texture and grassland management on soil organic C and N and rates of C and N mineralization. Soil Biol Biochem 26:1221–1231

Hassink J, Bouwman LA, Zwart KB, Bloem J, Brussaard L (1993) Relationships between soil texture, soil structure, physical protection of organic matter, soil biota, and C and N mineralization in grassland soils. Geoderma 57:105–128

Koerselman W, Meuleman AFM (1996) The vegetation N:P ratio: a new tool to detect the nature of nutrient limitation. J Appl Ecol 33:1441–1450

Kooijman AM, Besse M (2002) On the higher availability of N and P in lime-poor than in lime-rich coastal dunes in the Netherlands. J Ecol 90:394–403

Kooijman AM, Haan MWA de (1995) Grazing as a measure against grass-encroachment in Dutch dry dune grasslands: effects on vegetation and soil. J Coastal Conserv 1:127–134

Kooijman AM, Dopheide J, Sevink J, Takken I, Verstraten JM (1998) Nutrient limitation and their implications on the effects of atmospheric deposition in coastal dunes: lime-poor and lime-rich sites in the Netherlands. J Ecol 86:511–526

Kooijman AM, Besse M, Haak R (2000) Effectgerichte maatregelen tegen verzuring en eutrofiering in open droge duinen, Eindrapport fase 2, 1996–1999. Report Univ of Amsterdam, 120 pp

Lambers H, Chapin III FS, Pons TL (1998) Plant physiological ecology. Springer, Berlin Heidelberg New York

Lindsay WL, Moreno EC (1966) Phosphate phase equilibria in soils. SSSA Proc 24:177–182

Meulen F van der, Jungerius PD (1989) Landscape development in Dutch coastal dunes: the breakdown and restoration of geomorphological and geohydrological processes. In: Gimingham CH, Ritchie W, Willetts BB, Willis AJ (eds) Coastal sand dunes. Proc R Soc Edinb (Sect B) 96:219–229

Meulen F van der, Kooijman AM, Veer MAC, Boxel JH van (1996) Effectgerichte maatregelen tegen verzuring en eutrofiëring in open droge duinen; eindrapport fase 1. Fysisch Geografisch en Bodemkundig Laboratorium, Univ of Amsterdam

Neitzke M (1998) Changes in nitrogen supply along transects from farmland to calcareous grassland. Z Pflanzenernähr Bodenk 161:639–646

Olff H, Ritchie ME (1998) Effects of herbivores on grassland plant diversity. Tree 13:261–265

Pavlik BM (1983) Nutrient and productivity relations of the dune grasses *Ammophila arenaria* and *Elymus mollis* I. Blade photosynthesis and nitrogen use efficiency in the laboratory and field. Oecologia 57:227–232

Pluis JLA (1993) Algae in the spontaneous stabilization of blow-outs. PhD Thesis, Univ of Amsterdam

Sjöberg RM, Persson T (1998) Turnover of carbon and nitrogen in coniferous forest soils of different N-status and under different $^{15}NH_4$-N application rate. Environ Pollut 102:385–393

Scheffer F (1982) Scheffer/Schachtschabel; Lehrbuch der Bodenkunde. Enke Verlag, Stuttgart

Swift MJ, Heal OW, Anderson JM (1979) Decomposition in terrestrial ecosystems. Blackwell, Oxford

Veer MAC (1997) Nitrogen availability in relation to vegetation changes resulting from grass-encroachment in Dutch dry dunes. J Coastal Conserv 3:41–48

Veer MAC (1998) Effects of grass-encroachment and management measures on vegetation and soil of coastal dry dune grasslands. PhD Thesis, Univ of Amsterdam

Veer MAC, Kooijman AM (1997) Effects of grass-encroachment on vegetation and soil in Dutch dry dune grasslands. Plant Soil 192:119–128

Verhoeven JTA, Kooijman AM, Wirdum G van (1988) Mineralization of N and P along a trophic gradient in a freshwater mire. Biogeochemistry 6:31–43

Verhoeven JTA, Maltby E, Schmitz MB (1990) Nitrogen and phosphorus mineralization in fens and bogs. J Ecol 78:713–726

Weeda EJ, Westra J, Westra Ch, Westra T (1994) Nederlandse oecologische flora, wilde planten en hun relaties 5. IVN, Amsterdam

Westhoff V, Bakker PA, Leeuwen CG van, Voo EE van der (1970) Wilde planten; flora en vegetatie in onze natuurgebieden. Deel 1. Vereniging tot behoud van natuurmonumenten in Nederland. 's-Graveland

16 The Costs of Our Coasts: Examples of Dynamic Dune Management from Western Europe

F. van der Meulen, T.W.M. Bakker, and J.A. Houston

16.1 Introduction

Coastal dunes are highly diverse and valuable, but vulnerable ecosystems. In certain regions of the densely populated western part of Europe, coastal dunes are multifunctional, serving several, often conflicting functions at the same time, such as nature conservation, recreation, drinking water catchment and coastal defence. Yet nature management has been accepted politically and socially. Who are the managers, what are the costs of this management and what does society gain from it? An example is given from the Sefton Coast in north-west England near Liverpool and from Meijendel, near The Hague, The Netherlands.

16.2 Coastal Dunes: Dynamic Systems and Management

Coastal dunes are extensively described elsewhere in this book. For a recent overview of dunes as a dynamic system in management and conservation, the reader is referred to Arens et al. (2001). For this chapter, some essential concepts are summarised.

A useful model of a dune system is given by Bakker et al. (1979; Fig. 16.1). This is a hierarchical model, showing, from top to bottom, the landscape-forming factors (climate, coastline processes, geological substratum, relief, groundwater, soil and vegetation) and their associated processes (respectively: changes in temperature, precipitation, etc.; coastal erosion and accretion, etc.). Two kinds of process are distinguished: natural (left) and human induced (right). In the latter, the role of man as a landscape-forming factor is given. The hierarchical arrangement indicates that factors (and processes) towards the top of the model dominate more over those at lower levels than

Ecological Studies, Vol. 171
M.L. Martínez, N.P. Psuty (Eds.)
Coastal Dunes, Ecology and Conservation

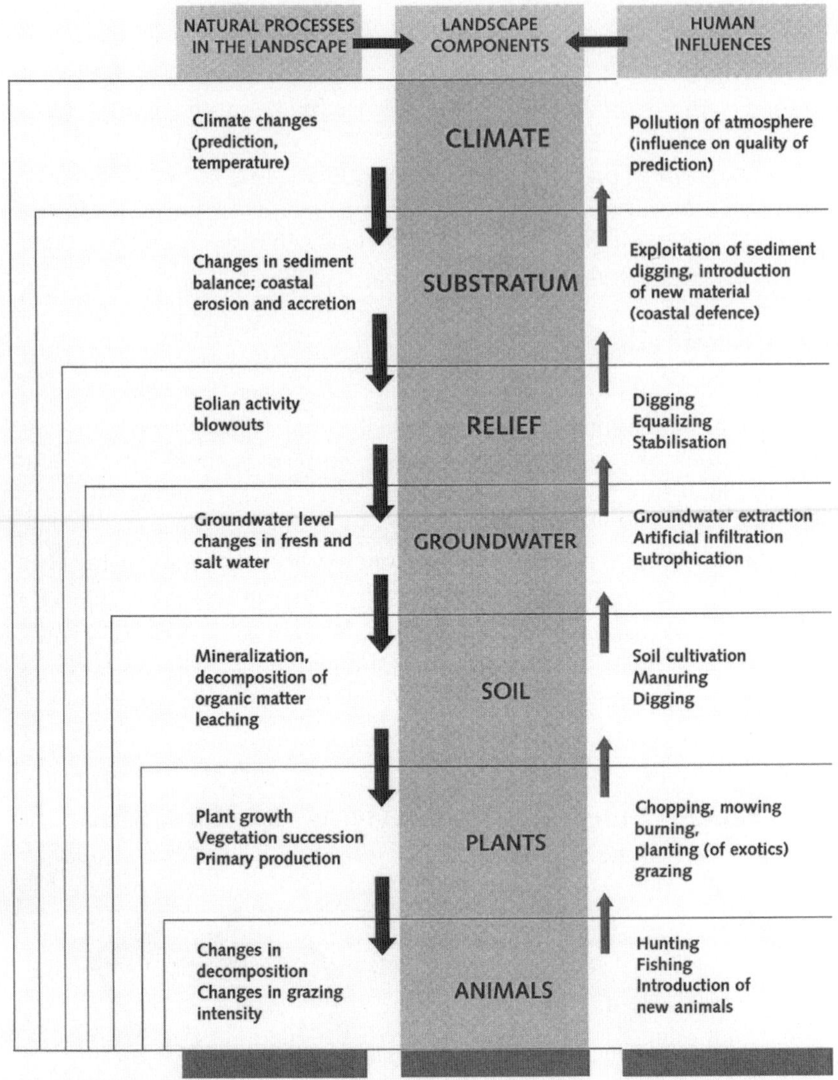

Fig. 16.1. Hierarchical model of landscape forming factors. (Bakker et al. 1979)

the other way around. For example, relief dominates over vegetation and soil (compare north-facing and south-facing slopes and their different soil and vegetation as a result of microclimate differences); in final instance, climate dominates over all factors. Not to be seen in the model is the importance of scale (temporal as well as spatial): processes on top of the model take more time and space than those near the bottom (coastal erosion acts over kilometres, while soil leaching acts over centimetres or decimetres). It is clear that changing of climate will ultimately affect all processes and may change a dune completely. Some signs of this can already been seen in the effects of (acid and eutrophic) atmospheric deposition (Harkel 1998).

What is the use of this model for the manager? At first, it presents the dune landscape in an orderly way, a frame for thinking and managing. Every management plan eventually results in a measure, in a physical intervention in the landscape. Mowing, grazing, cutting of sods, the reintroduction or stabilisation of mobile dunes, all these measures influence the existing processes and introduce new processes. Looking at the model, the manager can assess at which level to intervene in the dune system and how this intervention might influence processes at other levels. Dynamic management primarily is working with natural processes. The more managers are in line with the natural processes at higher hierarchical levels, the more they are managing towards a "natural" system. This is what is the objective in a so-called natural core area (see later this chapter).

16.3 Examples from Western Europe: England and The Netherlands

Dunes, because of their very position at the border between land and sea, are attractive for many other functions than nature conservation alone. This is especially the case in intensively used dunes of western Europe, but it holds for many places over the world, where dunes border densely populated areas. Dunes become multifunctional landscapes, and the manager has to deal with several, often conflicting, uses. How does this work and what are the costs of these coasts?

We give two examples of managers taking care of important dune systems adjacent to large industrial and populated areas: Sefton Coast dunes, near the city of Liverpool, north-west England and Meijendel dunes, near the city of The Hague, The Netherlands. These dunes are under great pressure, but nature management has been accepted politically and socially. Despite great pressure, it was decided to maintain nature conservation areas in the middle of densely populated land, both for the benefit of nature itself and for other uses, such as recreation, production of drinking water and coastal defence. Who are the managers, how is the management organised and what does it cost? What does society have to pay to maintain these dune areas and what does it gain from it?

The management of the Sefton Coast Dunes and the Meijendel Dunes will give more insight into these questions. To understand management in its social-cultural context, it is interesting to see how the attitude towards nature management has changed in the past 50 years in The Netherlands.

Towards the middle of the 20th century, nature conservation concentrated on (rare) species, species habitats and inventories of nature areas. This was followed by more attention for the physical and biological processes, which, in fact, "produced" these habitats, and for whole landscape ecosystems, consisting of aggregates of different habitats. As a consequence, more attention focused on process management. At the onset of the 21st century, more and more managers became members of society, offering a "product" to be used by that society. The general public commented on the management and the managers were explaining to the public why certain measures were necessary. Managers and politicians asked for the quality of the "product", the appreciation by the general public and the costs and benefits of all this work.

16.4 The Sefton Coast (England)

16.4.1 Area

The Sefton Coast in north-west England is a 14-km arc of dunes lying between the city of Liverpool and the resort town of Southport (Fig. 16.2). The coastal landscape is formed by a wide foreshore, a dune system of high (20–30 m) dunes and a dune backland of older low dunes. In the early 18th century, the dune system had a wilderness character and was used mainly agriculturally with extensive areas of rabbit warren. In the late 19th century, rapid changes took place with the construction of railways, urbanisation, the laying out of golf courses, military use, the planting of pinewoods and growth in tourism and recreation. In the 20th century, increasing development pressures and increased tourism led to the loss of many dune areas and severe damage to others.

The three primary functions of the area are coastal defence (the dune system helps to protect over 150 km^2 of low-lying land from flooding), nature conservation and recreation. Other issues include military training, golf and agriculture. It is estimated that the dunes and beaches (excluding the resort town of Southport) are visited by 1.3 million people a year (Atkins 2001).

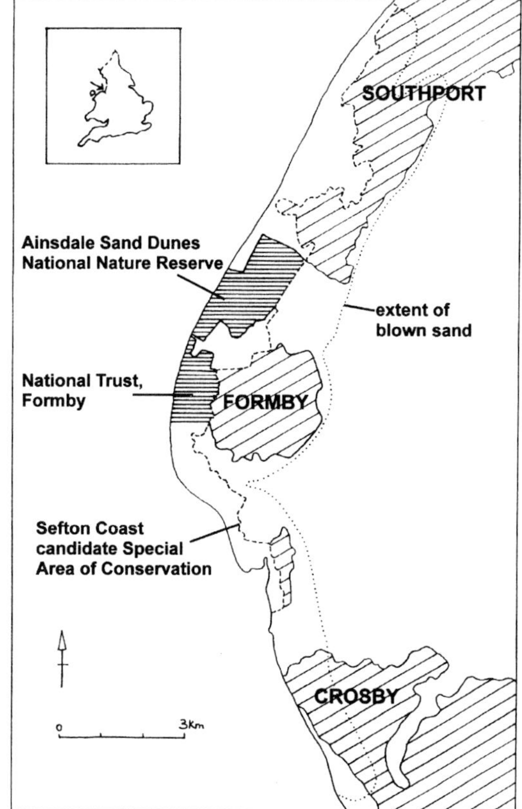

Fig. 16.2. The Sefton Coast in north-west England showing the extent of blown sand, the main urban areas, the boundary of the Sefton Coast candidate Special Area of Conservation (covering most of the remaining area of open dunes) and the location of the sites referred to in the text

16.4.2 Management

The breakup of private estates in the early 20th century led to piecemeal land acquisition with currently 12 managers within an area of 2000 ha. Despite attempts to establish a co-ordinated management framework in the 1960s, it was not until 1978 that a proposal by the County Council, the District Council and the Government's countryside advisory agency led to the establishment of the Sefton Coast Management Scheme (see Houston and Jones 1987).

The management scheme brought together statutory agencies, principal land managers and conservation and recreation interests in a collaborative partnership to restore and enhance the dune coast. The three principal managers are Sefton Council (the local authority), English Nature (a government body) and the National Trust (a land-owning charity). At first, coast defence issues took priority over conservation and recreation interests and a programme of dune building and dune repair was initiated. Even looking back today from the perspective of a dynamic approach to dune management, the dunes of the Sefton Coast had suffered unacceptable damage from uncontrolled recreation pressure, vehicular damage and loss of natural foredunes.

The restoration works took a decade and were only possible through central government funding through unemployment relief programmes. The focus of the work was to control recreation pressures whilst repairing the extensive damage to the dune landscape and planning for the future with the designation of nature reserves, employment of a ranger service and encouragement of volunteer and community support. By the end of the 1980s, the Sefton Coast Management Scheme had developed effective techniques for dune stabilisation and recreation management. Experience demonstrated the high costs that can result from a lack of management and emphasised the need for appropriate long-term management. The total costs of this work between 1977–1985 were over 2 million Euros at 1985 prices with the employment of over 600 labourers (Wood 1985).

With many of the earlier problems, such as the trampling damage to the foredunes, under control, work in the 1990s focused on working with natural processes and developing a greater understanding of habitat and species management requirements. However, managers' adoption of new approaches to dune management was not always sufficiently supported by public information and education. Consequently, the managers sometimes found that their ideas differed from the views of sectors of the local community. There is a similar shared experience throughout north-west Europe where dune managers have found it increasingly important to listen to the views of local people (Edmondson and Velmans 2001; Geelen 2001; Zwart 2001). These authors advocate that it is no longer sufficient for managers to simply provide access or undertake conservation work without public consultation. The need for public support and accountability is becoming increasingly important. This has a direct bearing on the cost of management and, in some instances, community officers have been appointed to improve liaison with local communities.

The Sefton Coast, with several land-owning interests, multiple uses and social constraints imposed by a combination of local needs and local politics, requires a special form of management. Here, as in The Netherlands, the 'ideal' ecological or geomorphological position cannot be realistically achieved and management objectives must aim to satisfy a balance of interests.

16.4.3 The Sefton Coast in the 21st Century

A pattern of land use and land management has become well established on the Sefton Coast. The last remaining substantial area of private duneland was purchased by the local authority in 1995 and most of the dune area falls within the Sefton Coast candidate Special Area of Conservation (a Natura 2000 site, based on the UK system of Sites of Special Scientific Interest). The co-ordination of management is guided by a non-statutory Coast Management Plan prepared by the Sefton Coast Partnership (formerly the Sefton Coast Management Scheme). The plan supports the statutory planning and sea defence functions of the local authority and develops more detailed strategies for nature conservation, woodland management and visitor management.

Recreation has changed considerably. In the early 20th century, visitors came primarily to the resort towns and beaches. Photographs from the 1930s onwards show beaches packed with cars (Fig. 16.3). These activities remain popular, but in recent decades the use of dunes themselves as a recreation area has significantly increased. Whilst beach activities are very seasonal, the use of the dune area (walking) is year-round. Part of this popularity is the general trend towards more active pursuits and gentle exercise, but also a growing local population who use the dunes as their open space. The town of Formby, e.g., grew from a pre-World War II population of 3000 to its current 30,000. Managers responded to the increasing pressure with greater use of paths, boardwalks, fencing and signposting.

Fig. 16.3. Car parking on Ainsdale beach, Sefton Coast 1938. Photograph by R.K. Gresswell, English Nature archive, Ainsdale Sand Dunes NNR

16.4.4 Recreation

16.4.4.1 Visitor Research

The pattern of landownership on the Sefton Coast helps to create a number of zones with differing levels of access. The area, as a whole, can provide for the four different kinds of nature identified by Korf (1997): semi-park nature, accessible nature, 'wild' nature and strict reserve.

In a situation similar to Meijendel, visitors to the Sefton Coast put a greater value on an introduced feature, the pinewoods and their associated red squirrels, than on the natural dune landscape. Dune managers, whilst naturally wishing to promote the conservation value of open dune landscapes, need to be aware of these popular opinions if they are to successfully work with their local communities and visitors.

16.4.4.2 Visitor Typology

On the Sefton Coast, a comparison can be made between the dune area managed as a National Nature Reserve by English Nature and an adjacent area managed primarily as an amenity by The National Trust (Fig. 16.2, Table 16.1). The National Nature Reserve is a relatively large area of 'wild' nature, whereas the National Trust's property would fall into the semi-park category. Visitor surveys, however, reveal some surprising results. Most visitors to the nature reserve come to walk, relax and enjoy the scenery (only one of five visitors mentions 'nature' as a reason for their visit), whereas on the much busier

Table 16.1. Comparative costs of management for two sites on the Sefton Coast

Property	Estimated visitors/year[a]	Area[b]	Cost per ha/year[c]	Cost per visitor (Euros)
National Trust, Formby	340,000	345 (170 dunes)	880 Euros (1800 if only dune area considered)	0.9
Ainsdale Sand Dunes National Nature Reserve	55,000	495 (340 dunes)	285 Euros (415 if only dune area considered)	4.1

The National Trust figures are gross costs; income from admissions, members and other sources can achieve a surplus

[a] Atkins (2001)

[b] Areas include intertidal land

[c] All costs associated with site management

National Trust property over 40 % of visitors cited 'nature' as a reason for their visit.

Such results are interesting for the management of multi-use or multi-ownership sites. Nature conservation is no longer the sole preserve of nature reserves. The promotion of biodiversity has shown that nature is everywhere, yet 'nature reserves' still have a place for society, and perhaps an increasing role as places of tranquility.

16.4.5 Costs of Management

Dune management encompasses elements of coast protection, nature conservation, landscape and amenity, access and recreation and sustainable commercial activities (e.g. forestry, water supply, golf). Studies attempting to measure the socio-economic costs and benefits of integrated coastal management (Firn Crichton Roberts and University of Strathclyde 2000) look at both qualitative and quantitative impacts. Qualitative benefits of multiple-use management can include better partner understanding, stronger community feeling, more sustainable activities and improvements to landscapes and habitats. Quantitative benefits included habitat protection (the value of safeguarding habitats from deterioration), local infrastructure and business and tourism benefits. To simplify comparisons, the costs in Table 16.1 do not include those for coast defence, the management of tourist beaches or research and monitoring.

Meijendel (along with the dunes of the Amsterdam Water Works and the Provincial Water Company of North Holland to the north) is a large area under single management regime. In contrast, the Sefton Coast dunes are divided between several owners and so, to achieve the same level of management as the Dutch sites, there is a need for a partnership approach. The cost of maintaining an integrated coastal management Unit for the Sefton Coast is 170,000 Euros per year. The costs of this are largely borne by the local authority, although the coordination activity of the unit helps to secure considerable external funding for nature and recreation projects.

A dynamic approach to management is now integral to the overall management of the coast. Much less effort is now needed in dune restoration work or in fencing to reduce damage to habitats. Recreation provision has to be worked into the natural processes operating along the coast. The introduction of a Beach Management Plan by the local authority in 1993 has zoned the use of the beaches, reduced recreational damage to the foredunes and encouraged the formation of new embryo dunes and slacks. There has been little additional cost to the manager for substantial nature conservation and coast defence gains.

16.5 The Meijendel Dunes (The Netherlands)

16.5.1 Area

The Meijendel dunes are situated north of The Hague in one of the most densely populated areas along the mainland coast of The Netherlands (Fig. 16.4). They stretch for 6 km along the coast and are 3.5-km wide. In these 2000 ha of dunes a rich variety of animals and plants are found. Meijendel is part of the Ecological Main Framework of The Netherlands. This framework is defined by the national government and areas within this framework have the highest degree of protection (LNV 1992).

Meijendel is managed as a nature reserve and a drinking water catchment area (see, e.g. Bakker and Stuyfzand 1993). It is of great importance for the production of drinking water for the people of the surrounding cities (1.5 mil-

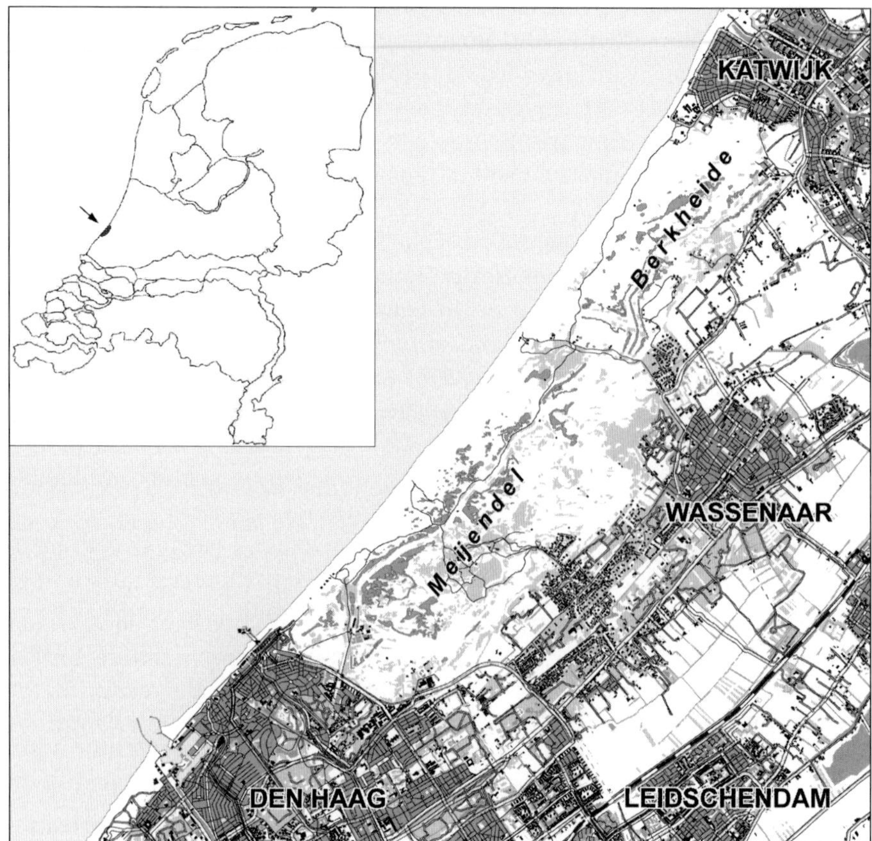

Fig. 16.4. Meijendel in west Netherlands

lion people). Further, the dunes play a major role in sea defence by protecting the low-lying western part of the country against the sea. Last, but not least, it is an important recreation area for about 3 million people living within 25 km distance. About 1 million people visit Meijendel yearly; this does not include people visiting the beach.

16.5.2 Management by the Dunewater Company

The management of the area is done by the Dunewater Company of South Holland. It deals with the more or less conflicting interests of nature-conservation, production of drinking water, coastal defence and recreation. In a management plan for the area, the company sets out the goals for the future. In 2000, a new management plan for the years 2000–2009 was completed (Vertegaal 2000).

Zonation is an important tool in dealing with conflicting interests (Fig. 16.5). In some parts of the area, the production of drinking water has priority, while in others the priorities are coastal defence, nature conservation or recreation. Most important is that in the entire area, solutions are sought for in which particular functions take care of the interests of other functions. In the new management plan three main aims are formulated:

1. Better possibilities for recreation, e.g. better possibilities for people who enjoy nature.
2. The development of a so-called natural core area of some 700 acres within the dune area.
3. Optimisation of the production of drinking water in combination with nature conservation.

16.5.3 Recreation: Better Possibilities for People to Enjoy Nature

Since World War II, recreational use of the dunes has been planned. In the 1950s, the general idea was that large parts of Meijendel should be made attractive for recreation through the planting of trees and shrubs. Small cosy corners were created, where families could picnic or play. Meijendel should look more like a city park than a nature-reserve. An important development at the end of the 1950s was the increasing number of people coming to Meijendel by car: on sunny days, the central dune area (where day recreation was concentrated) was turned into an enormous car park (Schoep and van der Toorn 2001).

In the 1960s, the awareness grew that, together with the increase in recreation, the natural value of the area decreased. A better zonation for recreation was realised. Parking of cars within the area was limited. Cycle and walking paths were made. In the following years, the number of visitors was more or

Primary functions
Nature and landscape
Recreation
Drinking water recovery
Coastal defence
KATWIJK
Berkheide
Meijendel
WASSENAAR
DEN HAAG
LEIDSCHENDAM

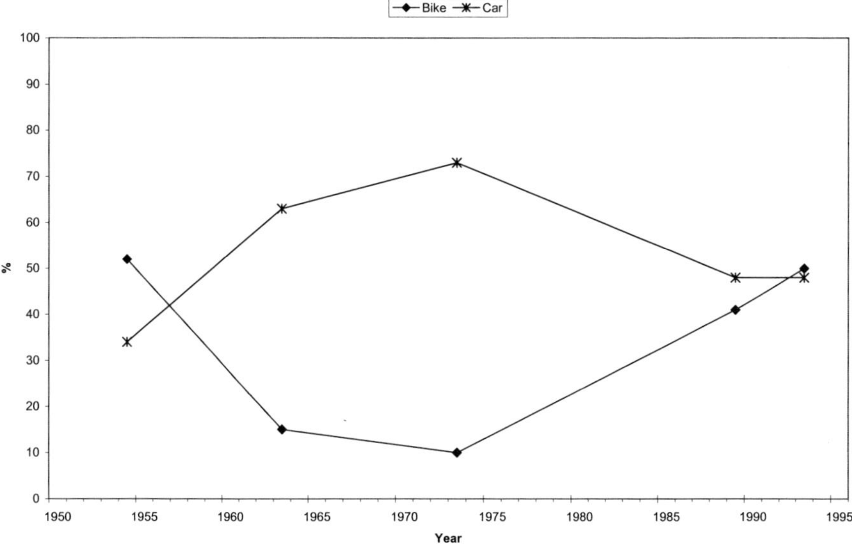

Fig. 16.6. People coming to Meijendel by car and by bike in 1954, 1963, 1974, 1990 and 2000

less stable, but the number of people who came to the dunes by bike increased (Fig. 16.6). The Dunewater Company did not stimulate recreational use but mainly dealt with nature conservation and production of drinking water. Recreation was "tolerated" and there was no clear vision on the way recreation should or could take place. Areas for the production of drinking water were not open to the public.

In the 1990s, this attitude changed. The company strived for more and better opportunities for people to enjoy nature. In the new management plan this is even one of the main goals. The general idea is that recreation concentrating on nature is far better to combine with the maintaining of natural values and the demands of the production of drinking water than was thought until now. Consequently, the public access to the area will increase and areas where the production of drinking water takes place are no longer strictly forbidden. In areas near the city, the infrastructure will improve, especially for walking, and more people will be allowed to enter and enjoy the dunes.

In 2000, a recreation study was carried out in Meijendel (Bakker 2001). Visitors gave their opinion on the management and on the recreational facilities.

Fig. 16.5. (*top*) Zonation of functions in Meijendel

Fig. 16.7. (*bottom*) Artificial infiltration lakes for the production of drinking water in Meijendel

The main reason to visit the area appeared to be walking and enjoying nature. In general (80 %), visitors were very content with the area. Cyclists, dogs and bustle and noise were the most irritating items people mentioned. A surprising result was that people highly appreciate the most artificial type of landscape: the artificial dune lakes for the production of drinking water. An explanation may be that these lakes have a natural appearance, because existing dune valleys have been inundated.

16.5.4 Meijendel and the Production of Drinking Water

Fresh groundwater is scarce in the western part of The Netherlands. Beneath the dunes, a limited amount of fresh groundwater is present. This is well suited as a source of drinking water. In 1874, the extraction of drinking water started in Meijendel. A canal was dug in the dunes and collected superficial groundwater, which was delivered to The Hague. After decades, a lowering of the groundwater table by several metres occurred, and wet slacks dried out. After about 50 years, the amount of extracted water decreased so much that in places brackish water, unsuited for drinking water, was pumped up. A new technique had to be found. Water from the river Rhine was brought to the dunes via a pipeline and became the major source of "artificial" groundwater. Via artificial lakes this water "flows" to the groundwater and then into the extraction wells. Since 1976, pre-treated water from the river Meuse is used for this artificial recharge. The benefits of producing drinking water in this way include:

1. Artificial groundwater is free from pathogenic organisms.
2. Artificial groundwater has, unlike river water, a more or less constant quality and temperature.
3. Beneath the dunes, a huge fresh groundwater dome is present. In case of an interruption of the delivery of water from the river to the dunes, this water body serves as a reservoir of drinking water for months.

Nowadays, the amount of drinking water produced in Meijendel is 50 million m^3/year (Fig. 16.7).

16.5.5 Development of a Natural Core Area

Because the production of drinking water takes place almost everywhere in Meijendel, hydrological circumstances are also artificial. Therefore, wet slacks and their vulnerable and rare biota are scarce. Another typical characteristic of dunes, the free mobility of sand, is restricted to a high degree, due to the planting of Marram grass for centuries.

As a consequence, man now steers two of the most important natural processes in dunes (cf. Fig. 16.1), the geomorphological and the hydrological processes. The National Government emphasises the great importance of areas where circumstances are completely or almost completely "natural". The creation of these "natural core areas" is an important item in the Ecological Main Framework of The Netherlands, to which all Dutch dunes belong. One of the objects in the new management plan is to create such an area in Meijendel . It will cover over 500 ha of the total of 2000.

To realise this, fairly big changes are needed in the way the production of drinking water now takes place. In parts of the area the production will have to stop. In other parts it will be intensified. Also, the way in which coastal defence, and more specifically the planting of Marram grass, take place, will have to change. Dunes, even foredunes directly at the seafront, should be allowed to be mobile again. During the last decade, nature conservationists showed to the public opinion and the public bodies that the systematic and complete fixation of dunes (which has been practice for over a century) prevents the free blowing of sand and damages the natural values, because rejuvenating of landscape and vegetation cannot take place. Nowadays, small-scale blowouts are tolerated and even stimulated at places. Figure 16.8 gives an idea of the amount of Marram grass planted since the 1950s. Because of a decrease in the amount of planted Marram grass, blowouts have increased, although only of small size and on a very small scale. There is no evidence that because of the stop of Marram planting, large-scale and uncontrolled blowouts develop. At present, the idea is that large-scale blowouts and parabolic dunes will only occur under specific climatic circumstances and perhaps even under specific human influence.

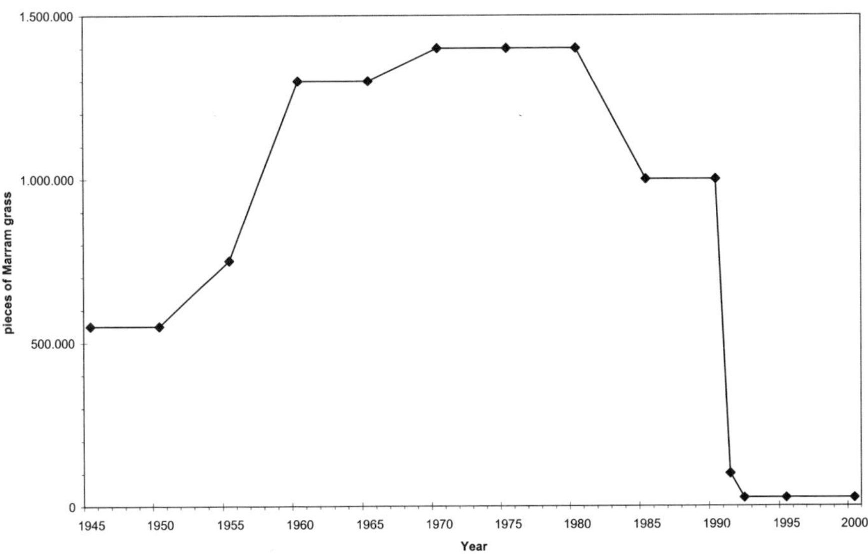

Fig. 16.8. Total amount of planted Marram grass for the years 1950–2000

Recreation in a nature core area will only be possible on foot. The costs of creating such an area are high. The changes in the way the production of drinking water is going to take place will cost some 10 million Euros, to mention just one aspect. Part of this amount will be paid by the drinking water company in the course of the regular investments.

To establish a nature core area, as mentioned above, it will be necessary to produce drinking water in a smaller area. To realise this, the drinking water company uses a concept called: "optimal artificial recharge". The area where drinking water is produced must be minimised, while the natural values of that area should increase at the same time. Artificial lakes and wells are constructed in such a way that natural values of the area are high. On the other hand, civil engineers will construct the lakes and wells in such a way that the amount of water produced per unit of space is higher than before.

16.5.6 Costs of Management

The total costs of management of Meijendel (including surveillance by some 15 wardens) amount to 3.2 million Euros per year. All this is paid by the company, although only a small part (0.65 million Euros) is a direct consequence of the production of drinking water: maintenance of infrastructure such as roads, wells and artificial lakes. The yearly income of the company is some 84 million Euros. All this money comes from the customers by selling drinking water to them. Compared to this amount, the costs of nature management (3.2 million Euros; only 4 %) are remarkably small. Especially, when we keep in mind the benefits of the dunes for the production of drinking water as mentioned above.

Nature management and recreation management cost 960 Euros ha^{-1} $year^{-1}$. As shown in Table 16.2, this is, compared to other Dutch organisations dealing with nature conservation, a large amount of money. To explain this big difference two facts are important:

1. Meijendel is situated in the middle of a very densely populated area, directly bordering large cities. Efforts to develop recreational areas are larger than at an average nature-reserve.
2. Meijendel is not a nature reserve pur sang. A huge amount of drinking water is also produced here. This has negative effects on the natural values of the dunes. Intensive management reduces these effects.

Table 16.2. Average costs of management of some Dutch nature reserves, according to annual figures of 1999. (Natuurmonumenten 2000; Staatsbosbeheer 2000). Natuurmonumenten is a more or less private organisation (comparable to the National Trust in England) and manages some 85,000 ha all over The Netherlands. The State Forestry Department is a former state organisation (comparable to English Nature as owner and manager of state land) and manages some 220,000 ha all over The Netherlands

	Average costs of management per ha/year (incl. costs of wardens)	
	All costs including drinking water, infrastructure (Euros)	Nature management infrastructure (Euros)
Dunewater Company	1200	960
Natuurmonumenten		460
State Forestry Department		390

16.5.7 Visitors Appraisal

We have explained that the management of Meijendel has changed considerably over the past 20 years and that still more changes will take place. The manager has asked the visitors about their opinion on all this. A majority (60–70 %) feels these changes are positive. For some aspects, such as the regeneration of artificial lakes into natural dune habitats, and the closing of parking areas, the views are less positive.

In 1997, research was done under the "customers" of the Dunewater Company (NIPO 1997). Of course, the production and distribution of reliable drinking water were regarded as the main task of the company. Good nature management was regarded as one of the other important tasks. About 70 % of the customers were willing to pay more for their drinking water, in order to make this good management possible. About 90 % supported the fact that the costs of nature management are paid as a part of the drinking water fees. In general, it appears that visitors have a very positive view on the activities of the company, both on the production of drinking water and on the way the dunes are managed.

16.6 Dune Management in a Changing Society

In modern western European society, more and more people have free time and money to spend. A considerable amount of time is spent in the open air walking, cycling, jogging and enjoying nature. At densely populated coasts, dunes meet an enormous need, as can be seen in the examples of Sefton Coast (England) and Meijendel (The Netherlands).

Dune managers aim at nature conservation and recreation (for nature). It is found that maintaining natural values of dunes is far better to combine with such forms of recreation and even with the demands of the production of drinking water than was thought until now. At the same time, an important economic use, such as the production of safe, healthy drinking water, is a strong legal form of protection, which is indispensable in densely populated areas.

More than 50 years of management have shown that negative effects of recreation and drinking water production can be minimised to a large degree so that the damage to the coastal ecosystem is relatively small. Active restoration of ecosystems (dune slacks and mobile sands) is also taking place. Management is done in adherence with the natural processes, rather than against them. This is best done in areas of at least several hundreds of hectares in size or more.

The overall costs are considerable, but for the time being, society is willing to bear these costs in turn for more access to the dunes and for more information on the "what" and "why" of the manager's activities. The future will tell us more about the sustainability of these efforts and if all this leads to a real win–win situation, both for society and for nature.

References

Arens SM, Jungerius PD, Meulen F van der (2001) Coastal dunes. In: Warren A, French JR (eds) Habitat conservation: managing the physical environment, Chap 9. Wiley, New York

Atkins WS (2001) Quality of coastal towns: sustainable tourism on Merseyside; assessment of coastal visitor facilities. Unpublished report for Sefton Council (available on www.seftoncoast.org.uk)

Bakker JG (2001) Recreatieonderzoek in duingebied Meijendel in Zuid Holland. Report, JG Bakker, Bennekom. 23 pp

Bakker TWM, Stuyfzand PJ (1993) Nature conservation and extraction of drinking water in coastal dunes: the Meijendel area, In: Vos CC, Opdam P (eds) Landscape ecology of a stressed environment. Chapman & Hall, New York, pp 244–262

Bakker TWM, Klijn JA, Zadelhoff FJ van (1979) Duinen en duinvalleien, een landschapsecologische studie van het Nederlandse duingebied. Pudoc, Wageningen

Doing H (1988) De landschapsoecologie van de Nederlandse kust. Stichting Duinbehoud Leiden, 228 pp plus kaartbijlagen

Edmondson SE, Velmans C (2001) Public perception of nature management on a sand dune system. In: Houston JA, Edmondson SE, Rooney PJ (eds) Coastal dune management: shared experience of European conservation practice. Liverpool University Press, Liverpool, pp 206–218

Firn Crichton Roberts/University of Strathclyde (2000) An assessment of the socio-economic costs and benefits of integrated coastal zone management. Report to the European Commission. Firn Crichton Roberts Ltd. and Graduate School of Environmental Studies, Univ Strathclyde

Geelen LHWT (2001) Habitat restoration and public relations: a restoration project in the Amsterdam water supply dunes. In: Houston JA, Edmondson SE, Rooney PJ (eds) Coastal dune management: shared experience of European conservation practice. Liverpool University Press, Liverpool, pp 171–176

Harkel MJ ten (1998) Nutrient pools and fluxes in dry coastal dune grasslands. PhD Thesis, Univ Amsterdam, 152 pp

Houston JA, Jones CR (1987) The Sefton Coast management scheme: project and process. Coastal Manager 15(4)267–297

Korf B (1997) Recreation in the North-Holland dune reserve. In: Drees JM (ed) Coastal dunes recreation and planning. EUCC, Leiden, pp 76–80

LNV (1992) Structuurschema Groene ruimte. Het landelijk gebied de moeite waard. Dutch Ministry of Agriculture, Nature Conservation and Fisheries, The Hague

Natuurmonumenten (2000) Samenwerken aan natuur. Jaarverslag 1999. Natuurmonumenten, 's Graveland, 91 pp

NIPO (1997) Imago-onderzoek waterbedrijven. Rapport DZH. 25 pp incl appendices. NIPO, Amsterdam

Schoep J, Toorn B van der (2001) Onderzoek naar waardering en recreatiewensen voor het Noordhollands Duinreservaat. Report for NV PWN

Staatsbosbeheer (2000) Jaarverslag 1999. Staatsbosbeheer, Driebergen, 53 pp

Vertegaal CTM (2000) Beheersplan Berkheide, Meijendel, Solleveld 2000–2009. Deel A en B. Duinwaterbedrijf Zuid-Holland, Voorburg/Katwijk; Staatsbosbeheer, Nieuwegein

Wood P (1985) Dune restoration at Formby Point: report of environmental improvement works carried out by Merseyside County Council 1977–1985. Report, Merseyside County Council, Liverpool

Zwart F (2001) Dune management and communication with local inhabitants. In: Houston JA, Edmondson SE, Rooney PJ (eds) Coastal dune management: shared experience of European conservation practice. Liverpool University Press, Liverpool, pp 219–222

17 Animal Life on Coastal Dunes: From Exploitation and Prosecution to Protection and Monitoring

G. Baeyens and M.L. Martínez

17.1 Introduction

The attitude of humans towards dune mammals and birds has changed throughout history. Initially, in the Middle Ages, wild animals and grazing stock provided food and fur. In Europe, for example, rabbits played a crucial role as meat, fur and felt (van Dam 2001a). Birds provided meat and eggs while deer were hunted for the excellent taste of their meat. Such animal exploitation lasted until the beginning of the 20th century, when bird protection incited habitat protection. The result was that animal groups other than birds were protected as well. Nowadays, dune management aims at the conservation of biodiversity and the stimulation of characteristic processes like dune rejuvenation. To integrate nature with other coastal socio-economic functions as tourism, coastal zone management (CZM) is coming into practice on a worldwide scale. Its implementation requires the knowledge and monitoring of biotic interactions between animals and their effects on the dune ecosystem. This chapter illustrates the change in attitude of humans to dune animals since the Middle Ages and the role biotic interactions have played in coastal dune conservation. Exploitation of economically profitable species and prosecution of predators were the essentials in the past; protection and preservation are the main purposes now and in the future. Do we have the adequate methods to reach this aim? We have gathered the existing information, mostly from European and North American dune reserves, in order to analyze the above.

Ecological Studies, Vol. 171
M.L. Martínez, N.P. Psuty (Eds.)
Coastal Dunes, Ecology and Conservation

17.2 Cropping Stock and Game: The Medieval Coastal Dunes as a Store of Animal Goods

Sand dunes were among the earliest sites that were colonized on the Atlantic coasts of Europe and North America as well as on the Mediterranean coasts (Ranwell 1972; Doody 2001). In these environments, dune animals, both stock and game, served as an important food source: sand dunes were used as meat stores. In the Dutch dunes, for example, human settlements of 3000 and 2000 B.C. revealed remnants of sheep, goats, swine and dogs, indicating the utilization of these animals in everyday life (Klijn and Bakker 1992). Since the Middle Ages until recently, Dutch sand dunes have been grazed by herds of cattle and sheep, that were used for their meat, fur and wool (Baeyens and Duyve 1992). In the Mediterranean, sheep herding has been known for as long as 8000 years (Doody 2001). This intense habitat exploitation occurred over many centuries and throughout Europe. In fact, because of this impact of grazing on community dynamics, Doody tempers the view of those that consider many European coastal habitats as purely natural.

In Europe, grazing by domestic stock has influenced dune formation, vegetation and landscape. Grazing impedes vegetation succession; the development from grassland into scrub and woodland is slackened because trampling can cause erosion that in turn can be enhanced by the wind. In fact, Ranwell (1972) stated that "the structure of sand dune communities in Europe prior myxomatosis (before 1955) was the product of intensive rabbit-grazing". Van Dam (2001b), who recently studied the economic and ecological role of the rabbit in the medieval sand dunes of The Netherlands, goes even further by defining the outer dunes (along the seaside) as a man-made "rabbit garden". Grazing rabbits (*Oryctolagus caniculus*) keep the vegetation short and prevent trees and shrubs from sprouting. Their frequent burrowing for digging up roots as well as for making underground nests and corridors can originate blowouts, when it is done on a large scale.

In the Middle Ages, the rabbit density was artificially enlarged. From the 13th to the 18th century, sand dunes in Britain, Ireland, The Netherlands, Belgium, France and Denmark were used extensively as rabbit warrens (Klijn and Bakker 1992; van Dam 2001b; Doody 2001). Rabbits were cherished by gourmet cooks and by skin sellers. Rabbit hunting was a privilege of the nobility, who were usually the owners of the dune wilderness. Counts and lords urged the warreners to pamper their rabbits by feeding them in wintertime, expelling cats and dogs, prosecuting wild predators and even by deterring shelduck (*Tadorna tadorna*) from nesting in rabbit burrows (Baeyens and Duyve 1992; van Dam 2001a). Such measures can be considered as an initial effort of conservation and management of animals in sand dunes. Thus, the animal harvest was protected and enlarged everywhere, by reducing predation risks. In The Netherlands and probably throughout western Europe, several terrestrial and avian predator species were prosecuted for being direct competitors of the hunting men: hen harriers (*Circus cyaneus*) (Dijksen 1992), foxes (*Vulpes vulpes*) (Klijn and Bakker 1992), stoat (*Mustela erminea*) and polecat (*Mustela putorius*) (Mulder 1990). Predator prosecution persisted until the 20th century and is locally still common practice, especially in Southern Europe, where illegal hunting is commonplace (Tucker and Evans 1997).

The improvement and creation of suitable habitats were other ways to entice palatable species. So, wintering waterfowl have been cropped since the Middle Ages in marshes and ponds in the USA and in Europe (Ranwell 1972). Habitat management of pools, by excavation and by controlling tall marsh growth, has increased the catch, in the past and still in the present. During the 16th, 17th and 18th centuries, the Dutch dunes and shores provided several bird products (Baeyens and Duyve 1992). The omelettes of gull eggs were so much in demand that in 1524 a ban was declared to stop the egg collection. Eggs of geese (Anserinae), lapwings (*Vanellus vanellus*), and plovers (Charadriidae) were also considered a delicacy. Dead gulls (Laridae) were plucked carefully and the feathers desalinated in order to stuff bed mattresses.

Introducing and breeding animal species, especially those that were of culinary value, regularly enlarged the animal crop. Partridges (*Perdix perdix*) and pheasants (*Phasianus colchicus*) were bred like poultry, set free and then shot in the following hunting season. The hunting of pheasants was often combined with corn feeding in the winter season, to lower the natural mortality. These practices were not specifically confined to the dunes but to any large part of land that was used as wilderness. Red deer (*Cervus elaphus*) and fallow deer (*Dama dama*) were intentionally introduced in the Dutch coastal dunes by Prince Maurice (1584–1625) in order to savour the delicious meat. Deer were bought elsewhere and then set free in the dunes so that they could breed and have young. Also roe deer (*Capreolus capreolus*) were hunted under his reign. However, the three species disappeared again during the 17th and 18th centuries (Klijn and Bakker 1992) and only the rabbits remained as permanent catch.

17.3 Nature Conservation Starts with Bird Protection

Hunting rabbits, birds and other animals remained an age-long practice as far as historical data reveal. In the 19th century, the shotgun was carried by quite a different type of men: the ornithologists. They shot birds to stuff them as artist's models. The most elaborate portraits were drawn with the finest detail. In Europe and the US several bird atlases were issued which in turn stimulated bird watching as a sport. Curiosity lead to admiration and admiration lead to greediness as well as to protection. In The Netherlands and elsewhere it came into vogue to attach stuffed birds, wings or tails on lady's hats. This infuriated some bird-lovers to such an extent that the Dutch Association for Bird Protection was founded in 1899. Soon thereafter other unions and associations for nature conservation followed. The British were pioneers in nature protection legislature and the first conservation acts were created: the Sea Birds Protection Act (1869), the Wild Birds Protection Act (1872) and the Wildlfowl Protection Act (1876) (Doody 2001). In 1903, President Roosevelt signed an act to protect egrets, herons and other species on Pelican Island, off the coast of Florida. In 1936 the first Dutch law for bird protection was issued.

Gradually, species protection lead to habitat protection. This triggered the set up of the first wildlife refuges on the coast and elsewhere which later developed into the National Wildlife Refuge System in North America. In Europe and in North America, the first transcontinental studies on bird migration revealed the importance of dunes (and shores) as flyway and stepping stone along the migratory routes. Hence, the increasing need to conserve sites for migratory animals led to several international conventions such as the Western Hemisphere Shorebird Network. This was launched in 1985, covering bird flyways from both North and South America, from the Arctic to Tierra del Fuego (Doody 2001).

In the European Low Countries, the numbers of several coastal bird species increased spectacularly during the 20th century. According to Spaans (1998), the booming numbers of gulls were brought about by the cessation of the heavy prosecution. Early in the 20th century, only the black-headed gull (*Larus ridibundus*) and the herring gull (*L. argentatus*) bred in The Netherlands and both were rather scarce. Between 1908 and 1993, six additional gull species colonized the Dutch coast. The total numbers increased from a few tens of thousands to just over 250,000 in 1996.

Another spectacular numerical increment occurred in waterfowl, following a change in dune management. In the second half of the 19th century, large parts of the mainland coast of The Netherlands were utilized for drinking water extraction. This public function safeguarded the dune reserves from being built over and enhanced nature conservation. When the original aquifers did not supply enough water for the expanding urban populations, the dunes were artificially recharged with river water. Therefore large open water basins were laid out in three dune reserves between 1940 and 1965. Even before then, however, mallard (*Anas platyrhynchos*), shoveler (*Anas clypeata*) and shelduck were already breeding in the dunes. Between 1955 and 1970, other species appeared and numbers increased fast, teal (*Anas crecca*), gadwall (*Anas strepera*), tufted duck (*Aythia fuligula*) and pochard (*Aythia ferina*) (Baeyens and Vader 1990).

The colonization of the dunes by so many new bird species incited many bird watchers to regular monitoring and habitat protection. As a consequence, and step by step, the habitat as a whole gained public interest (van der Meulen and Udo de Haes 1996). Since approximately 1920, the younger sand dunes were progressively opened to the public and an outing to the dunes became more and more fashionable. Getting acquainted with sand dunes and with nature in general aroused a gradual change in attitude.

Social and political events also played an important role in the conservation of sand dune fauna. In the beginning of the 20th century, the outer dunes in western Europe were barren and sandy because of the superabundant rabbits and also because of the desiccation due to ground water extraction. The farmers who built their farm houses in the inner dune slacks, between 1850 and 1930 (Fig. 17.1), were still hunting waterfowl, pheasants, partridges, woodcock (*Scolopax rusticola*), and rabbits. Nature associations as well as governmental institutions started to acquire the first estates, also in coastal dunes. During the Second World War, most European dunes became military grounds, forbidden for the inhabitants and general public. After that, from 1945 to 1970, the European population worked hard at the post-war reconstruction. When the standard of living increased and people had free time during the weekends, nature, including sand dunes, became more enjoyable and tourist centers were developed. From 1970 onwards, nature lovers faced with increasing apprehension the recreational pressure from leisure seekers, whose massive trampling had deleterious effects on the ecosystem. This con-

Fig. 17.1. The exploitation of dune slacks in Belgium in 1904. In between the barren and eroding dunes, the moist slacks were cultivated by fishermen, despite the relative low soil fertility. The soil was enriched with pomace. Grazing stock was confined to the salt marshes further down the coast. These fields were too vulnerable for erosion and therefore protected by hedgerows and sometimes covered with poplar twigs (Photograph by J. Massart, in Vanhecke et al. 1981)

cern resulted in the development of protective legislature in all coastal European countries so that the golden fringe of Europe was conserved (Udo de Haes and Wolters 1992)! The socio-political distinction brought about a physical segregation between leisure space and room for nature, which nowadays takes shape in coastal zone management (Van der Meulen and Udo de Haes 1996; Rigg 1999).

17.4 The Complexity of Biotic Interactions

Biotic interactions lead to additional environmental problems. In medieval times, the dunes were over-cropped by grazing stock as well as by rabbits. Rabbits regularly escaped from the warrens but could never develop a wild population due to predation and food shortage in winter. When deforestation (i.e., reduction of predators) and agriculture (i.e., more food) changed the dunes into more hospitable habitats, the rabbits spread in such large numbers that they became considered as pest (van Dam 2001a). In the 17th century, the Dutch tried to "depopulate" the dunes by hunting rabbits intensely. Domestic stock also brought both benefit (in dune slacks) and damage (on dune ridges) (Drees and Olff 2001). Since 1344, the Dutch authorities promulgated law after law to safeguard the marram grass plantations from being grazed. Herding sheep, goats and cattle in the sea barrier was strictly forbidden (Baeyens and Duyve 1992).

Pine plantations were discovered as the sovereign remedy to fix drifting sand in the Atlantic, Baltic and Mediterranean coasts of Europe (Ranwell 1972). Between 1850 and 1950 large parts (ca. 40 %) of the coastal dunes were afforested with mainly *Pinus nigra*, *Pinus sylvestris* and *Pinus maritima* (Tucker and Evans 1997). Later and even today, the planting of *Eucalyptus camaldulensis* became fashionable around the Mediterranean basin. Later, sea buckthorn (*Hippophae rhamnoides*) a native European dune stabilizer was introduced in Great Britain, Belgium and Ireland. However, its rapid expansion became a serious threat to dune slacks and open dunes which therefore was counteracted by physical removal and treatment with an appropriate herbicide (Doody 2001).

The fixation of European dunes with pine plantations also attracted animals that depend on conifers for nesting and feeding. In the second half of the 20th century, pinewoods and their inhabitants were designated as extraneous organisms in sand dunes but the complete removal of the alien conifers was not considered acceptable. In Great Britain (Houston 1995; Shuttleworth and Gurnell 2001) and in The Netherlands, they are preserved for the sake of red squirrels (*Sciurus vulgaris*), birds of prey and migrating crossbills (*Loxia curvirostra*). Furthermore, pine woods are particularly enjoyed by tourists and visitors since they offer amenity, nature and shade (Ranwell 1972).

In 1954, the fixation of sand dunes in Western Europe was boosted by a totally unexpected event: the outbreak of myxomatosis in the wild rabbit populations. In the Dutch dunes, the population drop was nearly 100 % (van Breukelen, unpublished data).

The cessation of rabbit grazing favoured the sprouting of a new generation of scrub species. According to the sequential analysis of aerial photographs of 1185 ha of the Amsterdam Water Supply Dunes, the scrub coverage increased from 17 % in 1958 to 38 % in 1979. A similar trend in scrub encroachment was measured in all dunes of The Netherlands and Flanders (Belgium) (van Til et al. 2002). Scrub encroachment was enhanced by an increase in annual precipitation in the same period. The rainwater contained more and more nitrogen (Stuyfzand 1991), so that three factors acted at the same time. In areas where the water table rose, coincidentally also from 1955–1960 onwards, landscape succession was further accelerated. In the dry parts of the dunes, scrub plant cover of ca 40 % was reached in 20 years, while in the moist parts, this was reached in ten. The rising ground water table stimulated plant growth and at the same time reduced rabbit grazing because rabbits are unable to dig burrows in wet sand.

The reduction of open and short vegetation resulted in the reduction of foraging habitat for bird species as wheatear (*Oenanthe oenanthe*) and sky-

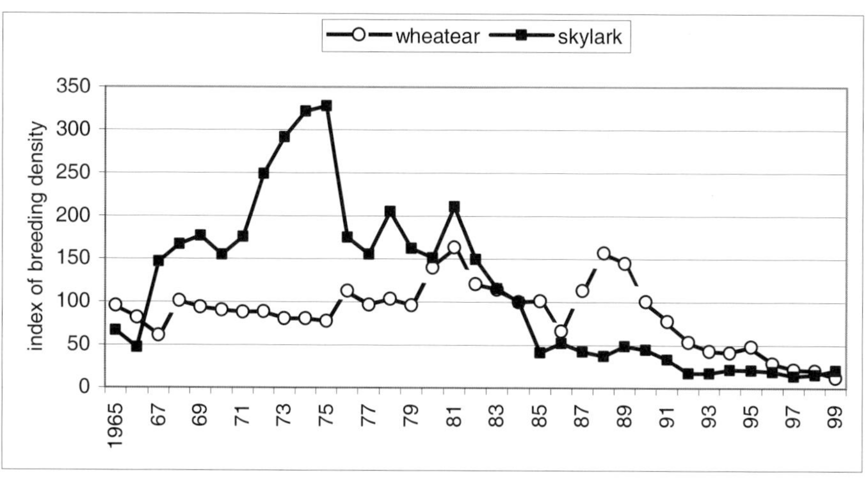

Fig. 17.2. Indexes of breeding densities of birds nesting in open dune vegetation in the mainland coast of Holland (the index of 1984 is set as 100). One would expect these breeding densities to decline to the same extent as the scrub expands and for the last 10 years this expectation meets the data. The increase in skylark (*Alauda arvensis)* numbers from the end of the 1960s is, however, in contradiction with this, and so are the fluctuations of the wheatear (*Oenanthe oenanthe)* index. The preference for nesting in open dunes and the avoidance of scrub are documented in various underlying local studies but the knowledge to interpret these trends as a whole is still lacking (Data from de Nobel et al. 2001)

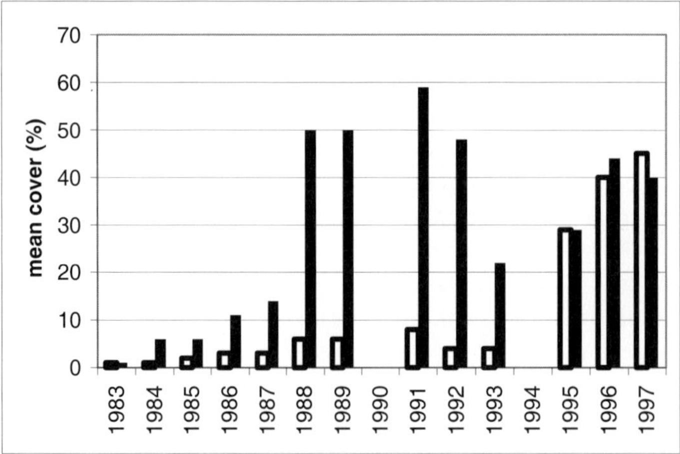

Fig. 17.3. Mean percentage of coverage by wood small reed (*Calamagrostis epigejos*) in a Dutch dune reserve, within a set of exclosures (*black bars*) and in unfenced reference areas (*open bars*). The dominance of wood small reed in the reference areas is now equal to the coverage within the exclosures since rabbit grazing has stopped in both situations. (Data from Snater 1999)

lark (*Alauda arvensis*). The wheatear showed a clear preference for short-grazed and moss vegetation, where it picked up food items more frequently: 24.3 pecks/5 min in moss and 8 or less in other vegetation types (NV PWN 2000). The breeding densities of bird species of open vegetation declined as the scrub coverage progressed (Verstrael and Van Dijk 1996; Fig. 17.2). Because of that, in countries such as Belgium, sea buckthorn is removed on purpose to optimize densities of birds breeding in open mosaic vegetation (Bonte and Hoffmann 2001). In fact, the response of birds breeding in open and short vegetation to scrub encroachment was a response to the fading away of the rabbit (Sierdsema, pers. comm.; Fig. 17.3). Before 1994, rabbit grazing restrained grass encroachment. Since 1994, rabbit populations have collapsed due to the epidemic impact of a new virus: VHD (viral haemorrhagic disease). The dominance of wood small reed in the reference areas is now equal to the coverage within the exclosures since in neither of both situations any grazing occurs (Fig. 17.3). Scrub encroachment was however favourable for bird species breeding in scrubs, like nightingale (*Luscinia megarhynchos*), whitethroat (*Sylvia communis*) and lesser whitethroat (*Sylvia curruca*); their numbers increased (Fig. 17.4).

Additionally, the reappearance of the red fox in the mainland dunes also affected meadow birds, and the breeding numbers decreased even (Van der Vliet and Baeyens 1995; van der Meer 1996; Veenstra and Geelhoed 1997). Other ground-breeding species like pheasant, partridge and woodcock also suffered from the increase of predation pressure (Koning and Baeyens 1990,

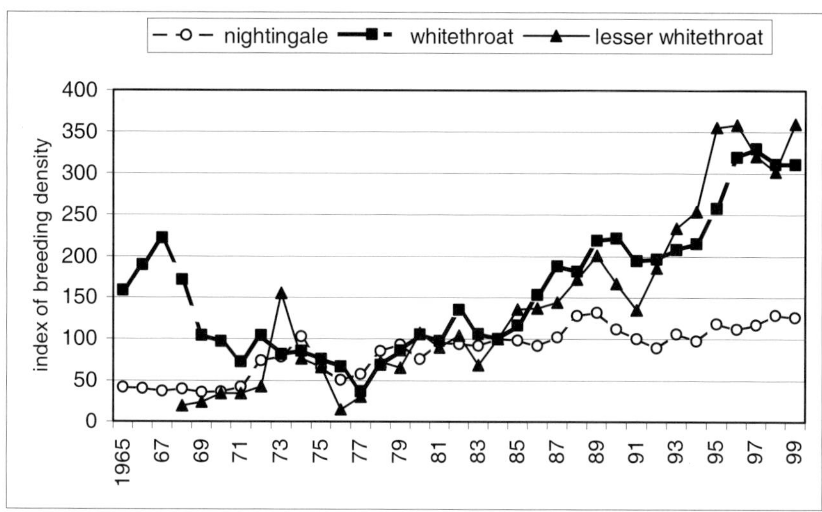

Fig. 17.4. Indexes of breeding densities of birds nesting in scrub in sand dunes of the mainland coast of Holland (the index of 1984 is set as 100). Scrub encroachment was profitable but for these migratory species but conditions in wintering areas always play an additional role. The decline in the whitethroat (*Sylvia communis*) before 1970 was probably influenced by the drought in the Sahel, which is the northern part of the whitethroats wintering grounds. (Data from de Nobel et al. 2001)

NV PWN 2001). In prior centuries, the fox had been exterminated, but between 1968 and 1980 the dunes were recolonized (Mulder 1992) and the population increased rapidly, also because there was (virtually) no hunting. Despite the fact that predation on eggs and chicks of these ground-breeding species was observed, it is not likely that predation was the *only* actual cause of the decline in numbers. In a large part of the Dutch dune reserves, the foxes are not strictly nocturnal and roam around in day time. They roused breeding gulls, ducks, curlews and other nesting ground-breeders so frequently that the birds just deserted their breeding grounds. The most probable explanation for the disappearance of meadow birds is that, the habitat already being deteriorated, nest-predation and disturbance acted as an extra stroke but it is impossible to determine the relative importance of all negative factors. A complex number of factors also affected the decline of some filter-feeding ducks in the Amsterdam Water Supply dunes. The impacting crash in the total number of the most common duck species (Fig. 17.5), frequently lead to instantly blaming the fox. However, a closer look at the species separately shows that shoveler and teal were already decreasing in numbers long before the fox arrived (Fig. 17.6). The banks where they used to forage were gradually overgrown by reed and cattail. The disappearance of foraging opportunities made the area unsuitable for breeding (Baeyens et al. 1997). In summary, the decline of numbers of meadow birds in the main-

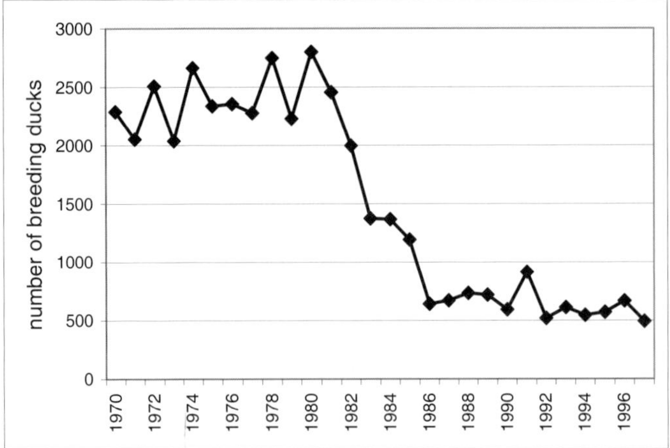

Fig. 17.5. Total number of breeding individuals of ducks in the Amsterdam Water Supply Dunes. The graph indicates an addition of numbers of mallard (*Anas platyrhynchos*), shoveler (*Anas clypeata*), teal (*Anas crecca*), gadwall (*Anas strepera*), shelduck (*Tadorna tadorna*), tufted duck (*Aythia fuligula*) and pochard (*Aythia ferina*), which are the most common breeding duck species in this area. In 1980, the first fox dens were discovered and the population increased very fast in 1981–1984. (Data from Vader, in Baeyens et al. 1997)

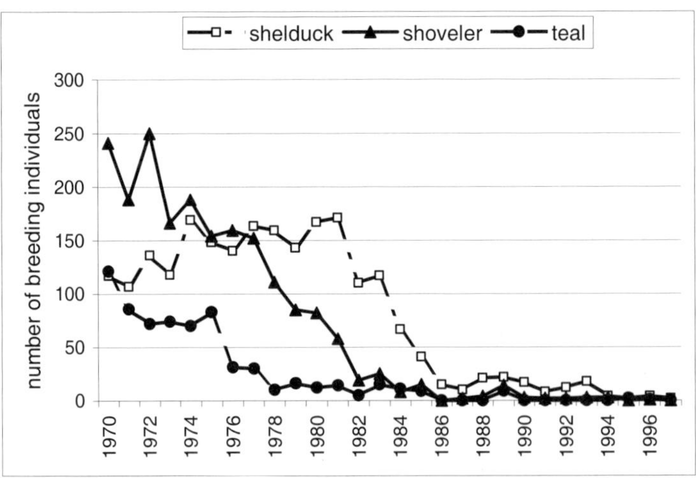

Fig. 17.6. Number of breeding individuals of three species of filter-feeding ducks in the Amsterdam Water Supply Dunes. The decrease in breeding density of shoveler and teal shows a more or less gradual course, starting before the increase in fox numbers since 1980. The filter-feeding species had less and less foraging opportunities as the seepage ponds were gradually overgrown with reed and cattail. The steep drop in shelduck numbers indicates that shelduck probably suffered more from fox predation because they breed in burrows where they are easily traced on smell (Data from Vader, in Baeyens et al. 1997)

land dunes of The Netherlands was thus caused by two main factors: an accelerated landscape succession paralleled by fox disturbance.

Ground-nesting birds and colonial breeders are also vulnerable to predation. In the German Wadden Sea, predation is the main cause of hatching failure for the black-headed gull, the oystercatcher (*Haematopus ostralegus*), the black-winged stilt (*Himantopus himantopus*) and several tern species (Thyen et al. 1998). Aerial predators like gulls and Corvids are sensitive to joint anti-predator defence. When these predators try to approach the nests, all the breeding birds respond together by chasing them away collectively and hence, the chance that eggs or chicks are taken away by gulls or corvids is minimal. Hatching failure is chiefly attributed to predation by night-active mammals like red fox, brown rat (*Rattus norvegicus*), and to a lesser extent to mustelids and hedgehogs (*Erinaceaus europaeus*). Foxes and brown rats have shown recent population increments in almost all European countries resulting in a greater impact on prey species than 30 years ago (Tucker and Evans 1997). For instance, in the mainland coast of The Netherlands, the fox was able to sweep away thousands of gulls in just a few years (Baeyens 1989; Bouman et al. 1991). The gulleries in a mainland dune reserve enlarged gradually in the 1970s: in the innerdunes close to the land side, in the middle dunes with a central position and in the sea barrier, directly behind the beach (Fig. 17.7). From approximately 1980, foxes colonized the dunes and increased fast in numbers. The colonies in the inner dunes were easy to disturb as the foxes had plenty of cover from scrub and shrub. Soon those gulleries were left and birds moved closer to the sea, where the landscape is more open. The effect of fox distur-

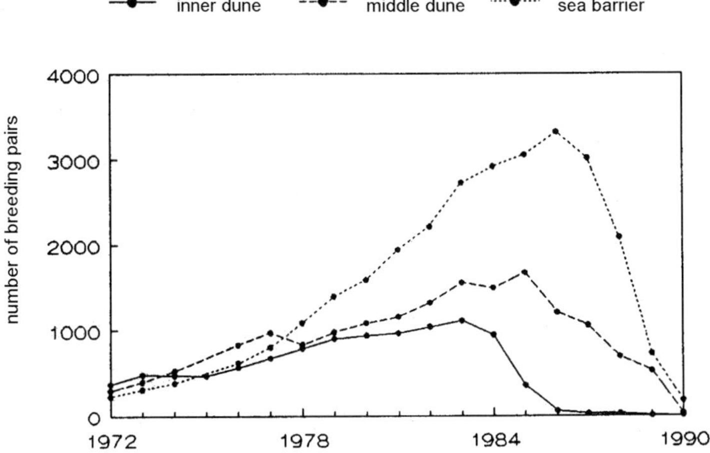

Fig. 17.7. Rise and fall of herring gull colonies in Meijendel, a dune reserve on the mainland coast of Holland. (Bouman et al. 1991)

bance "followed" the gulls and finally the gulleries in the sea barrier were ever so much disturbed that all gulls left the area.

On top of the above, an additional factor played a role: tourism. In The Netherlands, the reduction of working hours from 40 to 36 or even less per week resulted in more visitors having leisure time to spend in nature. In the breeding season, the curlew (*Numenius arquata*) is very sensitive to disturbance by walkers and cyclists (AWD 1981; Vos and Peltzer 1987; Mulder and Swaan 1988). Since approximately 1980, densities of curlew have decreased in all dunes where scrub encroachment and fox occurred concurrently. Despite these two detrimental factors, curlew density increased in two Dutch dune areas after recreation was strongly reduced by closing down foot paths (van Ommering and Verstrael 1987; Slings 1999). Also, in Latvia (Baltic coast), meadow birds disappeared because of the combination of two causes. Dune meadows were no longer mown and overgrown by ruderal vegetation while tourism increased at the same time (Opermanis 1995). Both factors were detrimental for breeding meadow birds. Again it is impossible to quantify how much each separate factor is responsible for the changes as a whole.

The interactions between breeding birds, predators and herbivores that greatly influence vegetation succession can obscure the results of dune management. Sudden changes in food webs can entail domino effects in all links of the chain. Even when the dose-effect relationship seems clear, it is not always possible to reverse a process by altering the dose. In The Netherlands for instance, several mitigating measures against grass and scrub encroachment, like mowing, sod cutting and domestic grazing, do not always restore the breeding densities of birds that disappeared. Earlier, Verstrael and van Dijk (1996) predicted that birds of open dune landscapes would benefit from the reintroduction of stock. However, a recent study in a Dutch dune reserve that has been grazed by cattle since 1990, revealed that the diversity in vegetation structure was improved but that the expected recovery of the bird population failed to occur (Van der Niet 2001). This will not exclude that such measures may be efficient in other dune areas or even benefit other animal groups.

17.5 Coastal Zone Management: Can Animals be Integrated?

When a dune area consists of a strand line with embryo dunes, mobile and fixed dunes, dune valleys and wet slacks, it can host a large number of breeding, migrating and wintering birds: gulls, waders, ducks, and specific passerines. However, when these dune systems are equally appreciated by tourists, they can come into conflict with the avian occupants. Tucker and Evans (1997) have made an inventory of sand dune loss and the adverse effect on birds. Tourism and recreation are by far the severest threat in dunes, especially in the Mediterranean basin. They affect 47 of the 75 priority bird species, according to the prioritization of SPECs (Species of European Conservation Concern). For example, the estimated loss of sand dunes to tourism over the last 100 years approximates or exceeds 50% of the total area in the Mediterranean coasts of France, Greece, Italy, Portugal, Spain and Turkey. In the Baltic region, losses from 35% up to more than 50% are caused by afforestation.

To segregate tourists and natural fauna, the integrated coastal zone management (ICZM) is now implemented on a worldwide scale (van der Meulen and Udo de Haes 1996; State of Queensland 1999; Texas General Land Office 2001). Coastal functions are relocated in such a way that the economic and leisure demands are restricted and concentrated to protect and conserve the ecological values. In various places in south Europe, Australia and the east coast of the USA, coastal settlements are removed and the building of new tourist resorts is being frozen (Arapis and Margaritoulis 1996; Cosijn et al. 1996). Within the protected dune areas, additional regulations should balance recreation pressure. An example in South Africa shows that conflicts can arise when recreationists show a preference for the most natural and sensitive coastal dunes (Avis 1995). In a comparison of three coastal dune systems, people preferred the one with the highest natural quality and the least commercial development. The recreational capacity must be carefully monitored to ensure the balance between increasing public access and the quality of the natural resources.

Not only birds, but also other vertebrates, have benefited by the conservation policies. The conservation of suitable and tourist-free habitat for smaller and less mobile vertebrates is relatively easy to achieve because their home ranges are relatively small. Rare species of snakes, lizards and toads are or can be successfully protected by isolation from human disturbance (Economidou 1996). In California, for example, habitat conservation and isolation is sustained by predator control (California Department of Parks and Recreation 1998; US Fish and Wildlife Service 1999). Enlarging suitable breeding habitat and relocating lizards at a greater distance from human settlements reduces predation by domestic cats on sand lizards (*Lacerta agilis*) at the Sefton coast in Great Britain (Larsen and Henshaw 2001). In the same British dune reserve, natterjack toads (*Bufo calamita*) are favored by removal of sea buckthorn, excavation of breeding pools and fencing to prevent trampling (Simpson et al. 2001). Zone management is thus applied on two scales: the regional conservation of the dunes as a whole and the local segregation of animals and humans by fencing off. A similar small-scale zonation serves the protection of nesting sea turtles (*Caretta caretta* and *Chelonia mydas*) on Greek beaches (Poland et al. 1995). Zone management on a larger scale can be effective for breeding birds. Breeding success of terns and plovers in New Jersey increased definitely since birds and recreationists have been separated by wardening, posting and fencing (Burger 1995).

For three animal groups, however, coastal zone management does not prevent the impact on dune ecosystems: seed dispersers, herbivores and predators. In South Africa, bushpocket biodiversity is endangered by numerous alien plant species, of which *Acacia cyclops* causes a widespread and ongoing problem (Kerley et al. 1996). Almost 30 % (10.4 % from *A. cyclops*) of all intact seeds dispersed by mammals and birds are invasive species. In the Dutch mainland dunes, a similar problem is apparent with the once imported American *Prunus serotina*, of which the cherry stones are spread by birds and foxes.

As has been shown, herbivores can upturn the vegetation succession completely, especially in the case of mammals with an apparently infinite reproduction rate, like rabbits and fallow deer. In a coastal dune reserve in San Rossoro (Italy), trapping and euthanizing fallow deer is the only possible method to prevent a complete deforestation (Van Breukelen, pers. comm.).

The abrupt increment of a predator population has its effects way beyond any zonation, as is illustrated by the reappearance of the red fox and the disappearance of gull colonies in the mainland coast of The Netherlands. The Dutch coast, however, extends northwards over the West Frisian islands in the Wadden Sea and southwards over the Zeeland islands. So far, these islands have not yet been reached by the fox. It is not surprising that large numbers of the herring gull and the lesser black-backed gull (*Larus graellsii*) have emigrated from the mainland dunes to these fox-free islands. These graphs

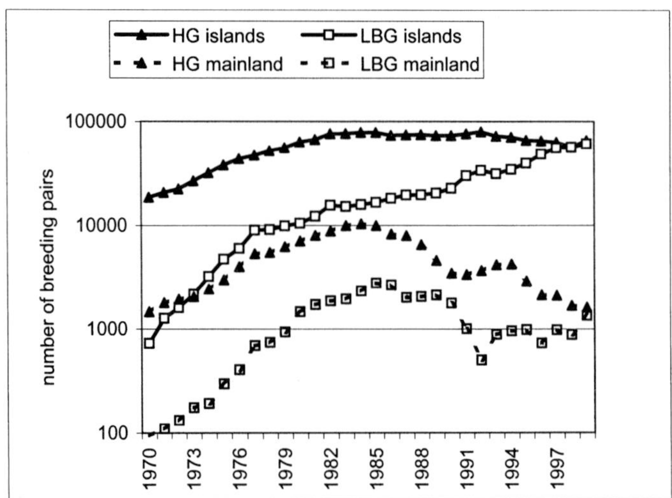

Fig. 17.8. Numbers of breeding pairs of Herring gulls (HG) (*Larus argentatus)* and lesser Blackbacked gulls (LBG) *(Larus graellsii)* on the mainland coastal dunes of Holland and on the dunes of the islands of Zeeland (south west of the mainland coast) and the Wadden Sea (north and north east of the mainland coast) (Data from Spaans 1998, completed and adapted by A.J. van Dijk, SOVON)

(Fig. 17.8) illustrate the impact of the reappearance and dispersion of the red fox on the mainland coast of Holland since the late 1960s. Until 1975–1980, gull numbers were increasing. Because foxes started to disturb the gulleries in the mainland dunes, one after the other disappeared. Some of them moved over to flat roofs of apartments and other large buildings at the seaside, which explains a great deal the remaining figures on the mainland coast. Others moved to the fox-free islands; many new gulleries were established just in the same period as they disappeared out of the mainland dunes.

So far, the increment in gull numbers has not produced unexpected effects in the existent island ecosystems but in theory the development of a predator population can indirectly affect dune areas that are miles away. With periodic monitoring of population dynamics and interspecific interactions, it is possible to assess the effect of such shifts in distribution. Eventually local management plans are needed as additional protective measures where new bird colonies are becoming established.

Coastal zone management is a useful instrument to protect sensitive and rare dune animals but it provides no safety measure against major changes in dynamics and against the sudden appearance or disappearance of essential links in the food chain. In that case, managers have to take specific additional measures, and still these measures are often not more than a continuous treatment of symptoms rather than of causes. Another choice is to accept the events and to monitor the impact on the ecosystem. In either situation, disturbed or not, monitoring of populations and their biotic interactions remains the crucial base for future decisions made by managers, NGO's and governmental institutions.

Acknowledgements. First of all, we thank Nico Penninkhof for collecting and screening a vast amount of information in libraries and on the Internet. We also thank Albert Salman and colleagues from the European Union of Coastal Conservation for their guidance in their inexhaustible coastal library. We are indebted to S.E van Wieren and J.J.M. van Alphen for their reviews of this chapter. The copyright for this chapter lies with the Amsterdam Water Supply Company.

References

Arapis T, Margaritoulis D (1996) Sea turtle conservation and sustainable tourism for the proposed marine park on Zakynthos island, Greece. In: Salman AHPM, Langeveld MJ, Bonazountas M (eds) Coastal management and habitat conservation. EUCC Leiden, The Netherlands, pp 25–28

Avis AM (1995) Recreational use of three urban beaches in South Africa and effects of coastal dune vegetation. In: Salman AHPM, Berends H, Bonazountas M (eds) Coastal management and habitat conservation. EUCC Leiden, The Netherlands, pp 467–486

Baeyens G (1989) Wildlife management in Dutch coastal dunes. In: Meulen F van der, Jungerius PD (eds) Perspectives in coastal dune management, SPB Academic Publ, The Hague, The Netherlands, pp 111–119

Baeyens G, Duyve J (1992) Lezen in het duin. Stadsuitgeverij Amsterdam/Gemeentewaterleidingen Amsterdam

Baeyens G, Vader H (1990) Aantalsontwikkelingen van eenden in drie duinwaterwingebieden. In: Koerselman W, Hoed MA den, Jansen AJM, Ernst WHO (eds): Natuurwaarden en waterwinning in de duinen; mogelijkheden voor behoud, herstel en ontwikkeling van natuurwaarden. KIWA-report 114, Nieuwegein, The Netherlands, pp 103–118

Baeyens G, Oosterbaan BWJ, Breukelen L van (1997) Restoration of wetland habitat in a Dutch dune reserve. In: Goss-Custard JD, Rufino R, Luis A (eds) Effect of habitat loss and change on waterbirds. ITE Symp No 30/Wetlands Intl Publ 42, London, pp 3–9

Bonte D, Hoffmann M (2001) A GIS study of breeding bird habitats in the Flemish coastal dunes and its implications for nature management. In: Houston JA, Edmondson SE, Rooney PJ (eds) Coastal dune management. Shared experience of European conservation practice. Liverpool University Press, Liverpool, pp 128–139

Bouman AE, Bruijn GJ de, Hinsberg A van, Sevenster P, Wanders EAJ, Wanders RM (1991) Meeuwen. Opkomst en ondergang van een meeuwenkolonie. Stichting Uitgeverij KNNV Utrecht/Duinwaterbedrijf Zuid-Holland

Burger J (1995) Beach recreation and nesting birds. In: Knight RL, Gutzwiller KJ (eds) Wildlife and recreationists. Coexistence through management and research. pp 281–295

California Dept Parks and Recreation (1998) Wildlife management plan for Torrey Pines State reserve: terrestrial vertebrates. San Diego. http://www.torreypine.org. Cited 30 Nov 2001

Cosijn R, Huisman B, Rens P, Salman AHPM (1996) Sea turtles and Dutch tourism in the Eastern Mediterranean. In: Salman AHPM, Langeveld MJ, Bonazountas M (eds) Coastal management and habitat conservation, EUCC Leiden, The Netherlands, pp 105–111

Dam P van (2001a) Status loss due to ecological success. Landscape change and the spread of the rabbit. Innovation: Eur J Social Sci 14:157–170

Dam P van (2001b) The Dutch rabbit garden. The creation of a man-made ecosystem in the dunes 1300–1600 (in Dutch with English abstract). Tijdschr Soc Gesch 27:322–335

Dijksen A (1992) Kiekendieven in de duinen. Duin 15:29–31

Doody JP (2001) Coastal conservation and management – an ecological perspective. Kluwer, Dordrecht

Drees M, Olff H (2001) Rabbit grazing and rabbit counting. In: Houston JA, Edmondson SE, Rooney PJ (eds) Coastal dune management. Shared experience of European conservation practice. Liverpool University Press, Liverpool, pp 86–94

Economidou E (1996) Ecology, conservation and management of Greek coastal biotopes and a case study in Greece. In: Salman AHPM, Langeveld MJ, Bonazountas M (eds) Coastal management and habitat conservation. EUCC Leiden, The Netherlands, pp 17–23

Houston J (1995) The sand dunes of the Sefton Coast, north west England, and their management. In: Van Dijk HWJ (ed) Management and preservation of coastal habitats. EUCC Leiden, The Netherlands, pp 109–119

Kerley GIH, McLachlan A, Castley JG (1996) Diversity and dynamics of bushpockets in the Alexandria Coastal Dunefield. Landsc Urban Plann 34:255–266

Klijn J, Bakker Th (1992) Vijfduizend jaar dieren in de duinen. Duin 15:3–5

Koning FJ, Baeyens G (1990) Uilen in de duinen. Stichting Uitgeverij KNNV Utrecht/Gemeentewaterleidingen Amsterdam

Larsen CT, Henshaw RE (2001) Predation of the sand lizard *Lacerta agilis* by the domestic cat *Felis catus* on the Sefton Coast. In: Houston JA, Edmondson SE, Rooney PJ (eds) Coastal dune management. Shared experience of European conservation practice. Liverpool University Press, Liverpool, pp 140–154

Meer HP van der (1996) Atlas van de broedvogels tussen Katwijk en Scheveningen. NV Duinwaterbedrijf Zuid-Holland, The Netherlands

Meulen F van der, Udo de Haes HA (1996) Nature conservation and integrated coastal zone management in Europe: present and future. Landsc Urban Plann 34:401–410

Mulder JL (1990) De hermelijn, flitsende konijnvanger van het duin. Duin 13:4–7

Mulder JL (1992) Vos. In: Broekhuizen S, Hoekstra B, Laar V van, Smeenk C, Thissen JBM (eds) Atlas van de Nederlandse zoogdieren. Stichting Uitgeverij KNNV Utrecht, The Netherlands

Mulder JL, Swaan AH (1988) De vos in het Noord-Hollands Duinreservaat. Deel 5: de wulpenpopulatie. RIN-rapport 88/45 Rijksinsitituut voor Natuurbeheer Arnhem en Provinciaal Waterleidingbedrijf Noord-Holland, The Netherlands

Niet T van der (2001) Begrazing en broedvogels: een gelukkig huwelijk? Duin 24:16–18

Nobel P de, Loos WB, Foppen R (2001) Biotoopspecifieke trends van 16 broedvogelsoorten in de Amsterdamse Waterleidingduinen en het Hollands duindistrict. SOVON, Beek-Ubbergen, The Netherlands

NV PWN Waterleidingbedrijf Noord-Holland (2000) Vossen in het Noord-Hollands Duinreservaat in de periode 1995–1998. Alterra rapport 197, Wageningen

Ommering G. van, Verstrael T. (1987) Vogels van Berkheide. Werkgroep Berkheide / Stichting Publikatiefonds Duinen, Leiden

Opermanis O (1995) Recent changes in breeding bird fauna at the seacoast of the Gulf of Riga. In: Healy MG, Doody JP (eds) Directions in European coastal management. Samara Publ, Cardigan, UK, pp 361–368

Poland R, Hall G, Venizelos L (1995) Sea turtles and tourists: the Loggerhead turtles of Zakynthos (Greece) In: Healy MG, Doody JP (eds.) Directions in European coastal management. Samara Publ, Cardigan, UK, pp 119–128

Ranwell DS (1972) Ecology of salt marshes and sand dunes. Chapman & Hall, London

Rigg K (1999) European Code of Conduct for the Coastal Zone. European Union for Coastal Conservation (EUCC) and the Council of Europe, Leiden. http://www.coastalguide.org/trends/tourism.html. Cited 1 Mar 2001

Shuttleworth CM, Gurnell J (2001) The management of coastal sand dune woodland for red squirrels (*Sciurus vulgaris L.*). In: Houston JA, Edmondson SE, Rooney PJ (eds) Coastal dune management – shared experience of European conservation practice. Liverpool University Press, Liverpool, pp 117–127

Simpson DE, Houston JA, Rooney PJ (2001) Towards best practice in the sustainable management of sand dune habitats: 2. Management of the Ainsdale Dunes on the Sefton Coast. In: Houston JA, Edmondson SE, Rooney PJ (eds) Coastal dune management – shared experience of European conservation practice. Liverpool University Press ,Liverpool, pp 262–270

Slings QL (1999) Het effect van natuurgerichte recreatie op de broedvogelstand van het duingebied bij Egmond. NV PWN Waterleidingbedrijf Noord-Holland, The Netherlands

Snater H (1999) Begrazingsonderzoek Noord-Hollands Duinreservaat. NV PWN Waterleidingbedrijf Noord-Holland, The Netherlands

Spaans AL (1998) Booming gulls in the Low Countries during the 20th century. Sula 12:121–128

State of Queensland, Environmental Protection Agency (1999) Cardwell-Hinchinbrook's coast: managing its future. Brisbane, Australia

Stuyfzand PJ (1991) De samenstelling van regenwater langs Hollands kust. KIWA report SWE 91.010, Nieuwegein, The Netherlands

Texas General Land Office. The Texas Coast, an owner's manual. http://www.glo.state.tx.us/coastal/ownersmanual/trust.html. Cited 30 Nov 2001

Thyen S, Becker PH, Exo KM, Hälterlein B, Hötker H, Südbeck P (1998) Monitoring Breeding Success of Coastal Birds. Final report of the pilot study 1996–1997 by the Joint Monitoring Group of Breeding Birds in the Wadden Sea

Til M van, Ketner P, Provoost S (2002) Duinstruwelen in de 20e eeuw (with English summary). De Lev Nat 103 pp 74–77

Tucker GM, Evans MI (1997) Habitats for birds in Europe: a conservation strategy for the wide environment. Birdlife International, Cambridge, UK

Udo de Haes HA, Wolters AR (1992) The golden fringe of Europe: ideas for an European coastal conservation strategy and action plan. In: Carter RWG, Curtis TGF, Sheehy-Skeffington MJ (eds) Coastal dunes: geomorphology, ecology and management for conservation. Balkema, Rotterdam/Brookfield, pp 525–532

US Fish and Wildlife Service (1999) Environmental assessment for the comprehensive management plan of Tijuana Slough National Wildlife Refuge. http://www.r1.fws.gov/planning/TJS/. Cited 30 Nov 2001

Vanhecke L, Charlier G, Verelst L (1981) Landschappen in Vlaanderen vroeger en nu. Nationale Plantentuin van België/Belgische Natuur- en Vogelreservaten, Brussels

Veenstra B, Geelhoed SCV (1997) Aantalsontwikkeling van broedvogels in het Nationaal Park Zuid-Kennemerland (1952–1996). NV PWN Waterleidingbedrijf Noord-Holland, The Netherlands

Verstrael TJ, Van Dijk AJ (1996) Trends in breeding birds in Dutch dune areas. In: Salman AHPM, Langeveld MJ, Bonazountas M (eds) Coastal management and habitat conservation, EUCC Leiden, The Netherlands, pp 403–416

Vliet F van der, Baeyens G (1995) Voedsel van vossen in de duinen: variatie in ruimte en tijd. Gemeentewaterleidingen Amsterdam / VZZ Med. 29, Utrecht

Vos P, Peltzer RHM (1987) Recreatie en broedvogels in heidegebieden, Strabrechtse en Groote Heide. State Forestry, Utrecht, The Netherlands

18 Coastal Vegetation as Indicators for Conservation

I. Espejel, B. Ahumada, Y. Cruz, and A. Heredia

18.1 Introduction

18.1.1 Environmental Indicators

According to Dale and English (1999), "environmental indicators are tools for different environmental decision-making situations. There are three types of tools aiding environmental decision-making: bits of information, or data; tools to gather data; and tools to organize and analyze data, including models to describe relationships among units of information. They may be quantitative or qualitative". According to the above, environmental indicators are direct and indirect forms to measure the environmental quality. They define the present state and the tendencies of the environmental capacity to sustain ecological and human health (EPA and SEMARNAP 1997). They are designed to quickly and easily inform decision makers about environmental information dealing with natural conditions. If used for monitoring, over time they may communicate information about ecosystem changes and trends. As management tools, they may provide awareness over developing problems and actions needed. There have been various efforts to design environmental indicators trying to make them analogous and comparable to economic or social indicators. For instance, an exotic species can be a variable to measure vegetation quality that is an environmental health indicator; income is a variable to measure poverty, a socio-economic indicator related to social health.

In the 1980s, environmental economists in developed countries (Canada and Europe) designed indicators of the pollution agents in air, water, and soils (Hammond et al. 1995). In the 1990s, several international organizations started to develop indicators to measure sustainability in other countries. Therefore, along with pollution as an environmental indicator, others were added, such as erosion, deforestation, and biodiversity. The main efforts refer to national scale indicators because they are used to evaluate a country's environmental performance (NRC 2000; SEMARNAP 2000).

Ecological Studies, Vol. 171
M.L. Martínez, N.P. Psuty (Eds.)
Coastal Dunes, Ecology and Conservation

In the USA, the National Research Council (2000) worked on ecological indicators for the national scale. They selected these based on general importance, conceptual basis, reliability, temporal and spatial scales, statistical properties, data requirements, skills required, data quality, data archiving, robustness, international compatibility, costs, benefits, and cost-effectiveness. The recommended indicators for the USA are indicators of the extent and status of the nation's ecosystems – land cover and land use. Indicators of the nation's ecological capital are total species diversity (measures the ecological capital actually present), native species diversity, nutrient runoff, and organic soil matter. Indicators of ecological functioning or performance are carbon storage, production capacity, net primary production, trophic status of lakes, stream oxygen, and for agricultural ecosystems, nutrient-use efficiency and nutrient balance.

In Mexico, the environmental ministry (SEMARNAP 2000) has developed an Environmental and Nature Resources Information System where 24 environmental indicators consist of 11 issues; water pollution in towns, biodiversity loss, global climatic change, county rubbish, erosion, forestry resources, fishery resources, air pollution, desertification, watershed pollution, and destruction of the ozone layer.

Until today, most of the biodiversity indicators measured the state of the ecosystem and the response to this state; endangered species lists, statistics of deforested land and abandoned fields, and statistics of land under any protection policy. Nevertheless, none of these clearly measures the pressure of human activities over natural ecosystems (Hammond et al. 1995). These indicators are useful for federal policies to evaluate the recovery of the national environment but cannot be compared at the regional or local level because local indicators need to be more specific.

Nevertheless, beginning efforts to select environmental indicators in smaller regions and sites, rather than countries, are developing. For example, in regional ordinances there are environmental impact assessments or risk and vulnerability studies (Villa and McLeod 2002).

18.1.2 Ecological Indicators

Ecological indicators can be used to assess the condition of the environment, to provide an early warning signal of changes in the environment, or to diagnose the cause of an environmental problem. Ideally, the suite of indicators should represent key information about structure, function, and composition of the ecological system (Dale and Beyeler 2001). These represent ecosystem characteristics related to or derived from a biotic or abiotic measurement. They can provide qualitative or quantitative information about the ecosystem composition, structure, and function (Noss 1997). The complexity of biotic systems suggests that the evaluation of the ecosystem can be by multiple organization levels and spatial and temporal scales (Table 18.1). In practice, the

Table 18.1. Ecological indicators for plant species in three complexity levels proposed by the authors

Authors	Ecological indicators		
	Landscape	Community-ecosystem	Population-species
Noss (1990, 1997)	**Composition:** To identify, distribution, richness, proportion of patch types, collective patterns of species distribution. **Structure:** Heterogeneity, connectivity, fragmentation, patch size frequency distribution. **Function:** Disturbance processes, energy flow rates, human land-use trends.	**Composition:** Life-form proportions, C_4:C_3 plants species ratios, proportion of endemics and exotics, threatened and endangered species, to identify relative abundance, frequency, richness, evenness, diversity of species, dominance-diversity curves. **Structure:** Physiognomy, foliage density and layering, horizontal patchiness, abundance, density. **Function:** Herbivory, parasitism and depredation rates, patch dynamics, human intrusion rates and intensities.	**Composition:** Absolute or relative abundance, frequency, importance or cover value, density. **Structure:** Dispersion, population structure. **Function:** Physiology, life history, phenology, demographic process, metapopulation dynamics, growth rate, adaptation.
Keddy et al. (1993)		Diversity, guild, plant biomass.	Exotics and rare species.
Argemeier and Karr (1994)	Fragmentation, number of communities, persistence.	Number of species, species evenness, number of trophic links.	Age or size structure, dispersal behavior.
Jone and Riddle (1996)	Clustering of habitat, habitat connectivity, status and trends data on vertical structure and species composition, configuration of habitats and ecosystems.	Representative species of each guild.	Number of species, species diversity indexes, number or percentage of federally-threatened or endangered species.
Cendrero (1997)	Fragmentation Spatial distribution of communities Persistence of habitats	Ecosystem diversity, productivity, percentage of natural cover, species diversity index.	Rare, threatened and endemic species
Dale & Beyeler (2001)		Species richness Species evenness Number of trophic levels	Age or size structure Dispersal behavior

elements are more frequently used as indicators than the processes, because elements are more sensitive to degradation and less expensively monitored (Angermeier and Karr 1994).

Every plant species is an indicator of environmental conditions. The traditional use of indicator species for evaluating and monitoring the environmental conditions is discussed. As Noss (1990, 1997) mentions, the use of these indicator species may present conceptual and procedure problems. For instance, scale errors occur if there is the usual supposition that an indicator species has the same response in a higher level of the biological organization. As well, (Landres 1992) mentions that communities, habitats, and landscapes may also indicate environmental qualities, but not by single species indicators but by a specific plant community or by a particular landscape.

The selection of indicators is a process that depends on the databases available (Fig. 18.1). These databases are broad and heterogeneous because they are products from divergent objectives driven by the needs of the decision-makers and public. These primary data contain all ecological variables. In this first phase of the indicator-generating process, the primary data are large (Hammond et al. 1995; Bergquist and Bergquist 1999) but only used by a small scientific population.

The second phase corresponds to the decision-maker sectors, which, in contact with scientific experts, select those variables that summarize landscape, ecosystem, community, or population performance. These variables need to represent standard situations or "universal" features to be used for comparisons around the world. For example, the composition indicator may

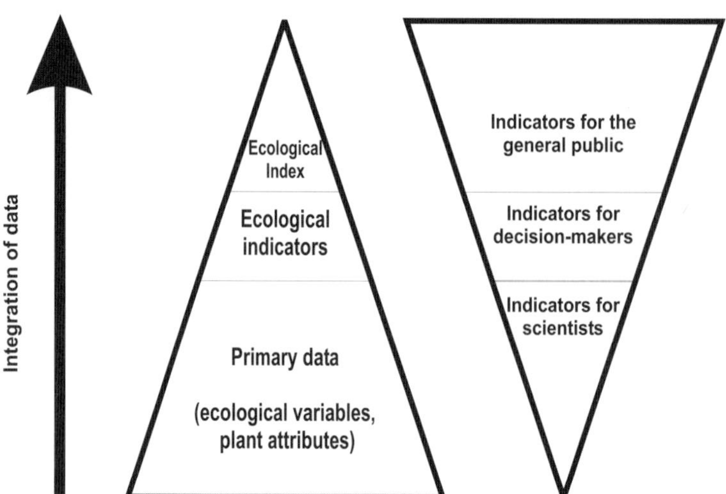

Fig. 18.1. Generation of the environmental index and the audience to whom they are focused

aggregate ecological variables such as species number and origin to explain human impact on plant communities. Structural indicators can combine species life forms and species life span to express natural impacts such as fire or hurricanes on a vegetation type. Functional indicators may join species physiological adaptations and productivity to select carbon-storage ecosystems (NRC 2000). The combined use of ecological variables, turn the variables themselves into ecological indicators (Table 18.1). For instance, they are useful to communicate information about environmental policy application, monitoring results, and environmental education successes.

The third phase corresponds to the integration of ecological indicators into ecological indices. For example, using compositional indicators together with structural indicators may evaluate the environment quality of an ecosystem. The number of species may mean a certain quality but if life form proportions are added, the description of the ecosystem to be evaluated is more accurate. Finally, the ecological index is a mathematical expression, which can be "translated" to be understandable by the public.

Our work builds on groundwork of the coastal plant community's ecology to propose the combinations of vegetation variables as ecological indicators to measure coastal ecosystem suitability for purposes of conservation and restoration. These indicators might be powerful tools for monitoring sand dunes of protected areas. Our work deals with more complex indices for conservation programs (SEMARNAP-INE 1995; Ahumada 2000; Espejel 2001; Espejel et al. 2002), but in this paper we refer to the ecological indicators of the sand dune vegetation that have been analyzed and incorporated to the coastal management indices and used in the northern coastal zone of Baja California, Mexico.

18.2 Methods

18.2.1 Ecological Indicator Selection

Higher-level scale indicators (landscape level) were generated by analyzing the coastal zone of northern Baja California with the Organization for Development and Economic Cooperation (ODEC) method suggested by Lourens et al. (1997) and used in Mexico (SEMARNAP 2000).

For the community-level scale indicators, we used our database which resulted from 212 "releves" (plots) collected between 1989 and 2001 along the Pacific coast of Baja California (Moreno-Casasola and Espejel 1986; Moreno-Casasola et al. 1998). Physically modified sites were sampled to add measures of the presence of invasive species, modifications of vegetation structure, and to detect functional changes.

The floristic list (Appendix) consists of 125 plant species. It shows the family and the species, to which we assigned the presence or absence of the 10 ecological variables or attributes used to build the ecological indicator. The list was classified to obtain the total number of species (region) and total per site (El Socorro, Punta Banda, Tijuana Estuary).

El Socorro and Punta Banda are part of our own generated data, but we used a species list from the literature of the Tijuana estuary to test the usefulness of literature data. We selected these three sites to test the indicators. Therefore, the three sites are different from one another. El Socorro is the more complex sand dune systems (sandy beaches, embryonic sand dunes, high mobile and fixed dunes, wet and dry slacks, slashed areas), more isolated and less modified by agriculture and urban areas than the other two sites. Punta Banda and the Tijuana estuary are sand spits with embryonic sand dunes and small mobile and fixed dunes (Fig. 18.2).

The criteria used to select ecological indicators were the importance of the plant species in these coastal sand dune communities (Table 18.2). We had in

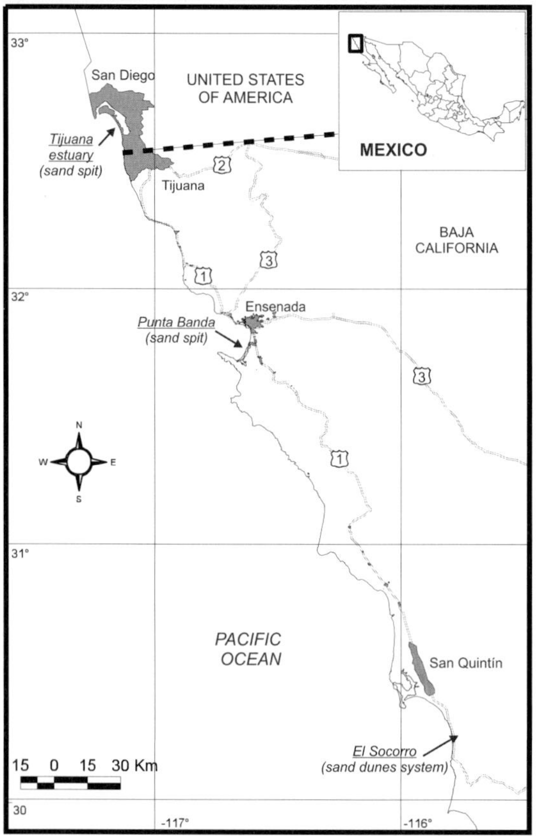

Fig. 18.2. Study area in which the ecological indicators for sand dune systems were analyzed

Table 18.2. Plant species variables or attributes. In ***bold*** are those selected for the ecological indicators (compositional, structural, and functional indicators)

Variables (attributes)	Ecological indicator types	Organization level
Species distribution		
Native species	Compositional	Species
Exotic species	Compositional	Species
Species abundance		
Absolute or relative abundance	Compositional	Population-species
Frequency	Compositional	Population-species
Species morphology		
Life forms	Structural	Species
Tree		
Shrub		
Sub shrub		
Desert forms (Cactaceae, Agavaceae, Crassulaceae and other succulent species of Aizoaceae, Chenopodiaceae, and Solanaceae)		
Perennial herb		
Annual herb		
Prostrate herb		
Erect herb		
Vines		
Vegetation associations	Structural and compositional	Community
Coastal succulent sage scrub		
Coastal chaparral		
Coastal dune		
Salt-marshes		
Riparian		
Introduced grassland		
Species functional features		
Pubescence	Functional	Species
Succulence		
Keystone species	Functional	Species
Soils fixer		
Nitrogen fixer		

mind the users of the ecological-indicator decision makers, who in this case are nonbiologists (Dale and Beyeler 2001). Therefore, we refused nonevident attributes such as physiological features (CAM, C_3, or C_4 strategies), which might be ecologically significant but difficult to visualize in the field. We consulted the literature on plant functional types of similar environments to contribute to our own selection (Barbour et al. 1985; Barbour 1992; Espejel 2001; Infante 2001; Garcia-Novo, this book). Native and exotic species are easily recognized using botanical catalogues; in this region Horn (1993), Whitson (1996), and Grennan (1999) are useful. The presence of exotics explains human impacts (roads, urban development); the presence of natives and desert forms indicate sand dune quality. Life forms may relate to continuous impacts, for example, grasses appear after fires or grazing. Shrubs and prostrated species stabilize sand, thus meaning more mature sand dunes. Vines (and lichens) may indicate microenvironmental humidity (Spjut 1996). Pubescence is an environmental adaptation to high temperatures and aridity. It improves water retention for plants in late successional stages. It can be considered an indicator of protection. Succulence is a plastic trait, induced by soil or air-borne salinity (Rozema et al. 1982) indicating early successional stages. It can be considered a stress tolerance indicator. Nitrogen fixers imply relations with richer soils and more stabilized older dunes. They can be consider an indicator of adaptation.

18.2.2 Calculation of Ecological Indicators

The ecological indicators were divided in compositional indicators based on the proportion of native species and the proportion of exotic species. It is expressed as

$$Ic = \frac{n_i / Nn}{e_i / Ne}$$

where
Ic =compositional indicator
n_i =native species in site or plot i,
e_i =exotic species in site or plot i,
Nn =total number of native species (of the region or of the site to be compared)
Ne =total number of exotic species (of the region or of the site to be compared)

The structural indicator is based on the proportion of life forms and is expressed by

$$Is = (s_i/Ns) + (d_i/Nd) + (p_i/Np) + (h_i/Nh) + (v_i/Nv)$$

where
Is =structural indicator
s_i =number of shrubs in the site or plot i.

Ns = total number of shrubs (of the region or of the site to be compared)
d_i = number of desert forms in the site or plot i
Nd = total number of desert forms (of the region or of the site to be compared)
p_i = number of prostrated herbs in the site or plot i
Np = total number of prostrated herbs (of the region or of the site to be compared)
h_i = number of erect herbs in the site or plot i
Nh = total number of erect herbs (of the region or of the site to be compared)
v_i = number of vines in the site or plot i
Nv = total number of vines (of the region or of the site to be compared)

The functional indicator is related to the proportion of legumes or nitrogen fixers, pubescence representing aridity resistance, and succulents meaning salinity adaptation. The indicator is calculated as

$$\mathrm{If} = (\mathrm{pu_i/Npu}) + (\mathrm{su_i/Nsu}) + (\mathrm{nf_i/Nnf})$$

where
If = functional indicator
pu_i = number of pubescent species in the site or plot i
Npu = total number of pubescent species (of the region or of the site to be compared)
su_i = number of succulent species in the site or plot i
Nsu = total number of succulent species (of the region or of the site to be compared)
nf_i = number of nitrogen fixers in the site or plot i
Nnf = total number of nitrogen fixers (of the region or of the site to be compared)

The ecological index was calculated adding all three ecological indicators. These values were normalized obtaining five classes; Very high (0.8–1.00), high (0.79–0.60), medium (0.59–0.40), low (0.39–0.20), and very low (0.19–0). For instance, a very high value of the compositional indicator means that the sample or site has more natives, thus it has a better value for conservation purposes. On the contrary, a very low indicator value means that the sample or site has more exotics, thus it has the lowest value for conservation purposes. If compared, in a decision-making process we can select the samples or sites with higher values for conservation and those with lower values can be used for other purposes.

18.3 Results

18.3.1 Environmental Indicators for the Region (Landscape-Scale Indicators)

At the higher level scale (landscape), environmental indicators were identified (Fig. 18.3), and ecological indicators, at the community level, were selected (bold). The columns of this figure show the potential threats, early warning and threat indicators in the northern Baja California coastal region where the sand dunes occur. This analysis allowed us to identify the pressure that causes impacts on the vegetation and helped us to select the elements, which could be the framework to search for ecological indicators. The state-response analysis allowed us to identify some environmental indicators but cannot measure them. Therefore, we selected community-scale indicators to be able to measure them quantitatively.

18.3.2 Ecological Indicators for Coastal Dunes (Plant–Community Scale)

The compositional indicator shows clear differences among sites, and for the site richness this indicator shows the quality of the site if compared to the region (Table 18.3). The Tijuana estuary and Punta Banda sand spits are rather poor sites, surrounded by urban and agriculture areas. Both showed very low and low compositional indicators. This means that for the five exotics, the proportion with native plants offers a site with less quality than El Socorro, which has a medium compositional indicator.

The structural indicator (Table 18.3) reflects the sand dune systems complexity, despite the compositional quality. This is shown in the El Socorro site where the structural indicator is high and the other two sites have a low and very low value. The proportion of life forms is more similar between El Socorro and the "region" than the other two sites, which have no desert forms or vines.

The functional indicator (Table 18.3) is clear in the nitrogen fixers when they disappear in the simpler sites. As with the structural indicator, the functional indicator reflects the functional complexity of the El Socorro sand dune system.

The sum of these three indicators forms the ecological index. If the value is closer to 1 it means that is similar to the "region" sand dune systems. In this case, El Socorro site reflects a high quality in terms of its composition, structure, and function. This site represents the better richness quality, the most structural complexity, and holds the highest key functional traits of the sand dunes of the Pacific coast of northern Baja California.

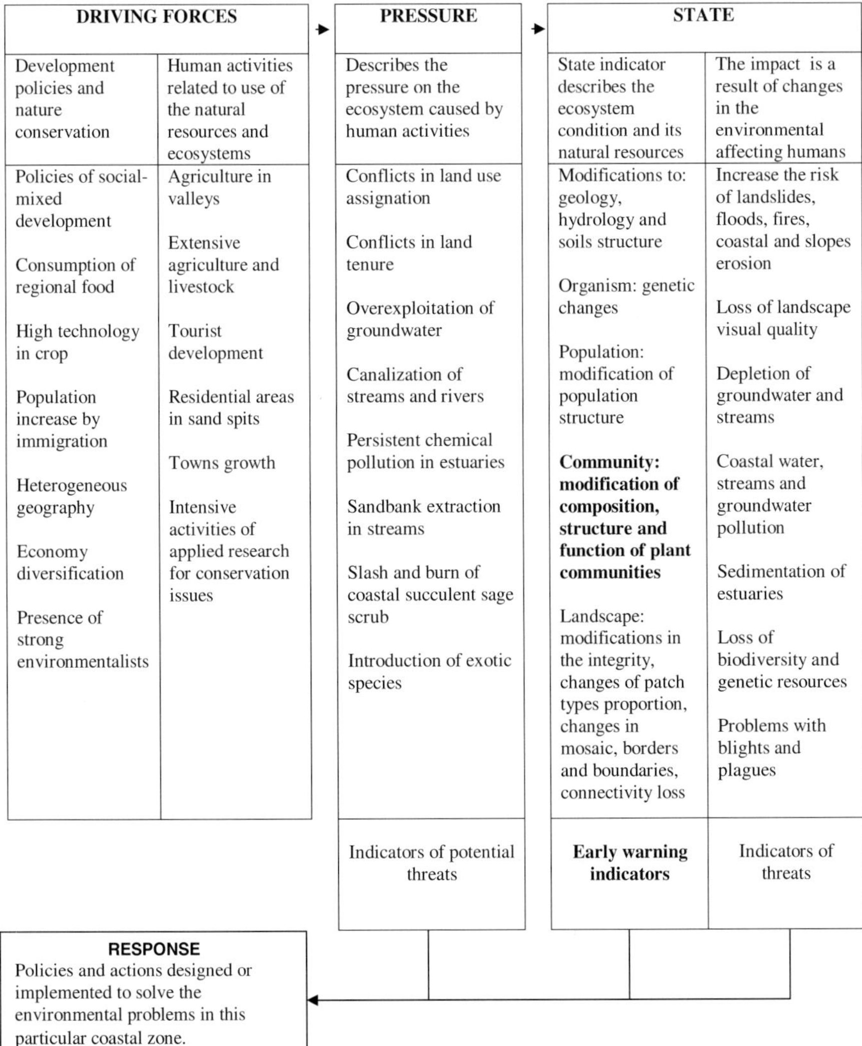

Fig. 18.3. Pressure-state-response scheme to identify environmental indicators in the Baja California coastal succulent sage scrub, sand dune, coastal chaparral, riparian wetland, agriculture, and urban field region. In *bold* are the early warning ecological indicators that we selected for this chapter. (After Laurens et al. 1997)

Table 18.3. Ecological index reflecting the sand dune plant species composition, structure, and function in a region (Pacific coast of northern Baja California) and three sites (El Socorro and Punta Banda (our own data) and the Tijuana Estuary. (Data from Delgadillo 1995)

Trait	Regional	El Socorro			Punta Banda			Tijuana Estuary		
	Number of species	Number of species	Indicator	Normalized	Number of species	Indicator	Normalized	Number of species	Indicator	Normalized
	125	85			32			21		
Native	109	78			27			16		
Exotic	16	7			5			5		
Compositional			1.64	0.56		0.79	0.27		0.47	0.16
				Medium			Low			Very low
Scrub	42	33	0.79		12	0.29		8	0.19	
Desert form	6	4	0.67		0	0.00		0	0.00	
Prostrate herb	36	22	0.61		12	0.33		11	0.31	
Erect herb	36	22	0.61		10	0.28		2	0.06	
Vine	5	4	0.80		1	0.20		0	0.00	
Structural			3.47	0.68		1.10	0.21		0.55	0.11
				High			Low			Very low
Pubescent	85	59	0.69		20	0.24		12	0.14	
Succulent	26	16	0.62		13	0.50		7	0.27	
Nitrogen fixer	10	6	0.60		0	0.00		0	0.00	
Functional			1.91	0.63		0.74	0.24		0.41	0.13
				High			Low			Very low
Ecological index				1.87			0.73			0.40
Normalized				0.62			0.24			0.13
				High			Low			Very low

18.4 Discussion and Conclusion

Most ecological indicators have been developed for aquatic ecosystems (Karr 1994, 1997a, b, 1998; Karr and Chu 1997; Done and Reichelt 1998), and in large-scale areas such as countries (NRC 2000; SEMARNAP 2000). We are searching for ecological indicators at the ecosystem and community level. We chose the sand dune vegetation to integrate ecological indicators "easy to see" to facilitate the daily work of the decision-makers and nonbiologists. Special concern was put into composition, structure, and function as suggested by Noss (1990, 1997). Dale and Beyeler (2001) suggest eight criteria to develop ecological indicators.

Our proposed ecological indicators meet several of these criteria, because they

1. are easily measured with a floristic list and little training in sand dune flora,
2. are sensitive to stress on sand dune systems such as the nitrogen fixers of the functional indicator,
3. respond to stress in a predictable manner. This is shown in the compositional indicator because perturbation allows predicting exotics invasion,
4. are anticipatory (signify an impeding change in the ecological system),
5. predict changes that can be averted by management actions,
6. are integrative (the full suite of indicators provides a measure of coverage of the key gradients across the ecological system). The ecological index is integrative itself,
7. All the ecological indicators selected have a known response to natural disturbances, anthropogenic stresses, and change over time.
8. All the indicators proposed reflect a specific floristic richness that shows the low variability in response to the particular stresses of most of the sand dunes in the world.

The examples that we present in this chapter show the present state of a beginning monitoring program for the sand dunes of northern Baja California. For a conservation or restoration monitoring aim, this is a useful tool to compare different sites or regions worldwide. With these 'present state' figures (Table 18.3), we can start monitoring as many sand dunes as possible as suggested for various ecosystems and communities (Noss 1990; Costanza et al. 1992; Keddy et al. 1993; Angermeier and Karr 1994; Jones and Riddle 1996; Cendrero 1997; Noss 1997). The compositional and the structural indicators are most effective to measure early changes of the landscape. Therefore, we suggest the variables of these ecological indicators be incorporated into the National Environmental Indicators System (SEMARNAP 2000). If changes are shown in one, five, or ten years, we can have a national response to the early warning indicators (Fig. 18.3) suggested to measure changes in the state of the coastal ecosystems (Lourens et al. 1997).

The ecological indicators applied to sand dune communities are useful for restoration and conservation issues. For ecological monitoring in the local context, the indicators provide essential information for the selection of the minimum number of variables to measure (Spellerberg 1991; Woodley et al. 1993; Dale and English 1999; Schulze 1999). The index also proved to be useful to choose representative areas in a regional context (El Socorro in this case).

If the index is calculated in several plots, quadrats or releves, it can also be useful in the selection of the most representative permanent plots to study continuously in long-term studies.

These types of studies are a priority to assess natural-area performance (Dale and English 1999; Schulze 1999). In addition, the indicators can help evaluate changes and trends caused by disturbance, especially if they are placed on protected or managed areas (Schmitt and Osenberg 1996; Wright 1996; Nelson and Serafin 1997). Applied research is mainly done using natural experiments (Connell 1975; Eberhardt and Thomas 1991) that often need to select the minimum data and the smallest area to minimize research efforts for time and budget (Spellerberg 1991; NRC 2000).

Because of the vulnerability of sand dunes to exotic species invasion (Brown and McLachlan 1990; Nordstrom 2000), the compositional indicator provides information for timing control maneuvers to stop or minimize the invasion and establishment of exotic species (Hiebert 1997).

The exercise presented in Table 18.3 seems to be useful to select representative sites on the regional scale. The ecological index, simultaneously analyzing ecological indicators, identifies the complexity of a dune system and provides a tool to select sites for conservation, restoration, and for monitoring programs. Furthermore, this analysis is useful for decision makers in the incorporation of literature data as shown by the Tijuana Estuary site. Managers often have low budget programs. Therefore, they rely on published data, which they then can compare with their own scarce field data.

Primary data (measured or literature-based) generated for scientific purposes other than management, conservation, or restoration can be used to select ecological indicators for decision makers in their local or regional programs. The ecological indicators of sand dunes seem to be a useful tool for monitoring conservation and restoration, impact and plant invasion assessment, and selecting protected natural areas. The public, from NGOs and local communities, can use these indicators for similar purposes. We propose these sand dunes variables as ecological indicators to be adapted and used in future monitoring conservation and restoration programs worldwide.

Appendix: Floristic List of Northern Baja California Coastal Sand Dune Systems

Family	Species	Life form	Distribution
Acanthaceae	*Justicia californica* (Benth.) Gibson	Shrub	
Aizoaceae	*Carpobrotus aequilaterus* (Haw.) N.E. Brown	Desert form	Introduced
Aizoaceae	*Mesembryanthemum crystallinum* L.	Desert form	Introduced
Aizoaceae	*Mesembryanthemum edulis* L.	Herb	Introduced
Aizoaceae	*Mesembryanthemum nodiflorum* L.	Desert form	Introduced
Anacardiaceae	*Rhus integrifolia* (Nutt.) Benth. & Hook. var *integrifolia*	Shrub	
Asclepiadaceae	*Asclepias subulata* Decne.	Herb	
Asclepidaceae	*Cynanchum peninsulare* S.F. Blake	Vine	Endemic
Asclepidaceae	*Sarcostemma arenarium* Decne.	Vine	Endemic
Asteraceae	*Ambrosia chamissonis* (Less.) Greene	Shrub	
Asteraceae	*Amblyopappus pusillus* Hook.& Arn.	Shrub	
Asteraceae	*Artemisia californica* Less.	Shrub	
Asteraceae	*Bebbia juncea* (Benth.) Greene var. *juncea*	Shrub	
Asteraceae	*Chaenactis glabriuscula* D.C. var. *glabriuscula*	Herb	
Asteraceae	*Chrysanthemum coronarium* L.	Herb	Introduced
Asteraceae	*Encelia californica* Nutt.	Shrub	
Asteraceae	*Encelia farinosa* A. Gray var. *farinosa*	Shrub	
Asteraceae	*Haplopappus berberidis* A.Gray	Shrub	Endemic
Asteraceae	*Haplopappus venetus* (H.B.K.) S.F. Blake ssp. *furfuraceus* (Greene) Hall	Shrub	
Asteraceae	*Haplopappus venetus* (H.B.K.) S.F. Blake ssp. *tridentatus* (Greene) Hall	Shrub	Endemic
Asteraceae	*Haploppapus squarrosus* H. & A. ssp. *grindelioides* (DC.) Keck.	Shrub	
Asteraceae	*Helianthus niveus* (Benth.) Brandegee ssp. *niveus*	Herb	
Asteraceae	*Heterotheca sessiliflora* (Nutt.) Shinn ssp. *sessiliflora*	Shrub	
Asteraceae	*Jaumea carnosa* (Less.) A. Gray	Herb	
Asteraceae	*Perityle emoryi* Torr.	Herb	
Asteraceae	*Pluchea serica* (Nutt.) Cov.	Shrub	
Asteraceae	*Porophyllum gracile* Benth.	Shrub	

Appendix: (*Continued*)

Family	Species	Life form	Distribution
Asteraceae	*Senecio californicus* DC.	Herb	
Asteraceae	*Senecio douglasii* DC. var. *monoensis* (Greene) Jepson	Shrub	
Asteraceae	*Stephanomeria virgata* Benth.	Herb	
Asteraceae	*Viguiera laciniata* A. Gray	Shrub	
Boraginaceae	*Cryptantha intermedia* (A. Gray) Greene	Herb	
Boraginaceae	*Heliotropium curassavicum* var. *oculatum* (Heller) I.M. Johnst.	Herb	
Boraginaceae	*Pectocarya linearis* DC var. *ferocula* I.M. Johnst.	Herb	
Cactaceae	*Echinocereus maritimus* (M.E. Jones) K. Schum.	Desert form	Endemic
Cactaceae	*Mammillaria dioica* K. Brandegeei	Desert form	
Cactaceae	*Opuntia cholla* Weber	Desert form	
Cactaceae	*Pachycereus pringlei* (S. Wats.) Britt. & Rose	Desert form	
Capparidaceae	*Isomeris arborea* Nutt. en Torr. & Gray	Tree	
Caryophyllaceae	*Cardionema ramosissima* (Weinm.) Nels. & Macbr.	Herb	
Caryophyllaceae	*Silene gallica* L.	Herb	Introduced
Chenopodiaceae	*Atriplex barclayana* (Benth.) D. Dietr. ssp. *palmeri* (S. Wats.) Hall & Clements	Shrub	
Chenopodiaceae	*Atriplex californica* Moq. in DC.	Herb	
Chenopodiaceae	*Atriplex canescens* (Pursh.) Nutt. ssp. *canescens*	Shrub	
Chenopodiaceae	*Atriplex canescens* (Pursh.) Nutt. ssp. *linearis* (S. Wats.) Hall & Clements.	Shrub	
Chenopodiaceae	*Atriplex coulteri* (Moq.) D. Dietr.	Shrub	
Chenopodiaceae	*Atriplex hastata* (L.) Hall & Clem.	Herb	Introduced
Chenopodiaceae	*Atriplex julaceae* S. Wats	Herb	Endemic
Chenopodiaceae	*Atriplex lindleyi* Moq.	Herb	Introduced
Chenopodiaceae	*Atriplex pacifica* Nels.	Herb	
Chenopodiaceae	*Atriplex semibaccata* R. Br.	Herb	Introduced
Chenopodiaceae	*Chenopodium murale* L.	Herb	Introduced
Chenopodiaceae	*Salsola kali* L. var. *tenuifolia* Tausch.	Shrub	Introduced
Chenopodiaceae	*Suaeda californica* S. Wats.	Desert form	

Appendix: (*Continued*)

Family	Species	Life form	Distribution
Convolvulaceae	*Calystegia macrostegia* (Greene) Brummit ssp. *tenuifolia* (Abrams) Brummitt	Vine	
Convolvulaceae	*Cuscuta salina* Engelm.	Vine	
Convolvulaceae	*Cuscuta veatchii* Brandegee	Vine	Endemic
Crassulaceae	*Dudleya attenuata* (S. Wats.) Moran ssp. *orcuttii* (Rose) Moran	Desert form	Endemic
Crassulaceae	*Dudleya lanceolata* (Nutt.) Britt. & Rose	Desert form	
Cruciferae	*Cakile maritima* Scop.	Herb	Introduced
Cruciferae	*Draba cuneifolia* Nutt. ex T. & G. var. *integrifolia* S. Wats	Herb	
Cruciferae	*Erysimum capitatum* (Dougl.) Greene	Herb	
Cruciferae	*Sisymbrium irio* L.	Herb	Introduced
Ephedraceae	*Ephedra californica* S.Wats.	Shrub	
Euphorbiaceae	*Croton californicus* Muell.-Arg.	Herb	
Euphorbiaceae	*Euphorbia micromera* Engelm.	Herb	
Euphorbiaceae	*Euphorbia misera* Benth.	Shrub	
Euphorbiaceae	*Euphorbia polycarpa* Benth.	Herb	
Euphorbiaceae	*Stillingia linearifolia* Wats.	Shrub	
Frankeniaceae	*Frankenia palmeri* Wats.	Shrub	Endemic
Frankeniaceae	*Frankenia salina* Jtn.	Shrub	
Hippocastanaceae	*Aesculus parryi* A. Gray	Shrub	Endemic
Hydrophyllaceae	*Phacelia distans* Benth.	Herb	
Hydrophyllaceae	*Phacelia hirtuosa* A. Gray	Herb	Endemic
Juncaceae	*Juncus acutus* L.	Herb	
Labiatae	*Hyptis emoryi* Torr.	Shrub	
Leguminosae	*Astragalus anemophilus* Greene	Herb	Endemic
Leguminosae	*Astragalus didymocarpus* Hook & Arn.	Herb	
Leguminosae	*Astragalus insularis* Kell var. *quintinensis* M. E. Jones	Shrub	Endemic
Leguminosae	*Astragalus sanctorum* Barneby	Herb	
Leguminosae	*Astragalus trichopodus* (Nutt.) A. Gray ssp. *leucopsis* (T & G) Thorne	Herb	

Appendix: (*Continued*)

Family	Species	Life form	Distribution
Leguminosae	*Lotus bryantii* (Brandegee) Ottley	Herb	Endemic
Leguminosae	*Lotus distichus* Greene	Shrub	Endemic
Leguminosae	*Lotus scoparius* (Nutt. ex T. & G.) var. *scoparius* Nutt.	Herb	
Leguminosae	*Trifolium gracilentum* Torr. & Gray	Shrub	
Malvaceae	*Sphaeralcea ambigua* A. Gray var. *ambigua*	Shrub	
Nyctaginaceae	*Abronia maritima* Nutt. ex Wats. ssp. *maritima*	Herb	
Nyctaginaceae	*Abronia umbellata* Lam.	Herb	
Nyctaginaceae	*Mirabilis californica* Gray	Shrub	
Onagraceae	*Camissonia bistorta* (Nutt. ex T. & G.) Raven	Herb	
Onagraceae	*Camissonia californica* (Nutt. ex T. & G.) Raven	Herb	
Onagraceae	*Camissonia cheiranthifolia* (Hornem es Spreng) Raimann ssp. *suffruticosa* (S. Wats) Raven	Herb	
Onagraceae	*Camissonia cheiranthifolia* (Sprengel) Raimann ssp. *cheiranthifolia*	Herb	
Onagraceae	*Camissonia crassifolia* (Greene) Raven	Herb	Endemic
Onagraceae	*Camissonia hirtella* (Greene) Raven	Herb	
Onagraceae	*Oenothera californica* (S. Watson) S. Watson	Herb	
Papaveraceae	*Eschscholzia californica* Cham.	Herb	
Plumbaginaceae	*Limonium californicum* (Boiss) Heller	Herb	
Plumbaginaceae	*Limonium californicum* (Boiss.) Heller var. *mexicanum* (Blake) Munz	Herb	
Poaceae	*Bromus rubens* L.	Herb	Introduced
Poaceae	*Distichlis spicata* (L.) Greene	Herb	
Poaceae	*Melica imperfecta* Trin.	Herb	
Poaceae	*Monanthochloe littoralis* Engelm.	Herb	
Poaceae	*Polypogon monspeliensis* (L.) Desf.	Herb	Introduced
Polygonaceae	*Eriogonum fasciculatum* Benth. var. *fasciculatum*	Shrub	Endemic
Polygonaceae	*Nemacaulis denudata* Nutt.	Herb	
Polypodiaceae	*Cheilanthes newberryi* (D. Eaton) Domin	Herb	

Appendix: (*Continued*)

Family	Species	Life form	Distribution
Polypodiaceae	*Polypodium californicum* Kaulf.	Herb	
Portulacaceae	*Calandrinia ciliata* (R. & P.) DC. var. *menziesii* (Hook) Macbr.	Herb	
Portulacaceae	*Calandrinia maritima* Nutt.	Herb	
Primulaceae	*Anagallis arvensis* (L.) Krause	Herb	Introduced
Resedaceae	*Oligomeris linifolia* (Vahl.) Macbr.	Herb	
Saururaceae	*Anemopsis californica* (Nutt.) Hook. & Arn.	Herb	
Scrophulariaceae	*Cordylanthus maritimus* A. Gray spp. *maritimus*	Herb	
Scrophulariaceae	*Cordylanthus orcuttianus* A. Gray	Herb	Endemic
Scrophulariaceae	*Galvezia juncea* (Benth.) Ball. var. *juncea*	Shrub	Endemic
Simmondsiaceeae	*Simmondsia chinensis* (Link.) C.K. Schneid.	Shrub	
Solanaceae	*Lycium andersonii* A. Gray	Shrub	
Solanaceae	*Lycium brevipes* Benth.	Shrub	
Solanaceae	*Lycium californicum* Nutt. ex Gray spp. *californicum*	Shrub	
Solanaceae	*Physalis crassifolia* Benth. var. *crassifolia*	Desert form	
Solanaceae	*Solanum hindsianum* Benth.	Shrub	

References

Ahumada CB (2000) Indices ecológicos para la evaluación y la gestión ambiental: aplicación en un estudio de caso (Punta Banda, Ensenada, Baja California, México). MSc Thesis. Universidad Autónoma de Baja California, Ensenada

Angermeier PL, Karr JR (1994) Biological integrity versus biological diversity as policy directives: protecting biotic resources. BioSciencie 44:690–697

Barbour MG (1992) Life at the leading edge: the beach plant syndrome. In: Seeliger U (ed) Coastal plant communities of Latin America, Academic Press, New York, pp 291–307

Barbour MG, De Jong TM, Pavlik BM (1985) Marine beach and dune plant communities. In: Chabot BF, Mooney HA (eds) Physiological ecology of North American plant communities. Chapman and Hall, New York, pp 296–322

Bergquist G, Bergquist C (1999) Post-decision assessment. In: Dale V, English M (eds) Tools to aid environmental decision making. Springer, Berlin Heidelberg New York

Brown AC, McLachlan A (1990) Ecology of sandy shores. Elsevier Science, Tokyo

Cendrero A (1997) Indicadores de desarrollo sostenible para la toma de decisiones. Naturale 12:5–25

Costanza R, Norton B, Haskell B (eds) (1992) Ecosystem health: new goals for environment management. Island Press, California

Connell JH (1975) Some mechanisms producing structure in natural communities: a model and evidence from field experiments. In: Cody ML, JM Diamond (eds) Ecology and evolution of communities. Harvard University Press, Cambridge, pp 460–490

Dale V, Beyeler CS (2001) Challenges in the development and use of ecological indicators. Ecol Indicators (1):3–10

Dale V, English M (1999) Tools to aid environmental decision making. Springer, Berlin Heidelberg New York

Delgadillo J (1995) Introducción al conocimiento bioclimático, fitogeográfico y fitosociológico del Suroeste de Norteamérica (USA and México). PhD Thesis, Alcalá de Henares, Spain

Done TJ, Reichelt RE (1998) Integrated coastal zone and fisheries ecosystem management; generic goals and performance indices. Ecol Appl (8)1:110–118

Eberhardt LL, Thomas JM (1991) Designing environmental field studies. Ecol Monogr 61:53–73

EPA (Environmental Protection Agency), SEMARNAP (Secretaría del Medio Ambiente, Recursos Naturales y Pesca) (1997) Indicadores ambientales para la Región Fronteriza/Environmental Indicators for the Bordeland Region. Programa Frontera XXI. México-Estados Unidos

Espejel I (ed) (2001) Selección de fragmentos de comunidades de matorral rosetófilo costero para su conservación en Baja California. Final Technical Report. Fondo Mexicano para la Conservación de la Naturaleza- Universidad Autónoma de Baja California. Ensenada, CD

Espejel I (Coord) (2002) Ordenamiento ecológico de la región de la escalera náutica. Techn Report. Instituto Nacional de Ecología (INE)-Secretaria de Medio Ambiente y Recursos Naturales (SEMARNAT). México, DF CD

Grennan JA (1999) Coastal sage scrub plants. San Diego Mesa College. San Diego California Pacific Botanical Press, San Diego, 168 pp

Hammond A, Adriaanse A, Rodenburg E, Bryant D, Woodward R (1995) Environmental indicators: a systematic approach to measuring and reporting on environmental policy performance in the context of sustainable development. World Resources Institute, 42 pp

Hiebert DR (1997) Prioritizing invasive plants and planning for management. In: Luken, OJ, Thieret JW (eds) Assessment and management of plant invasions. Springer, Berlin Heidelberg New York, pp 195–212

Horn E (1993) Coastal wildflowers of the Pacific Northwest. Mountain Press, Missoula, Montana. pp 179

Infante MD (2001) Diversidad morfológica y funcional de la vegetación de las dunas costeras de San Benito, Yucatán, México. Bachelor's Thesis, Benemérita Universidad Autónoma de Puebla, Puebla

Jones B, Riddle B (1996) Regional scale monitoring of biodiversity. In: Szaro R, Johnston D (eds) Biodiversity in managed landscape: theory and practice. Oxford University Press, New York, pp 195–209

Karr JR (1994) Landscapes and management for ecological integrity. In: Kim KC, Weaver RD (eds) Biodiversity and landscape: a paradox of humanity. Cambridge University Press, New York, pp 229–251

Karr JR (1997a) Measuring biological integrity. In: Meffe GK, Carroll CR (eds) Principles of conservation biology, 2nd edn. Sinauer, Sunderland, pp 483–485

Karr JR (1997b) The future is now: biological monitoring to ensure healthy waters. In: Streamkeepers: aquatic insects as biomonitors. The Xerces Society. Oregon, pp 31–36

Karr JR (1998) Biological integrity: a long neglected aspect of environmental program evaluation. In: Knap GJ, Kim TJ (eds) Environmental program evaluation: a primer. University of Illinois Press, Urbana, pp 148–175

Karr JR, Chu EW (1997) Biological monitoring: essential foundation for ecological risk assessment. Human Ecol Assess 3(6):993–1004

Keddy P, Lee H, Wisheu I (1993) Choosing indicators of ecosystem integrity: wetlands as a model system. In: Woodley S, Kay J, Francis G (eds) Ecological integrity and the management of ecosystems, St Lucie Press, pp 61–79

Landres PB (1992) Ecological indicators: panacea or liability? In: McKenzie DE, Hyatt DE, McDonald VJ (eds) Ecological indicators, vol 2. Elsevier, London

Lourens J, Van Zwol C, Kuperus J (1997) Indicators for environmental issues in the European coastal zone. Intercoast Network, pp 3–31

Moreno-Casasola P, Espejel I (1986) Classification and ordination of coastal sand dunes vegetation along the Gulf and the Caribbean Sea of Mexico. Vegetatio 66:147–182

Moreno-Casasola P, Espejel I, Castillo S, Castillo-Campos G, Durán R, Pérez J (1998) Flora de los ambientes arenosos y rocosos de las costas de México. In: Halffter G (ed) La Diversidad Biológica de Iberoamérica, Vol II. Acta Zoológica Mexicana, Instituto de Ecología A. C. – Ciencia y Tecnología para el Desarrollo, Veracruz, pp 177–258

NRC (National Research Council) (2000) Ecological indicators for the nation. National Academy Press, Washington, pp 180

Nelson JG, Serafin R (eds) (1997) National parks and protected areas: keystones to conservation and sustainable development. NATO. ASDI Series, Series G, Ecologicl Sciences 40. Springer, Berlin Heidelberg New York

Nordstrom FK (2000) Beaches and dunes of developed coasts. Cambridge University Press, Cambridge, 338 pp

Noss RF (1990) Indicators for monitoring biodiversity: a hierarchical approach. Conserv Biol 4(4): 355–364

Noss RF (1997) Hierarchical indicators for monitoring changes in biodiversity. In: Mcffe KG, Carroll CR (eds) Principles of conservation biology. Sinauer, New York, pp 88–92

Rozema J, Bijl F, Dueck T, Wesselman H (1982) Physiologia Pl 56:204–210

SEMARNAP (Secretaría del Medio Ambiente, Recursos Naturales y Pesca) (2000) Indicadores de desarrollo sustentable en México, Instituto Nacional de Estadística, Geografía e Información (INEGI), México

SEMARNAP-INE (Secretaría del Medio Ambiente, Recursos Naturales y Pesca-Instituto Nacional de Ecología) (1995) Ordenamiento ecológico de la Región de la Reserva de la Biosfera de la Mariposa Monarca. Final Technical Report CD

Schulze P (ed) (1999) Measures of environmental performance in ecosystem condition. National Academy Press, Washington

Schmitt R, Osenberg C (1996) Detecting ecological impacts: concepts and applications in coastal habitats. Academic Press, Toronto

Spellerberg IF (1991) Monitoring ecological change. Cambridge University Press, New York

Spjut WR (1996) Niebla and Vermilacinia (Ramalinaceae) from California and Baja California. SIDA, Botanical Miscellany No 14. Bot Res Inst Texas (BRIT), USA

Villa F, McLeod H (2002) Environmental vulnerability indicators for environmental planning and decision-making: guidelines and applications. Environ Manage 29(3): 335–348

Whitson TD (ed) (1996) Weeds of the West. Western Society of Weed Science, Wyoming, USA

Woodley S, Kay J, Francis G (1993) Ecological integrity and the management of ecosystems. St Lucie Press, Ottawa

Wright GR (1996) National parks and protected areas: their role in environmental protection. Blackwell, Cambridge

19 A Case Study of Conservation and Management of Tropical Sand Dune Systems: La Mancha–El Llano

P. Moreno-Casasola

19.1 Introduction

Coasts are typical environments in which human impacts have led to a whole range of changes with considerable variation in their degree of impact (French 1997). "Of all the coastal ecosystems, dunes have suffered the greatest degree of human pressure. Many dune systems have been irreversibly altered through the activities of man, both by accident and design" (Carter 1988). Most tropical countries have major economic problems and their economies still subsist on just one or two primary commodities. Worldwide tourism has become a very important economic force. Today tropical coasts – mainly beaches and sand dunes – are targeted for tourism of all kinds and are becoming a very important economic income for these countries.

Ecological features and dynamics of the diverse ecosystems, together with a distinctive coastal enterprise, make coastal zones unique. The coast should be recognized as a distinct region with resources that require special attention and environmental impacts that need to be controlled. There are a variety of available methods and necessarily the development of sustainable projects and the management of the coastal zone implies regaining an holistic perspective on coastal problems. Today, integrated coastal zone management schemes are being developed in numerous countries (Viles and Spencer 1995; Clark 1996; French 1997; Cicin-Sain and Knecht 1998).

19.2 Dune Conservation and Management

Sustainable development along the coast has to be based on an integrated management approach, which takes into account the dynamic nature of the littoral. In the following chapter, I would like to develop the idea that work has

Ecological Studies, Vol. 171
M.L. Martínez, N.P. Psuty (Eds.)
Coastal Dunes, Ecology and Conservation

to be done at several levels and with a variety of stakeholders, which are complementary and allow realistic planning for coastal activities. The first two approaches will be emphasized and a case study will be used to support the model.

1. A regional system of protected areas with management plans and economic resources that guarantee the preservation of the natural character of the coast and the protection of ecosystem functioning, habitats and species diversity and that includes the regional variation of beaches and dunes.
2. Community-based projects which create sustainable alternatives for development, promote participation of local people in the process and develop adequate structures for the inclusion of local governments in the management propositions and instrumentation. They are the grass root projects that will make environmental policies an everyday reality.
3. A view of the coast as a particular environment, which needs special legislation and an integrated coastal management program. The legal aspects that need to be addressed cannot be covered in this chapter.

Dune management has been visualized as constructing artificial dunes, dune stabilization, recovering dynamics reduced by over-management, beach nourishment (Clark 1996; Bird 1996; French 1997; Nordstrom 2000, among others). In tropical countries, it has to take into account a more integral perspective and learn from developed countries histories of errors and successes. Prevention rather than cure should be the everyday norm. It should include preservation of the natural character of the coast which refers to qualities that derive from natural as opposed to human influences on the coastal system. This means that sand dunes, their formation, dynamic and natural vegetation, are an essential part of the natural character of sandy beaches (Hesp 2000). This view is not perceived in most tropical countries, or indeed in many developing countries. Legislation for coastal areas exists in some tropical countries (Mexico lacks one), although very few have developed rules for dune management. Tourism is the force that is starting to develop management practices. In Quintana Roo, Mexico, a manual has been developed with guidelines for beach and dune protection (Molina et al 1998), although it is not compulsory.

19.3 Beach and Dune Biodiversity and Protected Areas in Mexico

Beaches and dunes occupy a very narrow strip along the coast, although floristically they represent between 5–7% of the flora in the country. The Pacific Mexican coast is on the edge of a tectonic plate, the coastal plain is very narrow or nonexistent except in the south, six climates are present from a

Mediterranean type in the north to a tropical wet climate in the south. Seasonal rivers and coastal lagoons are frequent along the rocky coast, as well as straight beaches and small bays. The Gulf and Caribbean coasts are located in the middle of a tectonic plate, the coastal plain is wide and sedimentary environments predominate. Climate is drier in the north Gulf, but the rest of the coast is warm and humid. There are many rivers along the Gulf and they provide significant sediment supply that littoral currents deliver to the beach and which form extensive sand dune fields. The Yucatan Peninsula is formed by a karstic platform, rivers are subterranean and calcareous sand is mainly of biogenic origin. The northeastern tip is very dry but the rest has a tropical humid climate.

The coast has been divided into different regions and subregions based on geomorphological, climatic, biogeographical, and floristic data (Moreno-Casasola 1999; Moreno-Casasola et al. 1998). I will use the regional classification described by Moreno-Casasola et al. (1998). The country has been divided into five regions (Fig. 19.1), three in the Pacific: the North Pacific (the oceanic coast of the Peninsula of Baja California), the Gulf of California (the states surrounding the Gulf of California or Sea of Cortes), and the South Pacific (south of parallel 21°). Along the Atlantic, there are two regions: the

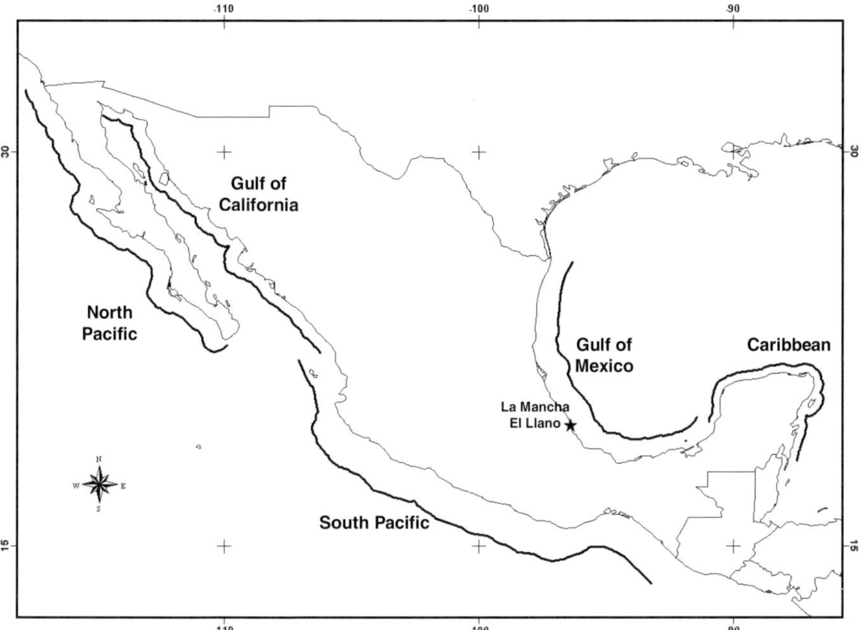

Fig. 19.1. Map of Mexico showing the different regions based on geomorphological, climatic, biogeographical and floristic beach and dune data (taken from Moreno-Casasola et al 1998). The case site of La Mancha-El Llano on the Gulf of Mexico is also indicated

Gulf of Mexico (states of Tamaulipas, Veracruz and Tabasco) and the Yucatan Peninsula (Campeche, Yucatan and Quintana Roo).

For both the Atlantic and Pacific coastlines, 1638 plant species have been registered belonging to 140 families (Moreno-Casasola et al. 1998). There are interesting floristic similarities and differences between regions. All of them share Leguminosae, Asteraceae, Poaceae, and to a lesser degree Euphorbiaceae and Cactaceae as the more abundant families. Other important families are Rubiaceae, Solanaceae, Chenopodiaceae, Cyperaceae, Boraginaceae, and Convolvulaceae.

From the phytogeographical perspective there are interesting differences (Moreno-Casasola et al. 1998). Distribution of tropical, temperate, cosmopolitan, and arid families varies among regions (Table 19.1). For example, the North Pacific coastal flora is influenced by the Sonoran Desert as well as the Mediterranean element it shares with California in the US, making it the more particular flora of all the regions. 171 species are found along both the Atlantic and the Pacific, but only 24 species are shared along the five regions. Growth forms also differ along regions (Fig. 19.2). Herbaceous forms predominate along the drier North Pacific and the Gulf of California. Woody species are the more abundant groups in the Gulf and Caribbean with high numbers in the South Pacific. There is a significant number of endemics: 64 species reported for the North Pacific (16 belonging to the Asteraceae, 7 Leguminosae, 6 Cactaceae, 6 Crassulaceae), 37 for the Gulf of California (9 Asteraceae, 4 Crassulaceae), 5 for the South Pacific, 10 for the Gulf of Mexico, and 34 for the Caribbean (6 Cactaceae, 3 Polygonaceae).

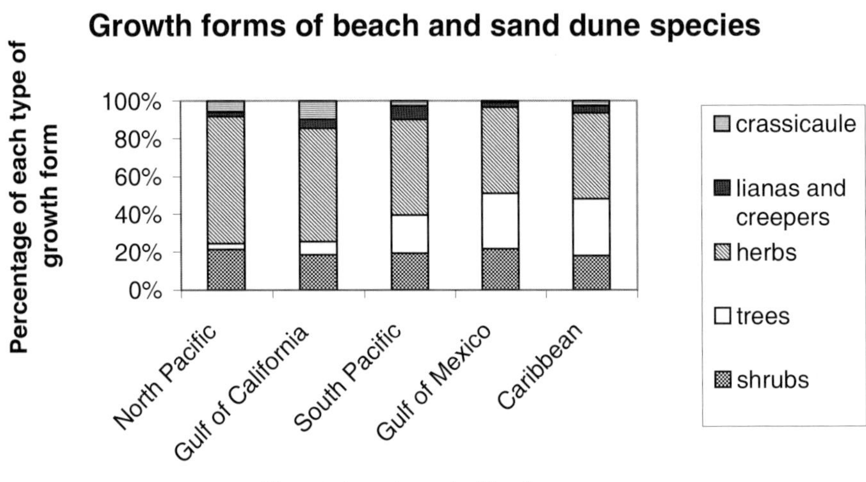

Fig. 19.2. Percentage of growth forms in the beaches and sand dunes of the five coastal regions in Mexico

Table 19.1. Diagnosis and degree of protection of beach and dune ecosystems in Mexico in established reserves. Data are based on (1) Moreno-Casasola et al (1998), (2) INEGI-SEMARNAP (1999), (3) degree of tourist development assigned using data on number of hotels and rooms offered, length of coast developed or under development schemes and number of tourists visiting the year annually

	North Pacific	Gulf of California	South Pacific	Gulf of Mexico		Yucatan Peninsula	
				Tamaulipas	Veracruz and Tabasco	Sisal	Yucatan and Quintana Roo
Plant families (1)	84 Families	59	92	87		78	
Tropical families	25.8 %	35.7	44.7	47		47.8	
Temperate	10.4	4.2	2.5	3		4.2	
Cosmopolitan	58.9	53.2	49.6	49.2		45.6	
Arid	4.7	6.8	3	0.7		2.4	
Plant species (1)	566 Species	235	555	427		456	
Endemics (1)	64 Endemics	37	5	2	10	34	
Protected surface (ha) (2)	2,546,790	3,996,223	820,967	0	52,327	59,130	1,077,956
Number of coastal protected areas (2)	3	8	6	0	2	1	11
Number of biosphere reserves (2)	2	6	3		1	1	5
Percent coastal population (2)[a]	19.4	25.4	21.3	5.3	18.3	10.2	
Percent population living in coastal counties (2)[b]	97.3	78.1	14	28.3	30.3	46.3	
Length of littoral (km) (1)	6107		2065	457	928	1730	
Tourist development (3)	Medium tending to high	Low tending to medium	Low	Very low	Very low tending to low	Low	Very high
Diagnosis and recommendation	Strong pressure from developments, more protected areas are needed (in the Mediterranean climatic region and tip of the Peninsula to ensure endemics habitats)	The northern Gulf coasts are well represented; there is need for protected areas in the south, both along the Peninsula and the continent	Strongly in need of further protected areas			Good, the area is small and the only Biosphere Reserve of Ría Celestún represents the coastal communities	Pressure from developments indicates more areas are needed i.e. there are plans to develop beach areas of Sian Ka'an

[a] Percentages were calculated with respect to Mexico's coastal population, 14,591,857 habitants in coastal counties

[b] Percentages of people living in coastal counties were calculated with respect to the total population for that region

Beaches and dunes are intensively used by visitors during holidays. Protected areas are a way of assuring that some landforms or areas will be protected from development, trampling, and other types of disturbance. In 1996, Mexico had under different forms of protection 5.8 % of the territory of the country (INEGI-SEMARNAP 1999). Among them the Biosphere Reserves (68.9 % of the protected area) receive funding, have a management plan, and monitor activities to ensure protection of species and habitats. Nineteen of the 59 coastal protected areas belong to this category. The highest numbers are found in the South Pacific region and the Caribbean (Fig. 19.3A). When one examines the area of these reserves in each region (Fig. 19.3B), the northwest of Mexico and the Caribbean have the biggest proportion of coastal land under protection, although only a small part of this area includes beaches and dunes. In these states, population numbers and settlements are not very high, making it possible to set aside larger protected areas (Table 19.1). The Caribbean has a good number of areas but they do not include a reasonable representation of the Caribbean beach and dune vegetation. The region shows the highest increase in tourist activities. Many of the South Pacific protected

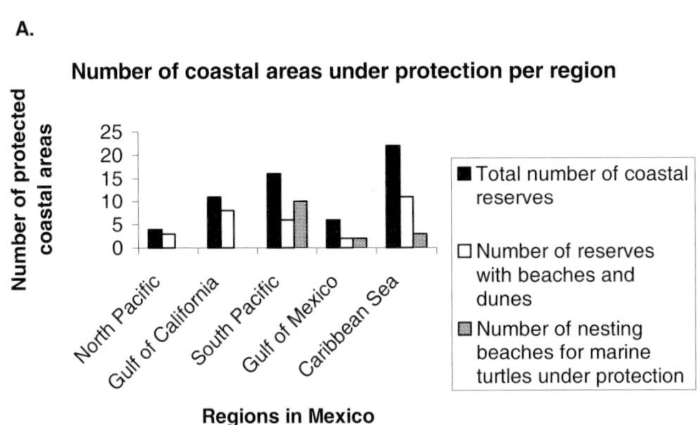

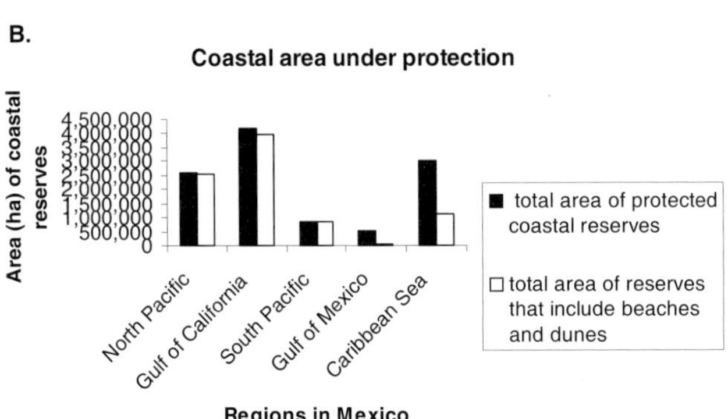

Fig. 19.3. Number *A* and area *B* of protected coastal areas in the five regions of Mexico. The area of nesting beaches for marine turtles under protection is always less than 1.5 % in the three regions (South Pacific, Gulf of Mexico and Caribbean)

areas belong to the category of nesting beaches for turtles. In these, only a small fringe of the beach is under protection, so that the total area is very small. The Gulf of Mexico has the lowest value (both in number of areas and the total hectares they cover) of all. It is also the most industrialized region.

Table 19.1 summarizes these results and shows that beach and dune communities still are in strong need of protection. There are floristically very rich regions with a high number of endemics, under strong pressure for tourism development. Beach protection should increase in these regions. Others are in strong need of protected areas such as the Gulf of Mexico. In this region, as well as in the South Pacific, federal and state authorities have to make a bigger effort for establishing natural reserves, but there should also be a strong community involvement and creation of local reserves.

19.4 Community Management for the Conservation of Coastal Resources: La Mancha-El Llano Case Study

Coasts are used by many different sectors of society. This has led to a proliferation of controlling and interest groups when it comes to coastal management. In the construction of a management plan, experts should incorporate as many views as possible from as many users as there exist. The plan should be based on the needs of the people where it is going to be applied (French 1997). Stakeholder participation is a relevant issue. Integrated Coastal Zone Management experiences in tropical countries are just beginning. Several authors have described various case studies (Clark 1996; French 1997; Cicin-Sain and Knecht 1998). Mexico's coastal areas, as well as those in many other tropical countries, give us a great opportunity to develop strategies within the framework of sustainable development. In most of these regions, local people see their world changing at a speed they cannot cope with. Government authorities in these rural counties are farmers and ranchers, elementary or high school teachers without access to many of today's information systems or environmental concerns. For them, ecology is a synonym of pollution and garbage; there is no comprehension of the need for an integrated and holistic view of the coastal zone; dunes are perceived as wastelands that should be put to use, for example, for cattle ranching.

Mangroves and estuaries are very much appreciated by local people and by governments. Many stakeholders will participate in activities related with these habitats. Other ecosystems such as freshwater wetlands are recently being recognized as providing important services and having ecological values. In tropical regions they are still associated with unhealthy conditions and government based programs to drain them, still exist. Beaches and dunes are poorly understood. Rural people see them as areas with nutrient-poor soils where low-intensity cattle ranching can take place. It is still very difficult to involve them in conservation schemes for these environments. As stakeholders, their interest is focused on transforming these environments to suit their productive activities.

A sustainable project would need to: (1) involve the local people and the local authorities, (2) create the space for interaction with other levels of government, (3) incorporate women as active members of society, (4) give children and youngsters information and education on the environment that surrounds them, and from which their parents are making an income, and (5) incorporate the concept of nature protection in their everyday lives. A case study is described to show how we are trying to create a sustainable way of living on a tropical coast. In the following, I will give a general overview of the project and develop with more detail the activities around the beach and sand dune system.

The study area of La Mancha-El Llano, located along the central coast of Veracruz (Fig. 19.1), is characterized by an extraordinary array of coastal ecosystems in an area of 22,484 ha. The beach and dune systems can be described and classified according to natural characteristics and actual use (Table 19.2). A detailed description of the beaches and dunes in the area appears in Moreno-Casasola (1997). Dunes reach 25 m in height and show different degrees of stabilization, with areas of erosion and sand accretion. In general, dunes are considered very fragile systems (French 1997; van der Maarel, 1997; Carter 1988), with degrees of transformation that range from low to very high. All (except the reserve at La Mancha) are cattle grazed. Both dune lakes (Cansa Burros and La Mancha) are highly modified by human activities and are considered to have a high fragility. They are invaded by introduced tilapia fish; *Pistia stratiotes* and *Eichhornea crassipes* have become plagues. Tectonic hills are part of the coastal landforms in the area and have been modified by cattle ranching. There are numerous stakeholders in the area: local people who have lived on the local resources for many years, others are just arriving and see the potential the area has for tourist development. Along the border of La Mancha Lagoon is the field station and protected area of the Institute of Ecology, a research institution.

Problems such as multi-users generating social conflicts, low quality education, rural economy where jobs are scarce and wages are very low, silted and polluted lagoons, extensive cattle ranching that increases deforestation and soil erosion, and the variety of stakeholders, indicate that action has to be taken before the area deteriorates still more. A management plan for the coastal are of La Mancha needs to take into account the environmental conditions of its ecosystems. There is a need for a strategic approach to coastline access management, a need to develop a tourist program and the strategies to change a rural county into an ecotourist resort, and all have to be integrated into an overall coastal zone management policy. In the neighboring Institute of Ecology a research project on conservation and management of coastal ecosystems has been in progress since 1998. The researchers and students working there believe that local people have to be involved in decisions of what is taking place in the area where they live. The project was designed trying to use a total catchment management approach as the means to develop conservation and good management practices in the area. To promote the restoration, conservation, and sustainable use of natural resources along the Gulf coast of Mexico, we are working in La Mancha–El Llano as a case study (Moreno-Casasola 2000; Moreno-Casasola et al. 2000). The reserve of the Institute of Ecology A.C., is the radiating point for many of the actions taking place.

We are using four instruments to develop the management plan, which will help us reach our objectives of sustainable use, conservation, and restoration in different ways:

1. The Watershed Management Committee is formed by the different stakeholders in the area: productive and social sectors, governmental authorities, and institutions for research and education. It is conceived as a forum promoting interaction between the different social sectors in the area and promoting environmental solutions to the watershed and coastal problems.
2. The productive projects allow local people to obtain a better income through the use of natural resources. It gives us the opportunity to link a particular group of people with the conservation and/or restoration of habitats. Table 19.3 shows the productive projects, their objectives, who is involved and what they do for the beaches and dunes.
3. Conservation and restoration projects are based on community action and environmental education. They are coupled with the productive projects to help people understand the ecosystems they live in and the importance that a healthy environment has on their everyday life. It allows us to incorporate children, youngsters, and women into activities and increase their awareness on the importance of sustainability for their future.
4. Land-use planning is a legal instrument in the Mexican legislation (Ordenamiento Territorial). It can be used as a planning exercise that allows local people, technical groups, and government to discuss and decide on ecological principles for land use. Table 19.2 is such an exercise that is being discussed at the community level. The process allows the discussion of areas that should be declared as reserves, with a strong community participation and involvement.

Other activities are also taking place, which promote the conservation of the beach and dunes: weekly beach clean-up, making compost with organic garbage, promoting the use of dry latrines, annual Beach Bird Festival, environmental education (talks, conferences, games, videos, visits to ecosystems, among others), citizen committees for protection and vigilance of resources, training courses for local people and county authorities.

We would like to exemplify how La Mancha project and the Institute of Ecology developed a series of agreements with the "palaperos" (restaurant

Table 19.2. Coastal description for the area of La Mancha and El Llano, Veracruz. Activity proposed: *C* conservation, *T* tourism development of low density, ecotourism, *R* recreation, infrastructure, *A* agriculture and cattle ranching

		San Isidro beach dunes	Los Amarillos beach and dunes	Cansa Burros beach – dunes	Dune lake Cansa Burros	Tectoni hill
Beach	Long or Curved, Narrow/Wide	L, W	L, W	L, N		
	Barrier island/ adjacent Dune fields	D	D	D		
	Subject to high flooding. No/Yes	Y	N	N		
	Abundant embryodunes/Scarce	A	S	S		
	Erosion/Accretion	A/E	A/E	A/E		E
Dunes	Highly mobile/Mix/Stabilized	HM	HM	M		S
	Pioneers, sand binding plants	x	x	x		
	Grassland/thickets			x		x
	Dry, low tropical forest		x			x
	Tropical semideciduous forest					
	Presence of endemics	x	x	x		
	Abundant beach birds, crabs				x	
	Invasive species	x		x		x
	Dunes slacks/dune lakes	x			x	
Activities	Recreation	x				
	Fishing				x	
	Cattle grazing		x	x		x
	Coconut cultivation					
	Sugar cane cultivation			x		
	Water extraction					
	Garbage dumps					
	Land partitioning for tourism					
	Public services provided					
	Nature Reserve					
	Buildings or structures					
Stakeholders	Only local stakeholders	x	x	x	x	
	External stakeholders present					
	Tourism developers					x
	Degree transformation (Low, Medium, High, Very High)	L	L	M	H	VH
	Fragility (L, M, H)	H	H	H	H	L
	Activity proposed	R	C	T, A	C	T

a Iancha each nd unes	La Mancha dune lake	Fosil dune hill Jicacos	Old dunes	Farallón beach and dunes	Barrier island/ beach	Villa Rica beach	V.Rica tectonic hill	V. Rica dunes	Limón beach
, N				L, N	L, W	C, N		L, N	L, W
				D		BI		D	River plain
				N	Y	N		N	Y
				A	S	S			A
/E		E		A/E	E	E	E	A/E	A
		S	S	M	S		S	HM,M	M
				x				x	x
		x	x	x	x		x	x	
		x					x		
				x				x	x
	x	x		x					
		x	x	x	x	x	x		x
	x				x				
				x		x			
									x
		x	x	x	x				
								x	
			x						
								x	
					x				
				x	x				
						x			
	x								
						x		x	
		x	x						x
	x			x	x	x	x	x	
						x	x	x	
	H	M	M	M	H	VH	H	VH	L
	M	L	H	H	H	M	M	H	H
	C	T	C	R	T	R	C	T	C, A

Table 19.3. Description of community productive projects linked with beaches and dunes in La Mancha and El Llano, Veracruz

Productive project: what is it?	Who works in it?	How does it help beaches and dunes?	Function within the community management plan
Services in **beach restaurants** and information center for the program	Peasants and fishermen who sold food on the beach in rustic establishments	– Keeping the beach clean – Keeping crab poachers away – Giving information to visitors – Providing touristic services in the beach – Not allowing vehicles on the beach	– Demonstrating that conservation is an economic alternative – Managing and preserving the coast, specially the beaches
Ecotourism guides and services	Peasants and fishermen	– increasing people's awareness of the importance of beaches and dunes – Promote organization for the conservation of habitats and resources	– Developing ecotourism activities for visitors and for local people – Demonstrating that conservation is an economic alternative
Propagation of **native species** in a nursery	Women from the town of Palmas de Abajo	– Developing the first nursery with native sand dune fixing plants – Providing alternatives to the use of *Casuarina equisetifolia*, an exotic used for sand dune fixing	– starting a productive and sustain able activity for women – Developing the first nursery with native local plants and trees
Conservation and management of the **blue crab** (*Cardisoma guanhumi*)	Fishermen from two cooperatives	– Increase awareness of the importance of crab populations – Promote organization for the conservation of habitats and resources	– Recovering crab populations – Transforming them into an eco nomic benefit for the fishermen, – Demonstrate that an adequate management preserves the resource and the habitat

owners) and the County authorities, both for the benefit of their own businesses, and for the beach and dune protection.

First set of agreements: the Institute of Ecology, a neighboring research institution, has to give its agreement for the establishment of beach restaurants. The conditions set were that the palaperos accepted a set of commitments for beach management. Agreements included that only five people would receive the authorization to make sure that they could make a good income out of this productive activity, authorizations are non-transferable (only their direct family can take over the business as a guarantee that the same agreements will be kept); they built restaurants according to a model designed with traditional local designs and materials. They bought materials and paid for the construction and the Institute provided the wood and technical advice. The palaperos got organized to keep the one kilometer strip of beach clean. In addition, they function as an information center for the management plan and the ecotourism project. Through these schemes a productive project generating an income is strongly linked to a conservation program. Their everyday work depends on the adequate management of the beach.

Second set of agreements. Training courses and workshops were designed to help the palapero group advance within the project (have a self-sustainable organization, good management practices, conflict-solving mechanisms). Through the workshops they became organized: named a president, secretary, and treasurer, defined their functions and established a control notebook (for keeping track of agreements, meetings, etc.). They agreed to share equally work, expenses, and maintenance of installations (either by working directly themselves or paying somebody else to do their share). The Institute built ecological toilets (dry latrines) for beach visitors. The palaperos charge a small fee for using the toilets which is used for buying cleaning materials and keeping the baths clean. The County will reduce the monthly tax until the business gets going and the Institute meanwhile will help with water and electricity until the palaperos can provide their own. The County will help with materials and equipment to fix the road and have good access for tourism.

Third set of agreements currently taking place: training on restaurant services and developing an environmental education program with them, to increase their knowledge of the beach, dunes, beach birds, lagoon ecosystems, and conservation and restoration needs. Two other local groups working within the area have developed an ecotouristic alternative which includes trails and guided tours on the beach and both the mobile and stabilized dunes. After the visit the tourists end their day having dinner with the palaperos. In this way, three groups of stakeholders in the area are linked together.

Other productive projects on ecotourism have been developed. Two local peasant groups have been trained as ecoguides and one of them now has a series of cabins for lodging groups. They both give guided tours along the beach, mobile dunes and stabilized dunes. They have learnt about the value of preserving the natural character of the coast and during the tours, talk about sand deposition and erosion, sand-loving plants, how slacks and dunelakes are formed. They decided that their names were Duneadventures in San Isidro and Ecoguides in Movement. One of the groups has started to give guided tours to school children from their village.

These projects and agreements will help these groups receive an income by providing services to local tourists, increase their awareness that this income is in direct relation with beach quality and service as well as ecosystem conservation, increase their training and that of their families and develop organizational capacities as well as protecting the beach and dunes, its flora and fauna. A model of beach management is being demonstrated and is being extrapolated to other beaches in the watershed. Work on productive and conservation projects coupled with development of organized groups is turning them into organized producers with strong environmental responsibility. The joint work between academic groups, government authorities and local inhabitants has shown to be very productive. Results are slow to come, but they are grass root achievements. Only through involvement and responsible participation of all stakeholders will we be able to manage and protect very fragile ecosystems such as beaches and dunes.

Acknowledgements. My thanks are extended to the people of La Mancha and El Llano who worked on the project as well as all the members of the working group: G. Salinas, L. Amador, H.H. Cruz, A. Juarez, B. Gea, A.C. Travieso, L. Ruelas and E. López. This research was financed by SIGOLFO (99–06–010-V) and NAWCA.

References

Bird ECF (1996) Beach management. Wiley, New York, 281 pp

Carter RWG (1988) Coastal environments. An introduction to the physical, ecological and cultural systems of the coastlines. Academic Press, New York, 617 pp

Cicin-Sain B, Knecht RW (1998) Integrated coastal and ocean management. Concepts and practices. Island Press, Washington, DC, 516 pp

Clark JR (1996) Coastal zone management. Handbook. Lewis Publishers, New York, 694 pp

French PW (1997) Coastal and estuarine management. Routledge, London, 251 pp

Hesp P (2000) Coastal sand dunes. Form and function. Forest Research, Rotorua, New Zealand. CDVN Tech Bull No 4, 29 pp.

INEGI-SEMARNAP (1999) Estadísticas del Medio Ambiente, vols I/II. INEGI, México DF.

Molina C, Rubinoff P, Carranza J (1998) Normas prácticas para el desarrollo turístico de la zona costera de Quintana Roo, México. Amigos de Sian Ka´an A.C. y Centro de Recursos Costeros, Univ. de Rhode Island. Kromagraphics Impresor, Q. Roo, 93 pp

Moreno-Casasola P (1997) Vegetation differentiation and environmental dynamics along the Mexican Gulf coast. A case study: Morro de la Mancha. In: van der Maarel E (ed) Dry coastal ecosystems, vol 2C. Elsevier, Amsterdam, pp 469–482

Moreno-Casasola P (1999) Dune vegetation and its biodiversity along the Gulf of Mexico, a large marine ecosystem. In: Kumpf H, Steidinger K, Sherman K (eds) The Gulf of Mexico large marine ecosystems - assessment, sustainability and management. Blackwell, New York, pp 593–612

Moreno-Casasola P (2000) Mangroves, an area of conflict between cattle ranchers and fishermen. Proc Intl Workshop. Asia-Pacific cooperation on research for the conservation of mangroves. United Nations University, Tokyo, pp 155–170

Moreno-Casasola P, Espejel I, Castillo S, Castillo-Campos G, Durán R, Pérez-Navarro JJ, León JL, Olmsted I, Trejo-Torres J (1998) Flora de los ambientes arenosos y rocosos de las costas de México. In: Halffter G (ed) Biodiversidad en Iberoamérica, vol 2. CYTED- Instituto de Ecología A.C., pp 177–258

Moreno-Casasola P, Saavedra T, Zárate D, Amador L, Infante DM (2000) La Mancha-el Llano: un estudio de caso sobre posibles métodos regulatorios voluntarios para la conservación y uso sustentable de las zonas costeras de México. In: Bañuelos M (coord) Sociedad, derecho y medio ambiente. Primer Informe del Programa de Investigación sobre Aplicación y Cumplimiento de la Legislación Ambiental en México. Coedición Conacyt-UAM-PROFEPA. México DF, pp 181–216

Nordstrom KF (2000) Beaches and dunes of developed coasts. Cambridge University Press, Cambridge, 338 pp

Van der Maarel E (1997) (ed) Dry coastal ecosystems. General aspects. Elsevier, Amsterdam, 707 pp

Viles H, Spencer T (1995) Coastal problems. Geomorphology, ecology and society at the coast. Arnold, London, 350 pp

20 European Coastal Dunes: Ecological Values, Threats, Opportunities and Policy Development

P. HESLENFELD, P.D. JUNGERIUS, and J.A. KLIJN

20.1 Introduction

This chapter describes international policy to conserve coastal dunes. Because it is not possible to describe all policy instruments of the world, the authors selected those of Europe as a case study. Perhaps more than other continents, Europe aims at a common and consistent international nature policy for the whole of its coastline. As a result, countries in the European Union have national as well as international legal instruments to protect dunes. This is of interest to other coastal regions of the world where attempts are made for an international cooperation in the management of the coastline.

Compared to other continents, Europe's winding and indented coastline is very long, roughly 3.5 times the Earth's perimeter (European Environmental Agency 1995). Europe's coastal landscapes are highly varied due to differences in geology, climate, coastal processes, geomorphology, biogeography, history of land use and actual human influence (Klijn 1990; Arens et al. 2001). Coastal habitats in Europe are therefore valued for their geological and geomorphological, historical, ecological and scenic properties (e.g. Bennet 1991; Doody 2000). Coasts, especially so-called soft coasts such as the coastal dunes, are very vulnerable to coastal erosion, and to a host of human influences. Coasts attracted man since prehistoric times. Settlement, fishing, hunting, agriculture and possibilities for easy transport were traditional attractors. Later, land reclamation, afforestation, water extraction, coastal defence works, urbanisation, industrial expansion, harbours, infrastructure and tourism became important. Well over one third of Europe's population of 200 million inhabitants lives within 50 km of the coastline. In many Mediterranean countries the great majority of urban areas (85 %) are situated near the coasts which attracts industries and transport of people and goods. Low-lying, fertile coastal plains are inductive to intensive agricultural land use.

Ecological Studies, Vol. 171
M.L. Martínez, N.P. Psuty (Eds.)
Coastal Dunes, Ecology and Conservation

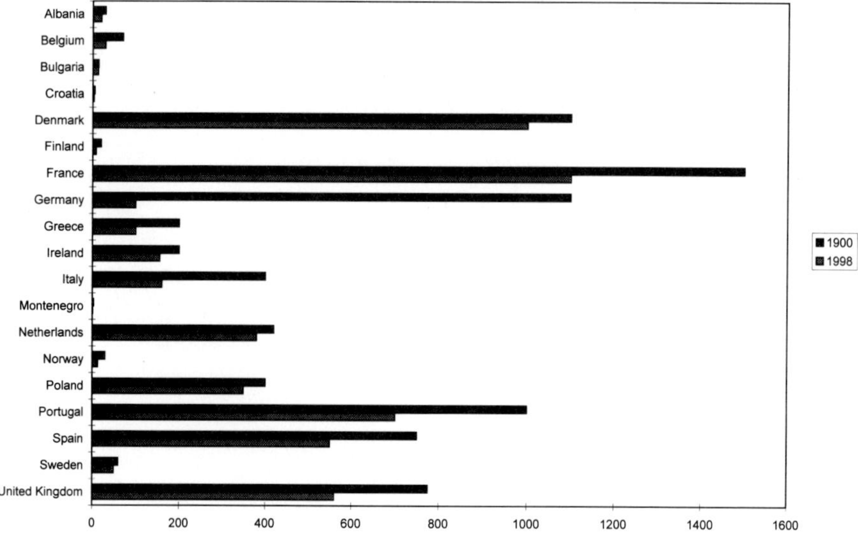

Fig. 20.1. Size of dune area in 1990 and 1998, in km². (Delbaere 1998)

All over Europe a net loss of coastal dunes of 25% has been reported since1900 (Fig. 20.1) and some 55% of the remaining coastal dune area has lost its natural character (Delbaere 1998). It has been estimated that roughly 85% of the present area is under threat (World Resources Institute, see Various Internet sites; European Environmental Agency 1999). Coastal erosion affects roughly 25% of Europe's coastline, much of which concerns soft coasts (European Environmental Agency 1999; Bird 1985). Increased erosion can be expected in the future due to sea level rise. To cope with all forthcoming changes in land use and coastal dynamics requires insight in the distribution and values of coastal dunes, expected trends, cause – effect relationships, resulting threats and opportunities, and possibilities to intervene by policy and adequate management.

20.2 Distribution of Coastal Dunes Along Europe's Coast: A Short Geography

The total area of Europe's coastal dunes has been estimated at over 5300 km² (Delbaere 1998). Coastal dunes are widespread along Europe's coastline, though their distribution is uneven (Fig. 20.2). Some rocky coasts harbour small and isolated dune areas in protected embayments, whereas an almost uninterrupted and kilometre-wide fringe of coastal dunes is found in sedimentary regions with favourable conditions (Klijn 1990; Doody 2000). Geo-

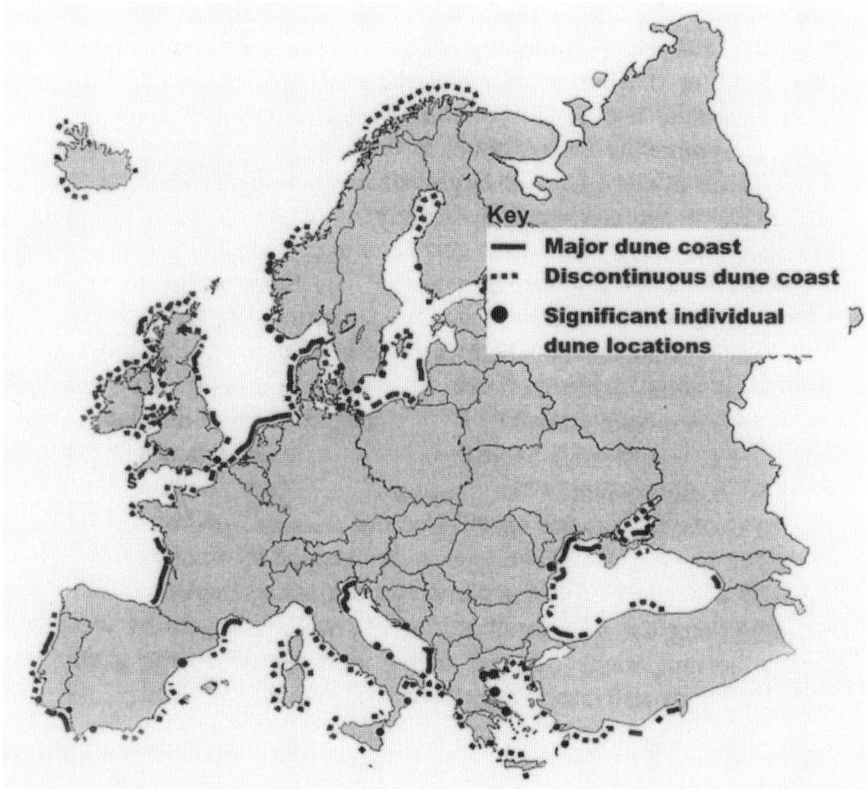

Fig. 20.2. Major dune areas along the European coasts. (Courtesy of J.P. Doody 2000)

graphically we can distinguish five major regions: the Baltic region, the North Sea region, the Atlantic region, the Mediterranean and the Black Sea region (European Union for Coastal Conservation 1997). Some characteristics of these regions are (Van der Maarel 1993):

1. Formerly glaciated mountainous Scandinavian coasts harbour small and isolated dune areas. The southern Baltic, however, is a sedimentary, periglacial region with ample material for beach barrier and dune formation (e.g. Poland, northern Germany).
2. The North Sea region includes the whole sweep from Calais to the tip of Jutland (northern France, Belgium, the Netherlands, Germany, Denmark) which is liberally beset with wide and almost uninterrupted dunes on barrier islands, along nearly straight coastlines and estuaries thanks to favourable sedimentary conditions, tidal range, wind and wave direction and energy.
3. The Atlantic region, Ireland and the UK have mostly small dune areas in sheltered bays, though larger areas occur both along the North Sea and the

western coasts. In France, south of the Normandy and Brittany promontories, the extensive sedimentary basin of Gascogne promoted extensive dune fields in Les Landes. North-western Spain is rather poor in dunes, whereas Atlantic Portugal and south-western Spain are again relatively rich.

4. The Mediterranean region has a complex and heterogeneous coastline. The larger dune fields are correlated with deltas and coastal plains, like in parts of Italy. Elsewhere, e.g. in Greece, dunes occur in smaller areas in embayments between promontories and cliffs. The latter relate to mountain ranges. The eastern Mediterranean littoral is a strip of varying width.
5. The coastline of the Black Sea region is very diverse. It includes deltas (Danube), estuaries (Dnjepr), systems of lagoons and sand-dune barriers and limans. Sand dunes are widespread along the north-western Black Sea. The coastline of Turkey is generally rocky. Dunes occur where lowlands border the sea.

Generally, it can be concluded that extensive coastal dune areas occur where an abundance of sediments from glacial, periglacial or fluviatile origin can be found and where tidal regimes, storm floods and prevailing winds stimulate a dynamic environment and a high aeolian transport capacity (Davies 1972; Klijn 1990; Doody 2000).

20.3 Ecological Values Related to Biodiversity

Landscape values of coastal dune areas comprise ecological, geomorphological, geological, historical, archaeological and scenic values. We restrict ourselves to ecology because it is mainly for ecological values that international agreements are presently being developed. Ecological values of coastal dunes are generally large. A good example are those of the Netherlands where more than half of all higher plant species occur in a coastal dune area representing no more than 1 % of the country (Bakker and Doing 1987).

We selected biodiversity to represent the ecological values of the European coastal dunes, because biodiversity, comprising the subspecies level, the species level, the ecosystem and the landscape level is presently an operational and politically acknowledged issue (Treaty on Biological Diversity, Rio de Janeiro 1992). Other ecological values, for instance life-support functions related to regulation processes (water purification, coastal protection among others, for a more comprehensive overview; see De Groot 1992) are not dealt with. Accepted criteria for nature protection are species data, species protection statuses and habitat requirements.

Species listed in various European directives include mammals, birds, reptiles, amphibians, fishes, butterflies and higher plants, and were selected on the basis of their international rareness, endemism, presence of so-called characteristic species and the rate of decline or possible extinction. National and regional policies have comparable or additional criteria and selections. Sustained ecological values depend first on species protection, e.g. by regulations on hunting or fishing, secondly on safeguarding basic environmental and spatial conditions such as sufficient area size and quality, and thirdly on apt management inside the areas and restrictions concerning land-use in adjacent areas (see Table 20.1).

Table 20.1. The main ecological conditions according to Adriani and Van der Maarel (1968) and Bakker et al. (1981)

Ia. Area and width of dunes

- These dimensions are relevant for (1) gradients from sea to inland (e.g. wind, salt spray), (2) buffering for outside influences (e.g. pollution or disturbance), (3) living area and carrying capacity of populations of a minimal size, (4) increase in age and successional stage related to natural zoning of dunes from the beach to the inner margin

Ib. Connectivity

- The extent to which dune areas are connected enables exchange of organisms (for migration, dispersal or other functions, such as foraging and resting)

IIa Parent material

- Parent material rich in lime and/or other minerals enables a variety of plants and plant communities (Grime 1979; Adriani and Van der Maarel 1968)

IIb Presence of water and moist environments

- Wet and moist dune slacks and open water (dune lakes) favour habitat diversity

IIc Variety of soil conditions

- The presence of various development stages of the soil (soil profile development, decalcification, humus formation, etc.) encourages variation in plant communities

IId Gradients towards other environments

- Gradients from dunes towards marshes, salt marshes, bogs and rocks add to habitat diversity

IIe Lack of adverse outside influences

- Absence of adverse external influences e.g. leading to lowering of groundwater tables, input of nutrients, pollution by air or water, and disturbance of fauna or flora by man, increases the vitality of the ecosystem

III Occurrence of a certain degree of natural dynamics,

- These dynamics (e.g. sand blowing) cause a renewed start of the succession and favour the presence of pioneer and successive communities. They also enable the dune system tot adapt to coastal erosion by natural retreat

IV Continuity of extensive management

- Extensive use (e.g. grazing), adds to ecosystem diversity by sustaining semi-natural habitats

20.4 Trends, Threats and Opportunities

Protection and possible restoration of dune ecosystem should focus on basic conditions that offer enough counterforces against adverse developments. It is crucial to translate this insight into effective, pro-active strategies. Trends or expected trends in Europe may vary from region to region. We present a rough and qualitative indication of trends, threats and opportunities grouped in large clusters of activities or processes, for the broad regions listed in Section 20.2 (largely after European Environmental Agency 1998; European Union for Coastal Conservation 2000).

20.4.1 Agriculture: Intensification as well as Marginalization

Agriculture was traditionally a common land use type in many coastal areas. Its effects vary from disastrous in case of overgrazing, excessive drainage and levelling for specific cultures (e.g. bulbs or onions), to real added ecological values where extensive grazing over centuries promoted rich plant communities. The decline or complete abandonment of grazing implies loss of ecological values as seen in the machairs along the west coast of Ireland and Scotland (Bakker et al. 1992). Changes in agricultural land-use are discussed by, e.g. the OECD (2000), explaining processes of marginalization (even land abandonment) in less-favourable areas and intensification elsewhere.

Expectations for the European dunes depend on the regions. The Baltic region will undergo both processes: intensification and land abandonment. The North Sea coasts are part of a region that will show a further decline of agriculture in the future, dunes making place for buildings and infrastructure. More pressure on dunes themselves is not expected unless from intensive horticulture. In the Atlantic region, the most remote areas will witness a loss of traditional extensive pastures in some places and intensification in other places. The coasts of the Mediterranean and Black Sea regions will probably experience more pressure from agriculture and horticulture in river valleys and coastal plains, due to the growing population and tourist industries, affecting coastal habitats through water availability, water quality and reclamation.

20.4.2 Urbanisation, Industries, Harbour Development

Urban sprawl and growth of industrial (power) plants and harbours are common features along coasts. There are several types of impact on the environment and on biodiversity, such as net loss of habitats, fragmentation of living areas, water and soil pollution by sewage and waste disposal, air pollution, water extraction, and sand quarrying. Investments in urban or industrial sites almost automatically lead to the construction of artificial coastal defence works.

The Baltic and Black Sea regions will undergo local developments, though spatial policies try to counteract. The North Sea countries show less urban development due to spatial policies as well as a decline of heavy industries. Ports are expanding. The Atlantic region seems to undergo further urban and industrial development in coastal areas. In the Mediterranean region the tendency of strong growth seems to continue due to population growth and immigration as well as touristic development affecting many areas. Pressure on coastal areas seems to go on.

20.4.3 Infrastructure

Railroads, highways and roads as well as airfields tend to increase in density and dimensions, thereby causing loss of habitats and fragmentation, disturbance and pollution of habitats. Coastal routes mostly require artificial constructions and sea defence works. The development of ports has the same effects and moreover often affects coastal dynamics by disturbing sediment flows by dredging and by constructing harbour entrances such as dams or piers.

Regional expectations indicate that increase of networks will occur everywhere, locally in the Baltic region and in other accession countries wanting to join the European Union. The North Sea region is already heavily developed. The Atlantic region will show a moderate increase. Most increase is expected in Mediterranean countries, notably near airports.

20.4.4 Tourism/Recreation

Coastal dunes, which often border attractive sand beaches, are favoured by tourists. The tourist industry is the world's fastest growing economic sector having major impacts on coastal regions. Construction of holiday houses and camping sites, hotels, marina's, transport routes, water supply, waste disposal, parking lots and golf courses consumed already many dune areas or added to their decline. Uncontrolled use of dune areas by tourists has other adverse effects such as trampling, disturbance of wildlife, and leaving litter.

Expectations indicate a general increase everywhere, although most pressure is to be expected in the Mediterranean countries. The Black Sea region seems to experience the same trends; assessments of growth between 1984 and 2025 show almost a doubling of the area consumed by tourist developments in coastal areas (European Union for Coastal Conservation 1997). The Atlantic and North Sea coasts are subject to growth as well, as are some coasts in the Baltic region.

20.4.5 Forestry

Traditionally, many coastal dunes have been planted with trees for reasons of stabilisation or wood production. Often, exotic species were used (pine, eucalyptus). Economic profits are limited. The effects on ecosystems are largely negative, though forest plantations created some new habitats and caused the arrival of new and sometimes valued species. Forest fires are a realistic and frequent threat endangering ecological values and causing large economic losses. Fixation of dunes prevents natural dynamics and brings a halt to succession processes.

Regional expectations do not foresee much expansion of forests. Natural expansion can take place where existing land use disappears. In some countries a return to more natural ecosystems with endemic tree species is promoted.

20.4.6 Coastal Processes, Climate Change and Sea Level Rise, Aeolian Processes

Coasts are dynamic systems responding to sea level movement, land subsidence or rise, tidal currents, wave action and, related to all these factors, to longitudinal or onshore or offshore movement of sediments. Sandy coasts respond quickly to changing conditions. On a geological time-scale, these coastlines tend to shift seaward and landward, preserving their characteristic though changing their position. Climate change and associated increased sea level rise or an increase in storm frequency can affect coastal dunes detrimentally (1) when losses cannot be compensated by sand input, or (2) when natural systems cannot shift landward because they are squeezed between the sea and a high or steep hinterland, or hard defence works (Klijn 1995). Sea levels rose 20–25 cm in the last century and will rise another 25–95 cm during this century due to global warming, possibly accompanied by a more vigorous wave attack exacerbating erosion rates (IPPC 1995). This results in net loss of dunes, estimated to amount to several hundred metres, with lowering of water tables among other impacts (Van der Meulen 1990; see also *Coastal defence works* below).

Regional differences are associated with geological conditions and climate zones. Especially countries in subsiding areas of the North Sea region are likely to be affected by an increased rate of sea level rise. Other regions however will also show more coastal erosion.

Aeolian processes are a natural phenomenon lying at the base of the history of coastal dunes. Both an excessive degree of sand blowing and an unnatural degree of sand fixation by man could result in loss of ecological values. Generally, dune managers show an overcautious attitude, leading to unnatural fixation of dunes and the disappearance of initial succession stages (Bakker et al.1981).

Management towards dynamic dunes should be promoted. A dynamic dune system is more resistant to erosion processes, is cheaper to maintain, has higher natural values and it is more sustainable than fixated dunes. A good example is the dynamic dune complex of the Polish Slowinski National Park. The dunes are over 40-m high, very dynamic and almost not influenced by humans. This dynamic system has already existed for thousands of years.

20.4.7 Coastal Defence Works

Artificial defence works comprise hard elements such as groynes, revetments, concrete walls, dikes and alike, or 'soft' measures such as (repeated) beach nourishment. The first group of measures affects the natural character of coastal dunes more than the second group. Artificial defence of retreating coasts logically helps to prevent a net loss of dune areas, but it has to be remembered that a shifting coastline is a natural phenomenon as such and dunes can move landwards and seawards under natural conditions by aeolian redeposition. A natural adaptation by retreat and landward shifting of dunes however is seldom accepted in regions where agriculture, infrastructure or built-up areas directly border dune areas or beaches. In fact, most countries have a very defensive attitude and prefer hard defence works. But some countries changed their strategies and adopted a more natural defence, using sand nourishment, either on the beach or on the shore face, which is even more natural and moreover much cheaper.

20.5 Policy Analysis of Dune Conservation in Europe

Many dune areas in Europe are under threat, as described above. However, almost every country has a number of instruments available that can be used to conserve the European dunes. It is impossible to evaluate all instruments of all countries in this chapter, so only the most relevant instruments are dealt with here.

There are three main instruments that are important for the protection of coastal dunes: the Bern Convention (including its Emerald Network), the EU Habitat Directive and national policies. Furthermore, three categories of countries can be distinguished: (1) the EU countries, (2) accession countries, which will enter in the (near or far) future, (3) the remaining non-EU countries.

20.5.1 Bern Convention

The legal instrument with the largest geographical (European) range of importance for coastal dunes is the Convention on the Conservation of European Wildlife and Natural Habitats. This 'Bern Convention' aims at the conservation of wild fauna and flora and their natural habitats, especially those species and habitats whose conservation requires the co-operation of several countries, and promotes such co-operation.

Over 40 European countries, including most of the member states of the Council of Europe have ratified the Convention or intend to do so. The member states are all EU and accession countries and most non-EU countries (the so-called contracting parties). Some countries are not (yet) member but have the status of observer states. However, the convention has little power, because it has almost no sanctions.

In response to an increasing need to protect natural habitats it was decided to establish a Pan-European network: the Emerald Network. This is a network of Areas of Special Conservation Interest (ASCIs), which is to be established in the territory of the contracting parties and observer states. The aim of this network is to protect specific flora and fauna species and endangered natural habitats. The responsibility for the designation of the ASCI's and their protection and management rests with the governments of the states concerned. There is, however, no precise recommendation to give legal protection to a designated site. Designation of ASCI's has started in 1997 (see "Various Internet Sites" after the references to this chapter).

20.5.2 EU Policy

Up to now, 15 European countries are united in the European Union. The EU has its own legal instruments. The most important EU instruments for the conservation of natural habitats are the Council Directive 92/43/EEC of 21 May 1992 on the conservation of natural habitats and of wild fauna and flora (the 'Habitat Directive', see "Various Internet Sites" below) and the Council Directive 79/409/EEC of 2 April 1979 on the conservation of wild birds (the 'Bird Directive').

The Habitat Directive aims to guarantee the biological diversity through the maintenance of natural habitats and the wild flora and fauna. It distinguishes the protection of areas ('natural habitats') and the protection of flora and fauna species. The protection of areas is the most important for the conservation of sand dunes. But also the presence of a priority species of the Habitat Directive can give a natural habitat a certain degree of protection. The Bird Directive aims to protect endangered European birds by protecting specific areas that are habitats of selected birds.

For every habitat type, the five most important areas (with a minimum surface of 100 ha) are designated within each EU country. These areas are selected by the national governments, based on criteria of the EU. The European Commission will make the definitive list of selected habitats (Bekker et al. 2001). Several types of sand dunes are included in the list of priority habitats of the Habitat Directive (see Table 20.2).

The Habitat and Bird Directives designate the establishment of an ecological network, Natura 2000, in which each member state has to nominate Special Areas of Conservation (SACs), also based on criteria of the EU. Natura 2000 is also the contribution of the EU to the EMERALD-Network under the Bern Convention.

Table 20.2. Coastal habitats of Community interest (sand dunes) which form the basis for the selection of Special Areas for Conservation (SACs). (Free after Doody 2000; from the Interpretation Manual of European Union Habitats, European Commission 1999)

Geographical area	Directive Name
Atlantic, North Sea and Baltic	Embryonic shifting dunes Shifting dunes with *Ammophila arenaria*
	Boreal Baltic dunes sandy beaches with perennial vegetation
	Fixed dunes with herbaceous vegetation (grey dune)
	Decalcified dunes with *Empetrum negrum* Eu-atlantic decalcified fixed dunes (*Calluno-Ulicetea*)
	Dunes with *Hippophae rhamnoides*
	Dunes with *Salix arenaria ssp. argentea* (*Salicion arenariea*)
	Wooded dunes of the Atlantic, Continental and Boreal region
	Humid dune slacks
	Machairs in Ireland only
Mediterranean coast	Crucianellion maritimae fixed beach dunes
	Dunes with *Euphorbia terracina*
	Malcolmietalia dune grasslands
	Brachypodietalia dune grasslands with annuals
	Coastal dunes with *Juniperus ssp.*
	Cisto-Lavanduletalia dune sclerophyllos scrubs
	Wooded dunes with *Pinus pinea* and/or *Pinus pinaster*

The EU countries are required to implement the regulations of the Habitat Directive in their national laws and policies. So, national laws have to be adapted. The consequence of this is that the Natura 2000 areas will attain a legal protection status. If the area is affected by human activities, sanctions will be implemented. These sanctions can vary from fines to large-scale compensation.

The Habitat Directive (and in some aspects also the Bird Directive) has a large impact also on countries outside the EU. The accession countries have to fulfil the requirements of the Habitat Directive before they can enter the EU (e.g. preparing to identify sites which later will be included in the Natura 2000 Network). Other countries which are not selected to enter the EU (mainly in central and eastern Europe, CEE) use the Habitat Directive as a guideline for their nature conservation policy. They do so to adapt their policy to EU standards to prepare the country for entering the EU in the future. In this way, the Habitat Directive has an influence on dune conservation not only in the EU, but also far beyond the borders of the EU.

The early approximation of nature conservation legislation of the accession and non-EU countries proves to be the best and most cost effective way to safeguard the existing natural areas of high conservation value and to ensure that they are not damaged or destroyed before accession (see Végh and Szücs 1999).

20.5.3 National Policies

Next to the above-mentioned international instruments, each European coastal country has its own coastal policy. There is a wide variety of protection, from almost none (like Albania and Georgia) to countries that have made special arrangements for the regulation of activities on the coast, especially for the dunes. Van Koningsveld et al. (1999) give three interesting examples of the last category.

In many countries, the coastal strip is protected. In Portugal, for example, this strip depends on the type of the coast. The coast is public domain. Outside urban areas, the protected belt on dunes is located 200 m inland from the land limit of the dune. Besides this protected belt, approximately 25 % of the coast in Portugal is protected. All planning instruments for this protection are submitted to the public for comments.

In Poland, 68 % of the open seacoast is protected for its natural values. Of this area, 10 % lies within national parks or nature reserves where all natural features are protected. This includes the Slowinski National Park, with its extended landscape of wandering dunes. The remaining area consists of landscape parks and protected landscape areas where the status of protection is less strict but new building and other developments are severely restricted. The shoreline consists of a technical belt (including beach, dune ridge and a zone up to 200 m behind the dune ridge) and a protected belt (a buffer of the technical belt which extends 2 km inward from the shoreline).

The Danish coast has a protection zone of 300 m. The protection of the coast is mainly regulated through environmental legislation (Van Koningsveld et al.1999).

Several countries use the principles of Integrated Coastal Zone Management (ICZM) in their policy to protect the coastal dunes. ICZM involves the comprehensive assessment, setting of objectives, planning and management of coastal systems and resources, taking into account traditional, cultural and historical perspectives and conflicting interests and uses. It is a continuous and evolutionary process for achieving sustainable development (Intergovernmental Panel on Climate Change 1994). ICZM can be an important instrument for the conservation and development of coastal dunes in Europe, including the connection between isolated dune areas (Van der Meulen and Udo de Haes 1996).

20.6 SWOT Analysis

A SWOT analysis can be used to evaluate the situation of the European dunes. It is a tool which originates from project management, but which is also very frequently used to give a good and brief overview of many other subjects, e.g. landscape or nature conservation (e.g. European Union for Coastal Conservation 1998). It helps to formulate strategies for nature conservation taking into account the strong and weak points of the values of landscapes in view of threats and opportunities set by changes in land use or natural conditions.

A SWOT analysis for each of the three categories of European countries is shown in Table 20.3. The valuation of each category is not absolute but relative to the other categories. Conclusions and recommendations can be derived from this overview. It must be made clear which actions are necessary to enlarge the strengths and opportunities and to decrease the weaknesses and threats.

20.7 Conclusions

20.7.1 EU Countries

Large dune areas in the EU countries are damaged or even destroyed, especially along the Mediterranean coast. Conservation of present dunes and restoration of damaged dunes is needed, which can be achieved by the imple-

Table 20.3. Overview of the SWOT-analysis. Relevant data of Table 20.1 are included

	Strong	**Weak**
EU countries	– Relatively good protection schemes – Awareness of the need to protect dune areas – Availability of funds and public support	– Limited area and width of dunes – Poor conductivity – Until now large damage to dunes due to large investments – Sharp transitions to surrounding areas – Few dynamic dune areas left – Fragmentation in isolated dune areas – Much pressure from surrounding areas
Accession countries	– Dunes relatively unspoilt – Almost complete dynamic dune networks (good connectivity) – Well-developed gradients to surrounding areas – Good future protection due to the need to implement the Habitat Directive before entering the EU – Availability of EU funds after entering the EU	– Present lack of good protection – Occasional lack of awareness of the need for dune conservation
Non-EU countries	– Dunes relatively unspoilt – Almost complete dynamic dune networks (good connectivity) – Well-developed gradients to surrounding areas	– Lack of good protection and funds – Lack of awareness of the need for dune conservation
	Opportunities	**Threats**
EU countries	– Conservation of remaining dunes – Funds and awareness available for development and restoration of dunes – Rehabilitation of affected dunes	– Damage to unprotected dunes (especially in the Mediterranean region)
Accession countries	– Safeguarding of unique extensive dune areas by quick acceptance of legislation	– Future flow of EU funding can negatively affect unprotected dunes by stimulating adverse land use
Non-EU countries	– Safeguarding of unique extensive dune areas by prompt introduction of protection measures (legal instruments, funds and awareness)	– Damage to relatively unspoilt dunes, due to lack of protection measures

mentation of the Habitat Directive. Compared to countries outside the EU, there is relatively large funding available for conservation and development of coastal dunes.

20.7.2 Accession Countries

It seems that accession countries are in the best position for sustainable development of their dune areas. The dunes are relatively unspoilt and they have to be protected according to the Habitat Directive before these countries can enter the EU. When they have entered the EU, these countries will have access to the EU funds that are available for the conservation and management of the dunes.

20.7.3 Non-EU Countries

The dunes in these countries are relatively unspoilt, but they are under heavy pressure. There is little awareness of the need of dune conservation: economical development is more important. The Habitat Directive is used to give some guidelines, but there is no large funding available to conserve the dunes.

20.8 Recommendations

1. Draw up an international long-term strategy for coastal protection and development with clear objectives and tasks for each of the participating countries. This should comprise an integral strategy for nature, natural processes and sustainable land use, with adequate spatial planning for activities in the surrounding land and sea areas, with active involvement of local communities.
2. Promote the inclusion of Integrated Coastal Zone Management (ICZM) in the national policy of all countries. With ICZM, a network of (dynamic) dunes can be developed in EU countries and conserved in the accession countries and the non-EU countries. An example can be the European Coastal Code of Conduct of the Council of Europe (see Various Internet sites, www.eucc.nl). In this Code, recommendations are given on how each actor should behave to achieve sustainable coastal development.
3. Involve all actors in the coastal zone, such as industry, transport, tourism, forestry, etc. in the implementation of ICZM. This is a very complex process. Good communication between the parties is the main condition for success.

4. Encourage a change of mentality for dune management. Dunes should not be fixed by afforestation. Instead, dynamic dune systems should be promoted.
5. Promote international exchange of knowledge and experience to learn from other countries. An example is the Coastal Guide on Dune Management of the European Union for Coastal Conservation (European Union for Coastal Conservation 2000). This provides a platform for dune managers all over Europe to exchange information.
6. Use part of the EU funds to finance dune conservation projects in non-EU countries. In that way it is possible to conserve the large dynamic dune areas that have disappeared from the EU. An example is the EECONET Action Fund (see Various Internet sites, www.eucc.nl) which finances the purchase of land with high natural values by local governmental and non-governmental organisations in central and eastern Europe.

Acknowledgements. The authors are indebted to Albert Salman for his support. The permission of Pat Doody to use Fig. 20.1 is gratefully acknowledged.

References

Adriani HI, Van der Maarel E (1968) Voorne in de branding. Stichting Wetenschappelijk Duinonderzoek, Oostvoorne (with English summary)
Arens SM, Jungerius PD, Van der Meulen F (2001) Coastal dunes. In: Warren A, French JR (eds) Habitat conservation: managing the physical environment. Wiley, London, pp 229–272
Bakker TWM, Doing H (1987) De Nederlandse kustduinen en hun internationale betekenis. Duin 10/1–2:34–41
Bakker TWM, Klijn JA, Van Zadelhoff FJ (1981) Nederlandse kustduinen: landschapsecologie. PUDOC, Wageningen
Bakker TWM, Beltman B, Jungerius PD, Klijn JA (1992) Irish machairs: valuation management and conservation. EUCC internal report 1, Leiden
Bekker HT, Verstrael T, De Vries H (eds) (2001) De Vogel- en Habitatrichtlijn. Rijkswaterstaat, Delft. Via Natura 7:1–5
Bennet G (ed) (1991) Towards a European Ecological network. IEEP, Arnhem, The Netherlands
Bird ECF(1985) Coastline changes: a global review. Wiley, New York
Davies JL (1972) Geographical variation in coastal development. Oliver & Boyd, Edinburgh
De Groot R (1992) Functions of nature. Thesis, Wageningen University, Wageningen
Delbaere BCW (1998) Facts and figures on European biodiversity; state and trends 1998–1999. European Centre for Nature Conservation, Tilburg, The Netherlands
Doody JP (2000) Coastal conservation and management: an ecological perspective. Kluwer, Dordrecht
European Environmental Agency (1995) Europe's environment. The Dóbøi_ Assessment, Copenhagen/Luxemburg

European Environmental Agency (1998) Second assessment Europe's environment. Copenhagen/Luxemburg

European Environmental Agency (1999) Coastal and marine zones: environment at the turn of the century. Chap 3/14, Copenhagen

European Union for Coastal Conservation (1997) Threats and opportunities in the coastal areas of the European Union. Report for the National Spatial Planning Agency, Leiden, The Netherlands

European Union for Coastal Conservation (1998) Threats and opportunities in the coastal areas of the European Union. A report for the National Spatial Planning Agency of the Ministry for Housing, Spatial Planning and Environment, The Hague, The Netherlands

European Union for Coastal Conservation (2000) Coastal guide on dune management. Leiden, The Netherlands

Grime JP (1979) Plant strategies and vegetation processes. Wiley, London

Intergovernmental Panel on Climate Change (1994) Preparing to meet the coastal challenges of the 21st century. Conference Report World Coast Conference held in 1993 in Noordwijk, The Netherlands, 1993

IPPC (1995) Second assessment report of Working Group 1. Cambridge University Press, Cambridge

Klijn J (1990) Dune forming factors in a geographical context. In: Bakker TWM, Jungerius PD, Klijn J (eds) Dunes of the European coasts: geomorphology – hydrology – soils. Catena Suppl 18, Cremlingen/Destedt, pp 1–13

Klijn J (1995) Scenarios for European coastal areas: a promising tool for decisions at various levels. In: Schoute JFTh, Finke PA, Veeneklaas FR, Wolfert HP (eds) Scenario studies for the rural environment. Kluwer, Dordrecht

OECD (2000) Environmental indicators for agriculture: methods and results. Executive summary. Paris, France, 53 pp

Van Koningsveld M, Marchand M, Heslenfeld P, Van Rijswijk L, Salman A (1999) Spatial planning in European coastal zones. Review of approaches in spatial planning, coastal policy and coastal defence. Delft Hydraulics, Delft and EUCC, Leiden, The Netherlands

Van der Maarel E (ed) (1993) Dry coastal ecosystems: polar regions and Europe. Ecosystems of the world 2A. Elsevier, Amsterdam

Van der Meulen F (1990) European dunes: consequences of climatic change and sea level rise. In: Bakker TWM, Jungerius PD, Klijn J (eds) Dunes of the European coasts: geomorphology – hydrology – soils. Catena Suppl 18, Cremlingen/Destedt, pp 209–223

Van der Meulen F, Udo de Haes HA (1996) Nature conservation and integrated coastal zone management in Europe: present and future. Lands Urban Plann 34:401–410

Végh M, Szücs D (eds) (1999) Establishing Natura 2000 in EU Accession Countries. Proc of intl seminars held in 1999. ECNC Tech Rep Ser, Tilburg, The Netherlands

Various Internet Sites:

http://europa.eu.int/eur-lex/en/lif/dat/1979/en_379L0409.html (text Bird Directive)

http://europa.eu.int/eur-lex/en/lif/dat/1992/en_392L0043.html (text Habitat Directive)

European Centre for Nature Conservation: http://www.ecnc.nl/doc/Europe/legislat/conveu.html (comments on Habitat and Bird Directives)

European Union for Coastal Conservation: http://www.eucc.nl/

World Resources Institute: http://www.wri.org/wr2000/coast

VI The Coastal Dune Paradox: Conservation vs. Exploitation?

21 The Fragility and Conservation of the World's Coastal Dunes: Geomorphological, Ecological, and Socioeconomic Perspectives

M.L. Martínez, M.A. Maun, and N.P. Psuty

21.1 Current Worldwide Status

Coastal dunes have a worldwide distribution and are comprised of a variety of forms going through successional changes in geomorphology and ecological association. Some changes are driven by natural processes whereas other changes are products of human endeavors. Coastal dunes are highly valuable multifunctional ecosystems that occupy a unique natural niche. Further, these same systems are ideal locations for recreation, replenishment of local aquifers, and coastal defense (van der Meulen et al., Chap. 16, this Vol.). However, in spite of their many valuable attributes, these ecosystems share a history of exploitation and mismanagement. The greatest threat to their survival is overuse, urban expansion, urban sprawl, mining, and pollution. Frequently, these systems and their functions are completely replaced by high rise buildings, residential development, cottages, tourist resorts, and recreation parks. The eventual consequence of such activities is the destruction of a fragile ecosystem. Tourism, in particular, has become an important activity with economic benefits that have resulted in severe environmental degradation of dunes. For example, hotels and houses are built directly on the dunes and trampling by people kills the stabilizing vegetation, thus exacerbating the process of sand movement and dune erosion. These changes in sediment supply and the regional sand budget produce extensive geomorphological and ecological damage with important economic consequences (Psuty, Chap. 2, this Vol.). As a result, there is the net loss of coastal dunes, their dynamics, and their natural biological diversity. Dune slacks have been similarly subjected to systematic alteration through overexploitation by grazing, hay making, and growth of crops.

Slowly, this alarming trend is reaching decision-makers, politicians, and the public in general, and, gradually, coastal dunes are becoming protected

Ecological Studies, Vol. 171
M.L. Martínez, N.P. Psuty (Eds.)
Coastal Dunes, Ecology and Conservation

ecosystems (mostly at mid-latitudes). Nevertheless, many important questions remain regarding management approaches and techniques for preserving form and function. For example, how much do we know about the methods of preservation of the form and function of coastal dunes? What remains to be studied? How should they be managed so that their evolving geomorphological and ecological systems are preserved for future generations? In the following sections we will address these questions and discuss whether dune conservation is a possibility. We will first consider current research trends and then examine different ways of managing and preserving fragile dune systems.

21.2 Current Research Trends (What Do We Know?)

Recent studies on coastal dune systems have focused on the spatial and temporal variations of morphologies, the sequences of biological succession in dune types, adaptations to the environmental stresses, and comparison between the patterns and processes in coastal dunes of the tropics and mid-latitudes. Different aspects of these topics were addressed widely in various sections of the book. This final chapter summarizes the complex topics covered in the book.

21.2.1 Variable Morphologies

Coastal dune systems are highly variable in form and dimension. Psuty (Chap. 2, this Vol.) and Hesp (Chap. 3, this Vol.) establish that the present-day morphological systems have been evolving through the most recent period of Holocene sea-level transgression and they have waxed and waned in their development as sediment supply has fluctuated and sequences of stabilization and destabilization have characterized regions. There is no one dune system, there are patterns, trends, and episodes that describe the creation of regional coastal dune forms and systems. As morphologies pass through sequences dependent on ambient sediment supply, wind conditions, stabilizing vegetation, and human interaction, they form the variable physical base that constitutes the coastal foredune in the beach profile and the inland extension of the dune fields. Whereas there is considerable progress accomplished in the mechanics of eolian sand transport, the specific association of quantities of transport and coastal dune form remains a fertile area of inquiry. Further, there is a very significant and difficult research topic regarding the dichotomy between the active dune-beach system (which may be eroding, accreting, or stable) and the highly variable equally dynamic dune topography inland of the beach.

21.2.2 Succession

Ever since the early studies performed by Henry Chandler Cowles, more than a century ago (1899), succession on coastal dunes has been the focus of many researchers. Importantly, the studies by Cowles were seminal for Frederic E. Clements who, almost 20 years later (1916), proposed the successional theory of plant communities, a key concept in current ecological theory.

Recent studies demonstrate that local environmental characteristics and heterogeneity affect successional trends (Lubke and Avis 1988; Grootjans et al. 1997;, Chap. 6, this Vol.). An array of complex interactions affects the successional process, namely: local hydrological and topographical conditions, nutrient availability, amount of precipitation, and freshwater discharge (Grootjans et al. 1991; Houle 1997; Vázquez et al. 1998; Lichter 2000; Martínez et al. 2001; Grootjans et al., Chap. 6, this Vol.; Vázquez, Chap. 12, this Vol.). There is evidence that the rate of vegetation succession is largely controlled by the productivity of the ecosystem, the decomposition of organic matter, and the recycling of nutrients within the ecosystem (Koerselman 1992). In addition, the increased atmospheric input of nitrogen, because of industrialization, automobile combustion, and increasing population size may accelerate the successional pathway (Maun and Sun 2002; Kooijman, Chap. 15, this Vol.).

More recently, it has become evident that biotic interactions play a key role during community succession. Facilitation of these interactions is especially important during the earlier successional stages (Vázquez et al. 1998; Lichter 2000; Shumway 2000; Martínez 2003; Lubke, Chap. 5, this Vol.; Grootjans et al. 1997 and Chap. 6, this Vol.; Martínez and García-Franco, Chap. 13, this Vol.; Vázquez, Chap. 12, this Vol.) when both phanerogamous species and algal mats ameliorate the environmental extremes and facilitate colonization by later successional species. However, biotic interactions are not restricted to a specific seral stage because facilitation and competition have both been observed during early and later stages of succession (Kellman and Kading 1992; Callaway and Walker 1997; Martínez and García-Franco, Chap. 13, this Vol.).

It is also important to note that the traditional explanation of dune succession usually overlooks the role of mutualistic fungi which facilitate the invasion of bare areas. Certainly, many dune-building plants are incapable of establishment in the absence of arbuscular mycorrhizal (AM) fungi (Koske and Polson 1984; Sylvia 1989; Corkidi and Rincón 1997; Koske et al., Chap. 11, this Vol.). This association is of particular ecological relevance since the fungi enhance nutrient uptake, soil aggregation, increase plant tolerance to drought and salt stress, and protect them from soil pathogens (Gemma and Koske 1997; Tadych and Blaszkowski 1999; Gemma et al. 2002). Thus, although overlooked, AM fungi certainly play a significant role in community dynamics. Further studies are needed to examine the role of AM fungi in ecological interactions during succession.

Vegetation succession is also known to affect animals and their interactions with plants (McLachlan 1991; McLachlan et al. 1987; van Aarde et al., Chap. 7, this Vol.). For example, in the case of birds, species richness and diversity are positively correlated with structural heterogeneity of vegetation and habitat (Kritzinger and van Aarde 1998). Alternatively, spatial and temporal variability in habitat, community structure, and plant height determine the occurrence of rodents in coastal dune forests (Ferreira and van Aarde 1999; Koekemoer and van Aarde 2000).

Different animal species also exhibit definite successional trends. They are dominant during the different seral stages. For example, the rodent, *Mastomys natalensis* was most abundant during the first few years of vegetation establishment and regeneration, while *Saccostomys campestris* was dominant in sites older than 15 years (van Aarde et al., Chap. 7, this Vol.). Similarly, birds were closely associated with different stages in succession because of changes in food sources, protection from predators, nesting sites, and plant community structure (van Aarde et al., Chap. 7, this Vol.; Kritzinger and van Aarde 1998).

Many sand dune systems around the globe are used for grazing by cattle and sheep. Such grazing is highly detrimental to dune systems because trampling by animals may arrest or revert succession by killing vegetation and enhancing erosion by wind action. Grazing by rabbits does not allow vegetation to increase in height and prevents shrubs and trees from sprouting and growing. When rabbit populations drop, shrubs and trees begin to flourish and bird populations change (Baeyens and Martínez, Chap. 17, this Vol.).

In brief, studies on successional changes on coastal dunes indicate that in addition to local abiotic conditions such as topography and hydrology (De Jong and Klinkhamer 1988; Grootjans et al. 1991 and Chap. 6, this Vol.; Houle 1997; Lammerts et al. 1999; Martínez et al. 2001), biotic interactions are integral to dune system development (Shumway 2000; Rico-Gray et al., Chap. 14, this Vol.; Vázquez, Chap. 12, this Vol.; Grootjans et al., Chap. 6, this Vol.; Martínez and García-Franco, Chap. 13, this Vol.). The understanding of the the concept of successional sequence (whether in accretional, erosional, or stable systems) and the interactive processes are of immense value in dynamic management plans aimed at maintaining the natural dynamics of dune systems.

21.2.3 Adaptations

Sand dune habitats are highly heterogeneous and offer a wide variety of microhabitats in which species utilize different strategies. The major environmental condition in foredunes is burial of plants by sand, initially supplied by wave action and then – episodically transported inland by onshore wind. This recurrent burial acts as a strong selective force that affects plant community composition and dynamics especially in the early stages of succession (Disraeli 1984; Davy and Figueroa 1993). Soon after burial of a plant its photosynthetic capacity may decrease but many dune species are able to withstand and even benefit from certain threshold levels of burial and the associated delivery of nutrients (Disraeli 1984; Maun 1994, 1998;, Chap. 8, this Vol.; Martínez and Moreno-Casasola 1996). Some foredune species are so well adapted that they require regular burial to maintain high vigor. In the absence of sand deposition, a marked decline in vigor and density of these populations has been observed after sand stabilization (Maun, Chap. 8). When this occurs, early pioneers may be eventually replaced by plants of later stages in succession.

Many factors limit the establishment and growth of plants in sand dune ecosystems. However, the major factors are high wind velocities, sand blasting, salt spray, salinity, and the scarcity of water and mineral nutrients. In particular, some dune plants are especially efficient in water utilization (Ripley and Pammenter, Chap. 9, this Vol.; Martínez et al. 1994). During dry spells, stomata are closed thus reducing transpiration and some plants have access to moisture in the form of internal dew. Some dune species also have the ability to re-cycle nutrients, suggesting that at the shoot meristem level, cell division and growth are not limited by nutrient availability (Ripley and Pammenter, Chap. 9, this Vol.). An efficient nutrient utilization has also been observed in tropical species (Valverde et al. 1997; Martínez and Rincón 1993).

In contrast, dune slacks have excessive moisture and may behave like temporary wetlands with cycles of flooding and dry periods. The species growing here are tolerant to both dry and wet (flooded) conditions (Grootjans et al., Chap. 6, this Vol.). Nevertheless, anoxic conditions during flooding may prevent the establishment of all but species with a well-developed aerenchyma (such as *Schoenus nigricans* and *Littorella uniflora*) which counteracts anoxia by active radial oxygen loss (Grootjans et al., Chap. 6, this Vol.). Many dune slack species also require very low amounts of nutrients.

The specific responses to environmental fluctuations generate functional groups (or types) (FTs) that contain species from different biogeographic and ecological zones but possess a common life strategy that enables them to thrive in a given environment (García-Novo et al., Chap. 10, this Vol.). The grouping of plants according to functional types is a useful approach that facilitates the recognition of patterns in species response to the environmental variability in coastal dune vegetation. The analyses of FTs not only helps to understand community functioning at the local scale but also facilitates the comparison between communities (with few species in common) exposed to similar environmental constraints at the regional scale.

21.2.4 Tropical vs. Mid-Latitude

There is a general scarcity of studies on tropical beaches and dunes; only recently are scientific publications emerging in Mexico, (Moreno-Casasola 1988, 1999) Cuba (Aguila et al 1996; Borhidi 1996) Brazil, Venezuela, and Chile

(Araujo 1992; Seeliger 1992). This is probably one of the reasons why earlier publications on coastal dunes stated that tropical coasts lacked extensive dunefields (Ranwell 1972; Carter 1988). However, current studies (Moreno-Casasola 1982) have demonstrated that coastal dunes are not only abundant along tropical latitudes but also have definite latitudinal variation among them.

The vegetation of coastal dunes has been thoroughly studied in temperate latitudes and less intensively in the tropics (van der Maarel 1993a, b). Although species and genera differ with latitude, in general, those occupying the upper beach and foredunes are tolerant of salt spray, substrate mobility, and strong winds. Farther inland, the vegetation varies depending on the microclimate and local conditions and there is less latitudinal similarity. Arboreal vegetation shows a closer resemblance to the regional flora.

Hesp (Chap. 3, this Vol.) suggests that there may be morphological differences between tropical and mid-latitude dunes, possibly the product of vegetation and weather conditions.

He indicates that, contrary to earlier publications (e.g., Jennings 1964), eolian coastal dunes are neither very poorly developed nor largely absent in the humid tropics. Certainly, dunes may be uncommon in many tropical coasts, probably due to frequent storm surges during typhoons and hurricanes, abundant rainfall coinciding with periods of strong winds, and a general lack of sediment supply. However, there are areas with large dunefields in the humid tropics, specifically in those regions with a positive sediment budget and strong winds during a marked dry season. Coastal dunes in these arid and semiarid tropical areas frequently extend into adjacent inland dunefields. Further, latitudinal coastal dune characteristics may vary because the many coralline shorelines and the abundant fine-grained deposits of the tropical rivers offer a different type and quantity of source material to the beach-dune system. In general, taller grasses and sedges are dominant on temperate foredunes whereas low creepers dominate the humid tropics. Creepers also occur in the dry tropics, but grassy species tend to be dominant along these coasts. In addition, colonization and stabilization rates are faster in the tropics which may lead to a reduced opportunity for changes in the original duneform. Given the above, coastal dunes in the tropics are apt to be more stable than their temperate counterparts, although outstanding exceptions probably occur in both regions.

Additional latitudinal convergences in dune vegetation and dynamics are (1) similar successional pathways (facilitation) (De Jong and Klinkhamer 1988; Kellman and Kading 1992; Shumway 2000; Martínez et al. 2001; Martínez 2003) and (2) a high susceptibility to invasion by exotic species (Castillo and Moreno Casasola 1996; Wiedemann and Pickart, Chap. 4, this Vol.; Lubke, Chap. 5, this Vol.). Furthermore, post-burial stimulation of plant growth has been observed in both temperate and tropical dune species alike (Maun 1994, 1998 and Chap. 8, this Vol.; Martínez and Moreno-Casasola 1996; Baskin and Baskin 1998).

21.3 Fragile Ecosystems?

Aspects of dune dynamics are driven by naturally occurring disturbances. However, even though recurrent disturbances are common disrupting forces in these environments, either a higher intensity or frequency of such disturbances or their absence may alter community dynamics. When too intense or highly frequent, mobility of sand may increase and stabilized dunes (such as forested areas) may be lost to erosion or sand deposition. When natural disturbances are absent, the early successional stages which are dependent on high substrate mobility may disappear. In brief, coastal dunes are dynamic and heterogeneous environments where naturally occurring patch dynamics maintain their diversity. They are considered fragile because only a slight disruption (either natural- or human-induced) may lead to change and long-term progressive alteration (Carter 1988) and their natural diversity might be compromised rather easily.

In addition to natural disturbances, the dune communities are exposed to different types of human-related disturbance events which cause both direct (trampling, grazing, sand and water extraction, leveling of dunes) and indirect damage (climate change, sea-level rise, and alterations in soil, moisture regimes, and sediment supply). The historical record of coastal Europe has a litany of coastal dune stabilization and destabilization conditions which are apparently related to a combination of natural and human-caused events (Sherman and Nordstrom 1994). The full range of these events has had a significant impact on natural dune dynamics and should be addressed in management practices.

21.4 Management Practices

On a worldwide basis, during the last two or three decades the public has become increasingly aware of their deteriorating and diminishing coastal recreational environment. Slowly, efforts at coastal protection and improved management are becoming priorities for the public and Government authorities, thus leading to a greater respect and urgent need for conservation of coastal dunes. Management practices have changed during the last decades depending on the use to which dune systems were subjected and their impact on human activities. For example, thirty years ago dunes were stabilized when they were encroaching on adjacent farm land. Bare sand surfaces were stabilized against deflation and inland engulfment. Because of its effectiveness in stabilizing active dunes, surface stabilization was achieved by introducing *Ammophila arenaria* (marram grass) into almost every temperate coast of the world. In time, this grass became invasive in some regions. For instance, following the introduction of marram grass along the western coast of the USA in 1869, the species flourished, spread quickly, and brought about major changes in the dunescape. Native dune-forming species were eliminated and the extensive grass cover stabilized the foredune, thus cutting off sand supply from the beaches to the back dunes. Early successional stages disappeared (Wiedemann and Pickart, Chap. 4, this Vol.) and indigenous taxa were suppressed (Richardson et al. 1992, 1997). The overall result was a reduction in biodiversity. Grass encroachment has also become an important environmental problem on European coasts (Veer and Kooijman 1997; Kooijman, Chap. 15, this Vol.; Grootjans et al., Chap. 6, this Vol.). At present, the pioneer stages are rare in most western European dunes, although large mobile systems can still be found in Spain (Coto Doñana); Poland (Slowinski National Park); France (Landes district); Denmark; the German Wadden Sea island of Sylt; and the Dutch Wadden Sea island of Texel (Kooijman, Chap. 15, this Vol.). Similar trends in spread by exotic species seem to occur in the tropics (for instance Mexico) where tall and dense stands of grasses (*Schizachyrium scoparium*) gradually replace local endemic vines and creepers (Martínez and García-Franco, Chap. 13, this Vol.; Martínez et al. 2001). Interestingly, *A. arenaria* has not become an invasive species in South African coastal dunes (Lubke and Hertling 2001; Lubke, Chap. 5, this Vol.).

Currently, numerous control programs exist on several continents that are aimed at eliminating *A. arenaria* and restoring the natural dune processes without sacrificing stability. Several activities such as (a) removal of aboveground biomass by annual mowing and grazing by cattle and (b) sod cutting and stimulation of eolian activity have been used to lower the vigor of alien species. Both measures seem effective, although their effectiveness depends on local soil characteristics (Kooijman, Chap. 15, this Vol.). However, these programs are costly. An eradication program of 10 ha of *A. arenaria* population cost US$ 350,000 over a four-year period (Wiedemann and Pickart, Chap. 4, this Vol.). Clearly, the correction of the problem of invasive species exceeds the resources available.

It is now evident that dune stabilization is not always a necessary and suitable management tool because it may create many new problems. First, artificial stabilization alters the eolian processes and may have a major impact farther inland or downwind. Second, stabilization can be very costly, and third, mobile dunes may be part of the natural successional landscape and should be allowed to function normally. In fact, recent management trends are directed towards an acceptance of these systems as valuable wilderness areas. Nature conservation policies have changed from a focus on rare and endangered species to biotic and abiotic processes and landscape functioning. That is, management is best done in accordance with the natural processes (Avis 1992; Avis and Lubke 1996), preferably in areas of at least several hundreds of ha in size (Bal et al. 1995). Given this scenario, when trying to conserve native dune species it should be considered whether stabilization is necessary and, if it is, under what circumstances and what are the species that should be used (preferably, native).

Based on the above emerging concerns, artificial stabilization measures in some parts of The Netherlands have been stopped, because they are no longer considered necessary. Dune fixation is now limited to special parts of the foredunes which are endangered by the sea, not the wind, and fixation is done mostly with hedges of branches or native marram grass without deleterious ecological consequences. This has saved the coastal management authorities millions of dollars, and has resulted in a much more beautiful and natural dune landscape (Heslenfeld et al., Chap. 20, this Vol.). Within the framework of the European Union of Coastal Conservation all countries except France are following the Dutch example. In other countries such as the USA, Canada, and South Africa, dune stabilization programs are selectively applied. Dune stabilization programs in other parts of the world are rare. In any case, any stabilization program should only be initiated when the need has been carefully ascertained through detailed studies. Whenever possible, natural dynamics should therefore be maintained and restored.

Ideally, management plans of coastal dunes should embrace the planting and maintenance of the natural vegetation, conservation of wildlife, control of visitors, the zonation of activities, and provision of amenities. When managing and preserving coastal dunes, it is important to bear in mind that species richness and habitat diversity in dune systems, which by their nature are highly dynamic, are greatest when a full sequence of successional stages is represented. Management should aim at maintaining this diversity and the natural dynamics of these ecosystems (Avis 1992; Davy and Figueroa 1993).

Management practices should also include animal populations. In a fashion similar to what occurred with coastal vegetation, the attitude of humans towards dune mammals and birds has changed throughout history (Baeyens and Martínez, Chap. 17). Initially, wild animals were used as sources of food and fur. Their populations were manipulated in several ways: (1) some populations (rabbits) were artificially fed and enlarged to facilitate hunting activities; (2) species of interest (deer) were protected from their natural predators; (3) new species were introduced and bred. However, at the beginning of the 20th century, birds started to receive special protection that led to habitat protection and the protection of wild animals in general. In addition, trends in both bird and rodent communities indicate that habitat rehabilitation is a management tool that could potentially reverse the negative impact of habitat loss and fragmentation that threatens the viability of species populations (van Aarde et al., Chap. 7, this Vol.). However, there is evidence that changes in food webs result in domino effects in many links of the food chain (Baeyens and Martínez, Chap. 17). In this case, mitigating measures aimed at controlling shrub and grass encroachment do not always restore the breeding densities of locally extirpated birds (Baeyens and Martínez, Chap. 17, this Vol.). It is thus evident that the relevance of biotic interactions should also be considered in restoration projects, because species reintroduction will not restore communities, given that interactions need to be restored as well.

Finally, it is important to bear in mind the importance of long-term monitoring. The systematic recording of the geomorphology, flora, and fauna over long periods of time is of particular value to detect ecosystem and community changes such as invasive plant species and their potential impact. Monitoring is important to establish trends, periodic variations, and aperiodic events in the history of change. Monitoring will lead to the establishment of a baseline which is essential to assess locally endangered species, and can only be achieved through long-term studies (Lubke, Chap. 5, this Vol.; van Aarde et al., Chap. 7, this Vol.; Grootjans et al., Chap. 6, this Vol.; Martínez et al. 2001).

21.5 Future Trends and Perspectives

Is it possible to preserve the form and the functions of coastal dunes? Based on current global trends and aspirations, the future looks bleak. For example, it is projected that (1) agriculture will intensify in these sandy, well-drained habitats; (2) ports will expand and urbanization will continue; (3) infrastructure will increase, especially near the airports and tourist resorts; (4) pressure from tourism will increase; (5) in some countries, a return to more natural ecosystems (forests) is likely to be promoted; and (6) sea level rise will displace the beach-dune system and narrow the protected dune zone; erosion rates will be exacerbated (Heslenfeld et al., Chap. 20, this Vol.). Thus, it seems that human development along the coast will continue, and it would be wise to do so in a proper and sustainable manner. This necessarily implies agreement with the philosophical supposition that conservation and development are not irreconcilable, and efforts must be made to strike a balance between them.

In light of the above, conservation efforts should be aimed at solving several problems of preservation, restoration, and management (which should include actions to prevent further species invasions) (Wiedemann and Pickart, Chap. 4, this Vol.). Because coastal defense mechanisms that fail to consider the natural processes are highly costly and ineffective, management towards maintaining dune dynamics should be promoted. All too often, however, management objectives focus on static solutions, such as securing the shoreline against erosion, maintaining a constant channel depth, or cultivating a continuous dune grass sward. Such solutions place an unwanted stress on many coastal environments and may, in some instances, promote a catastrophic event. In fact, it is known that the more diverse the systems are, the better able they are to withstand change. A dune system with diverse vegetation, morphology, and relief is more resilient to changing conditions. Indeed, the concepts raised in this book indicate that dune systems are often in a developmental stage of instability and migration as part of the natural system. Permitting some degree of mobility should be part of a holistic management strategy. Thus, the application of adequate standards of resource use, coupled with proper management and conservation policies will sustain these environments and maintain a wide variety of living organisms.

Heslenfeld et al. (Chap. 20, this Vol.) proposed the following recommendations to restore and preserve coastal dunes:

1. Generate an international long-term strategy for coastal management and development.
2. National policies of coastal countries should include an Integrated Coastal Zone Management (ICZM) approach.
3. Include all actors (local inhabitants and authorities) involved in the decision making process of ICZM.
4. Dunes should not be artificially fixed or afforested: they should be promoted as dynamic self-sustained systems.
5. Promote international exchange of knowledge and expertise.
6. Use international funds for dune conservation projects.

In addition to these recommendations, it is also important to bear in mind that:

7. Mobility and succession are dune formational processes and functions
8. Tropical dunes should be included in the above-described actions.
9. Finally, adequate legislation to protect coastal environments, generally lacking in tropical countries, needs to be addressed in the near future.

In particular, in the case of ICZM, zoning is a major management tool through which coastal functions are allocated to meet the economic and leisure demands without influencing their ecological well being (van der Meulen and Udo de Haes 1996). Segregated activities help to absorb the high numbers of visitors, and overall, the dune system is preserved. This is particularly important because the overall carrying capacity of dunes is relatively low (less than 100 persons per hectare) (van der Meulen et al., Chap. 16, this Vol.). Zoning operates on two embedded scales, one covering the entire coastline (and the countries involved) and the other within individual dune systems.

Biotic interactions: plant-plant, predators, herbivores, and seed dispersers should all be considered to make the ICZM effective. Because of their high relevance in community dynamics (Rico-Gray 2001; Rico-Gray et al., Chap. 14, this Vol.; Grootjans et al., Chap. 6, this Vol.; Vázquez, Chap. 12, this Vol.; Martínez and García-Franco, Chap. 13, this Vol.), coastal zone management should consider impact of the activity on biotic interactions as much as possible. However, when the impact of major changes in the food chain (appearance or disappearance of species), is not diminished, additional restoration measures will be required (Baeyens and Martínez, Chap. 17, this Vol.).

A complementary tool for ICZM plans is the use of different ecological indicators (compositional, structural, and functional) which are potentially helpful in measuring the environmental quality of dune systems (Espejel et al., Chap. 18, this Vol.). Such tools are particularly useful in conservation and restoration programs, since they provide quantitative values through which sites or regions can be compared and monitored. These ecological indicators are potentially very useful when planning to perform environmental quality assessment and regional coastal zone management. Furthermore, the adequate and sufficient knowledge of the functioning of coastal dunes will provide proper tools that will enable decision-makers to properly conserve and restore these ecosystems for future generations. Long-term monitoring studies should thus be promoted.

Coastal dune conservation provides a major challenge in densely populated areas, but it can be achieved. Two examples of extensive and protected dune systems lying next to large industrial and densely populated areas are the Sefton coast dunes near Liverpool (England) and Meijendel dunes near The Hague (The Netherlands) (van der Meulen et al., Chap. 16, this Vol.). In Mexico, conservation efforts are being planned by taking into confidence all stake-holders, local inhabitants, academic institutions, and Government authorities. The objective is to agree upon the necessary actions to utilize the natural resources without affecting their integrity and ecological balance (Moreno-Casasola, Chap. 19, this Vol.). Nevertheless, the total cost of dune management can be very high. For instance, in the Meijendel dunes, surveillance by 15 wardens plus management costs 3.2 million Euros per year (van der Meulen et al, Chap. 16, this Vol.).

Definitely, good design based on sound scientific research can contribute significantly to the management of dunes, but it must be emphasized that environmental education is an essential adjunct to planning and design. Public participation and awareness of the environmental problems are essential prerequisites to preserve important coastal features and processes for the present and future generations. At the onset of the 21st century, managers and society should no longer be considered separate entities. Managers are required to offer a product that is in accordance with nature and meets the needs of the society at large. Finally, the issues of enlightened conservation along with increasing coastal development should reflect on the fact that maintenance of the status quo of coastal dunes may not be attainable because of their natural dynamism and the changes wrought by humans.

References

Aguila CN, Moreno-Casasola P, Menéndez L, García-Cruz R, Chiappy C (1996) Vegetación de las dunas de Loma del Puerto (Cayo Coco, Ciego Avila, Cuba). Fontqueria 42:243–256

Araujo D (1992) Vegetation types of sandy coastal plains of tropical Brazil: a first approximation. In: Seeliger U (ed) Coastal communities of Latin America. Academic Press, New York, pp 337–348

Avis AM (1992) Coastal dune ecology and management in the Eastern Cape. PhD Thesis, Rhodes Univ, Grahamstown

Avis AM, Lubke RA (1996) Dynamics and succession of coastal dune vegetation in the Eastern Cape, South Africa. Landsc Urban Plann 34:247–254

Bal D, Beije HM, Hoogeveen YR, Jansen, Reest, van der PJ (1995) Handboek natuurdoeltypen in Nederland. IKC, Ministry of Agriculture, Nature Conservation and Fisheries, Wageningen

Baskin CC, Baskin JM (1998) Seeds. Ecology, biogeography, and evolution of dormancy and germination. Academic Press, New York

Borhidi A (1996) Phytogeography and vegetation ecology of Cuba. Akadémia Kiado, Budapest, 923 pp

Callaway RM, Walker LR (1997) Competition and facilitation: a synthetic approach to interactions in plant communities. Ecology 78:1958–1965

Carter RWG (1988) Coastal environments. An introduction to the physical, ecological and cultural systems of the coastlines. Academic Press, New York, 617 pp

Castillo SA, Moreno-Casasola P (1996) Coastal sand dune vegetation: an extreme case of species invasion. J Coastal Conserv 2:13–22

Clements FE (1916) Plant succession: analysis of the development of vegetation. Carnegie Inst Washington, Publ No 242, Washington, DC, 17.3.3, 17.4.5

Corkidi L, Rincón E (1997) Arbuscular mycorrhizae in a tropical sand dune ecosystem on the Gulf of Mexico. II. Effects of arbuscular mycorrhizal fungi on the growth of species distributed in different early successional stages. Mycorrhiza 7:17–23

Cowles HC (1899) The ecological relations of the vegetation on the sand dunes of Lake Michigan. Bot Gaz 27:95–117. In: Real LA, Brown JH (eds) Foundations of ecology (1991) The University of Chicago Press, Chicago, pp 28–58

Davy AJ, Figueroa ME (1993) The colonization of strandlines. In: Miles J, Walton DWH (eds) Primary succession on land. Blackwell, Oxford, pp 113–131

De Jong TJ, Klinkhamer PGL (1988) Seedling establishment of the biennials *Cirsium vulgare* and *Cynoglossum officinale* in a sand dune area: the importance of water for differential survival and growth. J Ecol 76:393–402

Disraeli DJ (1984) The effect of sand deposits on the growth and morphology of *Ammophila breviligulata*. J Ecol 72:145–154

Ferreira SM, van Aarde RJ (1999) Habitat associations and competition in *Mastomys-Saccostomus-Aethomys* assemblages on coastal dune forests. Afr J Ecol 37:121–136

Gemma JN, Koske RE (1997) Arbuscular mycorrhizae in sand dune plants of the North Atlantic coast of the U.S.: field and greenhouse studies. J Environ Manage 50:251–264

Gemma JN, Koske RE, Habte M (2002) Mycorrhizal dependency of some endemic and endangered Hawaiian plant species. Am J Bot 89:337–345

Grootjans AP, Hartog PS, Fresco LFM, Esslink H (1991) Succession and fluctuation in a wet dune slack in relation to hydrological changes. J Veg Sci 2:545-

Grootjans AP, van den Ende FP, Walsweer AF (1997) The role of microbial mats during primary succession in calcareous dune slacks: an experimental approach. J Coastal Conserv 3:95–102

Jennings JN (1964) The question of coastal dunes in tropical humid climates. Z Geomorph 8:150–154

Houle G (1997) No evidence for interspecific interactions between plants in the first stage of succession on coastal dunes in subarctic Quebec, Canada. Can J Bot 75:902–915

Kellman M, Kading M (1992) Facilitation of tree seedling establishment in a sand dune succession. J Veg Sci 3:679–688

Koekemoer AC, van Aarde RJ (2000) The influence of food supplementation on a coastal dune rodent community. Afr J Ecol 38:343–351

Koerselman W (1992) The nature of nutrient limitation in Dutch dune slacks. In: Carter RWG, Curtis TGF, Sheehy-Skeffington MJ (eds) Coastal dunes. Proc 3rd Eur Dune Congr, Balkema, Rotterdam, pp 189–199

Koske RE, Polson WR (1984) Are VA mycorrhizae required for sand dune stabilization? Bioscience 34:420–424

Kritzinger JJ, Van Aarde RJ (1998) The bird communities of rehabilitating coastal dunes at Richards Bay, KwaZulu-Natal. S Afr J Sci 94:71–78

Lammerts EJ, Pegtel DM, Grootjans AP, van der Veen A (1999) Nutrient limitation and vegetation changes in a coastal dune slack. J Veg Sci 10:111–122

Lichter J (2000) Colonization constraints during primary succession on coastal Lake Michigan sand dunes. J Ecol 88:825–839

Lubke RA, Avis AM (1988) Succession on the coastal dunes and dune slacks at Kleinemonde, Eastern Cape, South Africa. Monogr Syst Bot Mo Bot Gard 25:599–622

Lubke RA, Hertling UM (2001) The role of European marram grass in dune stabilization and succession near Cape Agulhas, South Africa. J Coastal Conserv 7:171–182

Martínez ML (2003) Facilitation of seedling establishment by an endemic shrub in tropical coastal sand dunes. Plant Ecol 168(2):333-345

Martínez ML, Moreno-Casasola P (1996) Effects of burial by sand on seedling growth and survival in six tropical sand dune species from the Gulf of Mexico. J Coastal Res 12(2):406–419

Martínez ML, Rincón E (1993) Growth analysis of *Chamaecrista chamaecristoides* (Leguminosae) under contrasting nutrient conditions. Acta Oecol 14:521–528

Martínez ML, Moreno-Casasola P, Rincón E (1994) Sobrevivencia de una especie endémica de dunas costeras ante condiciones de sequía. Acta Bot Mex 26:53–62

Martínez ML, Vázquez G, Sánchez-Colón S (2001) Spatial and temporal dynamics during primary succession on tropical coastal sand dunes. J Veg Sci 12:361–372

Maun MA (1994) Adaptations enhancing survival and establishment of seedlings on coastal dune systems. Vegetatio 111:59–70

Maun MA (1998) Adaptations of plants to burial in coastal sand dunes. Can J Bot 76:713–738

Maun MA, Sun D (2002) Nitrogen and phosphorous budgets in a lacustrine sand dune ecosystem. Ecoscience 9:364–374.

McLachlan A (1991) Ecology of coastal dunes. J Arid Environ 21:229–243

McLachlan A, Ascaray C, Du Toit P (1987) Sand movement, vegetation succession and biomass spectral in a coastal dune slack in Algoa Bay, South Africa. J Arid Environ 12:9–25

Moreno-Casasola P (1982) Ecología de la vegetación de dunas costeras: factores físicos. Biótica 7:577–602

Moreno-Casasola P (1988) Patterns of plant species distribution on Mexican coastal dunes along the Gulf of Mexico. J Biogeogr 15:787–806

Moreno-Casasola P (1999) Dune vegetation and its biodiversity along the Gulf of Mexico, a large marine ecosystem. In: Kumpf HK, Steidinger K, Sherman K (eds) The Gulf

of Mexico large marine ecosystems – Assessment, sustainability and management. Blackwell, Cambridge, pp 593–612
Ranwell DS (1972) Ecology of salt marshes and dunes. Chapman and Hall, London
Richardson DM, Macdonald IAW, Hoffmann JH, Henderson L (1997) Alien plant invasions. In: Cowling RM, Richardson DM, Pierce SM (eds). Vegetation of Southern Africa. Cambridge University Press, Cambridge, pp 535–570
Richardson DM, Macdonald IAW, Holmes PM, Cowling RM (1992) Plant and animal invasions. In: Cowling RM (ed) The ecology of fynbos: nutrients, fire and diversity. Oxford University Press, Cape Town, pp 271–308
Rico-Gray V (2001) Interspecific interactions. Encyclopedia of life sciences. Macmillan, Nature Publ Group. www.els.net
Seeliger U (1992) Coastal plant communities in Latin America. Academic Press, New York
Sherman DJ, Nordstrom KF (1994) Hazards of wind-blown sand and coastal sand drifts: a review. In: Finkl CW Jr (ed). Coastal hazards: perception, susceptibility and mitigation. J Coastal Res Spec Iss No. 12, pp 263–275
Shumway SW (2000) Facilitative effects of a sand dune shrub on species growing beneath the shrub canopy. Oecologia 124:138–148
Sylvia DM (1989) Nursery inoculation of sea oats with vesicular-arbuscular mycorrhizal fungi and outplanting performance on Florida beaches. J Coastal Res 5:747–754
Tadych M, Blaszkowski J (1999) Growth responses of maritime sand dune plant species to arbuscular mycorrhizal fungi. Acta Mycol 34:115–123
Valverde T, Pisanty I, Rincón E (1997) Growth response of six tropical dune plant species to different nutrient regimes. J Coastal Res 13:497–505
van der Maarel E (1993a) Dry coastal ecosystems: polar regions and Europe. Elsevier, Amsterdam
van der Maarel E (1993b) Dry coastal ecosystems: Africa, America, Asia and Oceania. Elsevier, Amsterdam
van der Meulen F, Udo de Haes HA (1996) Nature conservation and integrated coastal zone management in Europe: present and future. Landsc Urban Plann 34:401–410
Vázquez G, Moreno-Casasola P, Barrera O (1998) Interaction between algae and seed germination in tropical dune slack species: a facilitation process. Aquat Bot 60(4):409–416
Veer MAC, Kooijman AM (1997) Effects of grass encroachment on vegetation and soil in Dutch dry dune grasslands. Plant Soil 192:119–128

Taxonomic Index

S

T

U

V

Z

Subject Index

Ecological Studies

Volumes published since 1998

Volume 131
The Structuring Role of Submerged Macrophytes in Lakes (1998)
E. Jeppesen et al. (Eds.)

Volume 132
Vegetation of the Tropical Pacific Islands (1998)
D. Mueller-Dombois and F.R. Fosberg

Volume 133
Aquatic Humic Substances: Ecology and Biogeochemistry (1998)
D.O. Hessen and L.J. Tranvik (Eds.)

Volume 134
Oxidant Air Pollution Impacts in the Montane Forests of Southern California (1999)
P.R. Miller and J.R. McBride (Eds.)

Volume 135
Predation in Vertebrate Communities: The Białowieża Primeval Forest as a Case Study (1998)
B. Jędrzejewska and W. Jędrzejewski

Volume 136
Landscape Disturbance and Biodiversity in Mediterranean-Type Ecosystems (1998)
P.W. Rundel, G. Montenegro, and F.M. Jaksic (Eds.)

Volume 137
Ecology of Mediterranean Evergreen Oak Forests (1999)
F. Rodà et al. (Eds.)

Volume 138
Fire, Climate Change and Carbon Cycling in the North American Boreal Forest (2000)
E.S. Kasischke and B. Stocks (Eds.)

Volume 139
Responses of Northern U.S. Forests to Environmental Change (2000)
R. Mickler, R.A. Birdsey, and J. Hom (Eds.)

Volume 140
Rainforest Ecosystems of East Kalimantan: El Niño, Drought, Fire and Human Impacts (2000)
E. Guhardja et al. (Eds.)

Volume 141
Activity Patterns in Small Mammals: An Ecological Approach (2000)
S. Halle and N.C. Stenseth (Eds.)

Volume 142
Carbon and Nitrogen Cycling in European Forest Ecosystems (2000)
E.-D. Schulze (Ed.)

Volume 143
Global Climate Change and Human Impacts on Forest Ecosystems: Postglacial Development, Present Situation and Future Trends in Central Europe (2001)
J. Puhe and B. Ulrich

Volume 144
Coastal Marine Ecosystems of Latin America (2001)
U. Seeliger and B. Kjerfve (Eds.)

Volume 145
Ecology and Evolution of the Freshwater Mussels Unionoida (2001)
G. Bauer and K. Wächtler (Eds.)

Volume 146
Inselbergs: Biotic Diversity of Isolated Rock Outcrops in Tropical and Temperate Regions (2000)
S. Porembski and W. Barthlott (Eds.)

Volume 147
Ecosystem Approaches to Landscape Management in Central Europe (2001)
J.D. Tenhunen, R. Lenz, and R. Hantschel (Eds.)

Volume 148
A Systems Analysis of the Baltic Sea (2001)
F.V. Wulff, L.A. Rahm, and P. Larsson (Eds.)

Volume 149
Banded Vegetation Patterning in Arid and Semiarid Environments (2001)
D. Tongway and J. Seghieri (Eds.)

Volume 150
Biological Soil Crusts: Structure, Function, and Management (2001, 2003)
J. Belnap and O.L. Lange (Eds.)

Volume 151
Ecological Comparisons of Sedimentary Shores (2001)
K. Reise (Ed.)

Volume 152
Global Biodiversity in a Changing Environment: Scenarios for the 21st Century (2001)
F.S. Chapin, O. Sala, and E. Huber-Sannwald (Eds.)

Volume 153
UV Radiation and Arctic Ecosystems (2002)
D.O. Hessen (Ed.)

Volume 154
Geoecology of Antarctic Ice-Free Coastal Landscapes (2002)
L. Beyer and M. Bölter (Eds.)

Volume 155
Conserving Biological Diversity in East African Forests: A Study of the Eastern Arc Mountains (2002)
W.D. Newmark

Volume 156
Urban Air Pollution and Forests: Resources at Risk in the Mexico City Air Basin (2002)
M.E. Fenn, L. I. de Bauer, and T. Hernández-Tejeda (Eds.)

Volume 157
Mycorrhizal Ecology (2002, 2003)
M.G.A. van der Heijden and I.R. Sanders (Eds.)

Volume 158
Diversity and Interaction in a Temperate Forest Community: Ogawa Forest Reserve of Japan (2002)
T. Nakashizuka and Y. Matsumoto (Eds.)

Volume 159
Big-Leaf Mahogany: Genetic Resources, Ecology and Management (2003)
A. E. Lugo, J. C. Figueroa Colón, and M. Alayón (Eds.)

Volume 160
Fire and Climatic Change in Temperate Ecosystems of the Western Americas (2003)
T. T. Veblen et al. (Eds.)

Volume 161
Competition and Coexistence (2002)
U. Sommer and B. Worm (Eds.)

Volume 162
How Landscapes Change: Human Disturbance and Ecosystem Fragmentation in the Americas (2003)
G.A. Bradshaw and P.A. Marquet (Eds.)

Volume 163
Fluxes of Carbon, Water and Energy of European Forests (2003)
R. Valentini (Ed.)

Volume 164
Herbivory of Leaf-Cutting Ants: A Case Study on *Atta colombica* in the Tropical Rainforest of Panama (2003)
R. Wirth, H. Herz, R.J. Ryel, W. Beyschlag, B. Hölldobler

Volume 165
Population Viability in Plants: Conservation, Management, and Modeling of Rare Plants (2003)
C.A Brigham, M.W. Schwartz (Eds.)

Volume 166
North American Temperate Deciduous Forest Responses to Changing Precipitation Regimes (2003)
P. Hanson and S.D. Wullschleger (Eds.)

Volume 167
Alpine Biodiversity in Europe (2003)
L. Nagy, G. Grabherr, Ch. Körner, D. Thompson (Eds.)

Volume 168
Root Ecology (2003)
H. de Kroon and E.J.W. Visser (Eds.)

Volume 169
Fire in Tropical Savannas: The Kapalga Experiment (2003)
A.N. Andersen, G.D. Cook, and R.J. Williams (Eds.)

Volume 170
Molecular Ecotoxicology of Plants (2004)
H. Sandermann (Ed.)

Volume 171
Coastal Dunes: Ecology and Conservation (2004)
M.L. Martínez and N. Psuty (Eds.)

Printing: Saladruck, Berlin
Binding: Stein+Lehmann, Berlin